Withdrawn
University of Waterloo

D1267461

Waves Called Solitons

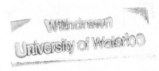

Withdrawn
University of Waterloo

Springer
Berlin
Heidelberg
New York
Barcelona
Hong Kong
London
Milan
Paris
Singapore
Tokyo

Michel Remoissenet

Waves
Called Solitons

Concepts and Experiments

3rd Revised and Enlarged Edition
With 170 Figures

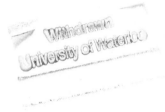

Withdrawn
University of Waterloo

 Springer

Professor Michel Remoissenet
Laboratoire de Physique
Université de Bourgogne
9, Avenue Alain Savary
F-21011 Dijon Cedex
France
E-mail: michelre@u-bourgogne.fr

ISBN 3-540-65919-6 3rd Edition Springer-Verlag Berlin Heidelberg New York
ISBN 3-540-60502-9 2nd Edition Springer-Verlag Berlin Heidelberg New York

Library of Congress Cataloging-in-Publication Data.
Remoissenet, M., Waves called solitons: concepts and experiments / Michel Remoissenet. –3rd rev. and enl. ed p. cm. Includes bibliographical references and index. ISBN 3-540-65919-6 1. Solitons. I. Title. QC174.26.W28R46 1999 531'.1133–dc21 99-23783 CIP

This work is subject to copyright. All rights are reserved, whether the whole or part of the material is concerned, specifically the rights of translation, reprinting, reuse of illustrations, recitation, broadcasting, reproduction on microfilm or in any other way, and storage in data banks. Duplication of this publication or parts thereof is permitted only under the provisions of the German Copyright Law of September 9, 1965, in its current version, and permission for use must always be obtained from Springer-Verlag. Violations are liable for prosecution under the German Copyright Law.

© Springer-Verlag Berlin Heidelberg 1993, 1996, 1999
Printed in Germany

The use of general descriptive names, registered names, trademarks, etc. in this publication does not imply, even in the absence of a specific statement, that such names are exempt from the relevant protective laws and regulations and therefore free for general use.

Dataconversion by perform GmbH, Heidelberg
Cover design: *design & production* GmbH, Heidelberg

SPIN: 10704787   55/3144/fm - 5 4 3 2 1 0 – Printed on acid-free paper

Preface

Since the printing of the second edition of this book the literature on nonlinear waves has grown greatly. The wide applicability of the subject to the physical, chemical and biological sciences generates a continuous supply of exciting new problems of practical and theoretical interest. Accordingly, this third edition has been thoroughly revised. I have made a number of changes in order to improve the presentation and have also brought some of the topics up to date with pertinent references. The new material includes a completely new chapter on solitary waves in diffusive systems. The overall objectives of the book are unchanged: to lead the reader to a thorough understanding of the basic nonlinear localization and soliton concepts, and to show how these concepts open doors to the advanced topics which they illuminate. The presentation of new topics has deliberately been kept simple for pedagogical purposes.

Chapter 1 was completed by references to the tidal bore and magnetic envelope solitons. Two new sections devoted to magnetic envelope solitons and signal processing with solitons have been added to Chap. 4. Short comments on models describing bloodpressure pulse propagation in terms of solitons were added to Chap. 5. A description of a new mechanical transmission line with two equilibrium states has been included in Chap. 6. Such an analog device is useful to illustrate the properties of kinksolitons and to observe solitary waves with a compact shape, called *compactons*. New references concerning recent advances in experimental techniques and lattice effects were added to Chap. 7. In Chap. 8 a short introduction to spatial optical solitons in continuous and discrete systems was included. Chapter 9 was completed by new sections devoted to nonlinear lattice models and energy localization. The concepts of *self-trapped states* and intrinsic *localized modes* or *discrete breathers* are discussed. Different analog chains, which make it possible to observe the characteristic features of discrete breathers, are described.

Contrary to previous chapters, where we have considered reversible or conservative systems where strict solitons can exist owing to the dynamical balance between linear dispersion and nonlinearity, Chap. 11, which is a new chapter, is devoted to irreversible systems where nonlinearity can balance the effects of dissipation leading to "diffusing solitary waves" or "diffusive solitons". We introduce this robust entity, which indeed is not a strict soliton, by considering a nonlinear electrical transmission line, where kink-shaped diffusive solitons can be predicted, and we present recent experiments. Then, we introduce reaction-diffusion processes by considering a chemical model for first-order nonequilibrium phase transitions. In this model the *power balanced solitary wave* or *diffusive soliton* represents a moving domain wall or interface layer between two phases. We then present an experimental electrical lattice which has been used as an analog model for the transmission of information along nerve fibers. In the next section we present a mechanical analog which makes it possible to observe diffusive solitons. The last section is devoted to a discussion of reaction-diffusion processes in lattices and propagation failure phenomenon.

I would like to express my gratitude to the many colleagues around the world who have, over the past few years, commented on various chapters of the book, made valuable suggestions and brought specific references to my attention. It is impossible to mention all of them by name and I hope they will forgive me. I would particularly like to thank G. Filatrella (University of Salerno, Italy), E. Jones (Proudman Oceanographic Laboratory, Birkenhead, UK), B. Kalinikos (St Petersburg University, Russia), Y. Kivshar (Optical Science Center, Australian National University, Canberra), J. Leon (University of Montpellier, France), C. Patton (Colorado State University, USA), P. Rosenau (Tel-Aviv University, Israel), M. Russell (Abingdon, Oxon, UK), A. Slavin (Oakland University, USA), A. Sievers (Cornell University, USA), S. Takeno (Osaka Institute of Technology, Japan), D. Vishwamittar (Panjab University, India), S. Watanabe (Yokohama National University, Japan).

I would like to thank H. Jauslin whose corrections and suggestions helped refine the manuscript of this third edition, and G. Millot for his comments on various chapters. I also take pleasure in thanking S. Dusuel, a talented young student whose research efforts are responsible for some of the new material added to this edition. I am particularly grateful to V. Boudon for computing and other facilities, to P. Michaux for his technical assistance in designing and performing new experiments, and to R. Chaux for preparing photographs.

Dijon   Michel Remoissenet
May 1999

Preface to the Second Edition

Encouraged by the friendly reception given to the first edition, I have preserved its basic form and most of the details. Apart from some corrections, minor changes, and addition of references where it was necessary, I have made the following changes.

Chapter 1 was expanded by a discussion of the discovery of solitons in the field of electromagnetic waves and optics. A new section devoted to nonlinear transmission lines and their applications in the microwave range has been added to Chap. 3. It seems to me that it was important to describe laboratory experiments on modulational instability, and subsequent generation of solitons, both in electrical transmission lines and in deep water in Chaps. 4 and 5. A description of a very simple experimental pocket version of the mechanical transmission line has been included in Chap. 6. Such a versatile and useful device should stimulate a practical approach to soliton physics. Chapter 7 was completed by a short presentation of some recent experimental results on discrete Josephson transmission lines. A discussion of the experimental modulational instability of coupled optical waves and a simple look at quantum solitons were added to Chap. 8 in order to introduce the reader to such remarkable topics.

Of the many people who made valuable comments on the first edition, I am particularly grateful to M. Dragoman, Y.S. Kivshar and A.W. Snyder.

I would like to thank R.S. MacKay whose corrections and suggestions helped refine the manuscript of this second edition.

I also take pleasure in thanking my Dijon colleague J.M. Bilbault and ex-student P. Marquié whose research efforts are responsible for some part of the new material added to this edition. I also wish to extend my appreciation to M. Pauty for a useful comment on the historical background.

Again, for this edition I have benefited from the technical assistance of B. Michaux and D. Arnoult in designing and performing new experiments, and in preparing a number of new diagrams and photographs.

Dijon   Michel Remoissenet
November 1995

Preface to the First Edition

Nonlinearity is a fascinating element of nature whose importance has been appreciated for many years when considering large-amplitude wave motions observed in various fields ranging from fluids and plasmas to solid-state, chemical, biological, and geological systems. Localized large-amplitude waves called *solitons*, which propagate without spreading and have particle-like properties, represent one of the most striking aspects of nonlinear phenomena. Although a wealth of literature on the subject, including theoretical and numerical studies, is available in good recent books and research journals, very little material has found its way into introductory texbooks and curricula. This is perhaps due to a belief that nonlinear physics is difficult and cannot be taught at an introductory level to undergraduate students and practitioners. Consequently, there is considerable interest in developing practical material suitable for students, at the lowest introductory level.

This book is intended to be an elementary introduction to the physics of solitons, for students, physicists, engineers and practitioners. We present the modeling of nonlinear phenomena where soliton-like waves are involved, together with applications to a wide variety of concrete systems and experiments. This book is designed as a book of physical ideas and basic methods and not as an up-to-the-minute book concerned with the latest research results. The background in physics and the amount of mathematical knowledge assumed of the reader is within that usually accumulated by junior or senior students in physics.

Much of the text of this book is an enlargement of a set of notes and descriptions of laboratory experiments developed over a period of years to supplement lectures on various aspects of wave motion. In spite of the diversity of the material, the book is not a collection of disconnected topics, written for specialists. Instead, I have tried to supply the practical and fundamental background in soliton physics, and to plan the book in order that it should be as much as possible *a self-contained and readable interdisciplinary whole*. Often, the important ideas or results are repeated several times, in different contexts. *Many of my choices of emphasis and examples have been made with experimental aspects in mind. Several experiments described in this book can be performed by the reader.* Although numerical studies play an important role in nonlinear science, I will not consider them in this book because they are described in a considerable body of literature.

In order to facilitate the use of this book, many illustrations have been included in the text and the details of theoretical calculations are relegated to appendices at the end of each chapter. A number of basic references are given as well as references intended to document the historical development of the subject. The referencing is not systematic; the bibliography listed at the end of the book serves only to advise the reader which sources could be used to fill in gaps in his or her basic knowledge and where he or she could turn for further reading.

The text is organized as follows. Our introduction in Chap.1 is devoted to the beautiful historical path of the soliton. The fundamental ideas of wave motion are then set forth in Chap.2 using simple electrical transmission lines and electrical networks as examples. At an elementary level, we review and illustrate the main properties of linear nondispersive and dispersive waves propagating in one spatial dimension. In Chap.3, we consider waves in transmission lines with nonlinearity. These simple physical systems are very useful for a pedagogical introduction to the soliton concept, and they are easy to construct and to model, allowing one to become quickly familiar with the essential aspects of solitary waves and solitons, and their properties. Specifically, we first examine the effect of nonlinearity on the shape of a wave propagating along a nonlinear dispersionless transmission line. Then we consider *the remarkable case where dispersion and nonlinearity can balance to produce a pulse-like wave with a permanent profile.* We describe simple experiments on pulse solitons, which illustrate the important features of such remarkable waves. In Chap.4 we consider the *lattice solitons*, which can propagate on an electrical network; then we examine periodic wavetrains, and modulated waves such as *envelope or hole solitons,* which can travel along electrical transmission lines.

Chapter 5 concentrates on such spectacular waves as the hydrodynamic pulse soliton, which was first observed in the nineteenth century, and the hydrodynamic envelope soliton. Simple water-tank experiments are described.

In Chap. 6, by using a chain of coupled pendulums, that is, a mechanical transmission line, we introduce a new class of large amplitude waves, known as *kink solitons* and *breather solitons* which present remarkable particle-like properties. Simple experiments that allow one to study qualitatively the properties of these solitons are presented.

Chapter 7 deals with a more sophisticated device: the superconductive Josephson junction. Here the physical quantity of interest is a quantum of magnetic flux, or *fluxon,* which behaves like a kink soliton and has properties remarkably similar to the mechanical solitons of Chap.5. In Chap.8 *bright and dark solitons* emerge, which correspond to the optical envelope or optical hole solitons, respectively. They can be observed in optical fibers where exploitation of the typical dispersive and nonlinear effects has stimulated theoretical and experimental studies on nonlinear guided waves.

Whereas the previous chapters are concerned with solitons in the macroworld, Chap.9 deals with nonlinear excitations in the microworld. Specifically, we consider the soliton concept in the study of nonlinear atomic lattices. The nonlinear equations that are encountered in the soliton story and models of several systems described in the text can be solved by using remarkable and powerful mathematical techniques, the main steps of which are given in the last chapter.

If a substantial fraction of users of this book feel that it helped them to approach the fascinating world of nonlinear waves or enlarge their outlook, its purpose will have been fulfilled. I hope the reader will feel encouraged to bring to my notice any remaining errors and other suggestions.

I have greatly benefited from frequent discussions with my colleagues and students. I am particularly grateful to Jean Marie Bilbault, who went over the entire manuscript and gave me invaluable comments.

I would also like to thank Alwyn Scott whose criticism and suggestions helped refine the manuscript.

I also wish to extend my appreciation to Patrick Marquié, Guy Millot, Jean François Paquerot, Michel Peyrard, and Claudine and Gérard Pierre for their comments on various chapters. Special thanks go to Bernard Michaux for his assistance in designing and performing experiments, and improving numerous illustrations throughout the book. Finally, it is a pleasure for me to acknowledge the technical assistance I have received from Dominique Arnoult and Claudine Jonon.

Dijon   Michel Remoissenet
October 1993

Contents

Solitons Blues

Music: Michel Remoissenet
Arrangement and transcription:
Michel Thibault

This book is dedicated to all the scientists
who have made the soliton concept a reality

On the occasion of a Conference on Nonlinear Coherent Structures in Physics and Biology organized by J. C. Eilbeck and D. B. Duncan, physicists and mathematicians solemnly gathered, on 13 July 1995, by an aqueduct of the Union Canal (see Nature, **376**, 373, 1995). This acqueduct, which was named *Scott Russell aqueduct* after a ceremony carried out by Alwyn Scott, is located at Hermiston near the site of the present Heriott-Watt University (Edinburgh). Martin Kruskal was one of the gathering, attempting, with the help of a borrowed boat to recreate *the solitary wave* of John Scott Russell (see Sect.1.2) which can be observed propagating a few meters in front of the boat. (Photo by K. Paterson, reproduced by kind permission of Heriot-Watt University).

1 Basic Concepts and the Discovery of Solitons

Today, many scientists see nonlinear science as the most important frontier for the fundamental understanding of Nature. The soliton concept is now firmly established after a gestation period of about one hundred and fifty years. Since then, different kinds of solitons have been observed experimentally in various real systems, and today they have captured the imagination of scientists in most physical discipline. They are widely accepted as a structural basis for viewing and understanding the dynamic behavior of complex nonlinear systems. Before introducing the soliton concept via its remarkable and beautiful historical path we compare briefly the linear and nonlinear behavior of a system.

1.1 A look at linear and nonlinear signatures

First, let us consider at time t the response R_1 of a linear system, an amplifier for example, to an input signal $E_1 = A \sin\omega t$ of angular frequency ω, as sketched in Fig. 1.1. In the low amplitude limit the output signal or the response of the system is linear, in other words it is proportional to the excitation

$$R_1 = a_1 E_1. \tag{1.1}$$

Here a_1 is a quantity that we assume to be constant (time independent) to simplify matters. If we double the amplitude of the input signal, the amplitude of the output signal is doubled and so on. The sum of two input signals E_1 and E_2 yields a response which is the superposition of the two output signals,

$$R = a_1(E_1 + E_2) = R_1 + R_2, \tag{1.2}$$

and a similar result holds for the superposition of several signals.

Next, if the amplitude of the input signal gets very large, distortion occurs as a manifestation of overloading. In this case, the response is no longer proportional to the excitation; one has

$$R = a_1 E_1 + a_2 E_1^2 + a_3 E_1^3 + \dots = a_1 E_1 \left(1 + \frac{a_2}{a_1} E_1 + \frac{a_3}{a_1} E_1^2 + \dots\right) \tag{1.3}$$

and signals at frequencies 2ω, 3ω, and so on, that is, *harmonics of the input signal are generated*. In some cases a chaotic response can occur: this phenomenon will not be considered in this book. Moreover, the sum of two signals at the input results not only in the sum of responses at the output but also in the product of sums and so on. *The superposition of states is no longer valid.*

1

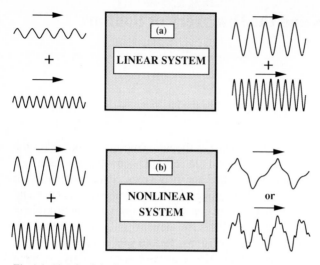

Fig.1.1. Sketch of the linear and nonlinear responses of a system to two input signals.

In equation (1.3) one has $a_1 >> a_2 >> a_3$... and we note that the nonlinear effects increase with the amplitude of the signal and with the coefficients a_2, a_3, ... If the quantities a_2/a_1, a_3/a_1, ... are not very small, say about 0.1, the system or material will be called *strongly nonlinear*.

Now, let us consider waves, that is signals which are not only time dependent but also depend on space, say one space dimension x, only, as will be assumed throughout this book. In this case any arbitrary pulse or disturbance can be regarded as a linear superposition of sinusoidal wave trains with different frequencies. If each of these linear waves propagates with the same velocity, the system is called *nondispersive* and the pulse travels without deforming its shape, as represented in Fig.1.2. If the velocities of each wave train are different, the pulse spreads out (see Fig.1.3) when propagating and the system is *dispersive*.

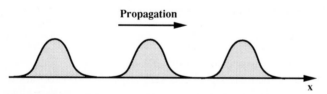

Fig.1.2. Undeformed propagation of a wave pulse in a linear and non dispersive system.

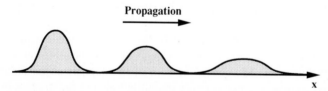

Fig.1.3. Propagation and spreading out of a wave pulse in a linear and dispersive system.

2

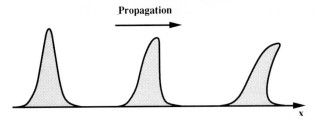

Fig.1.4. Points of large amplitude overtake points of small amplitude for a nonlinear wave pulse.

For many waves, nonlinearity introduces a new feature which is due to the generation of harmonics: the crest of the wave moves faster than the rest, in other words points of large amplitude overtake points of small amplitude, and the wave shocks and ultimately breaks (see Fig.1.4).

Having outlined the typical features of linear and nonlinear systems and media, let us now look into the beautiful history of nonlinear dispersive waves known as solitary waves and solitons.

1.2 Discovery of the solitary wave

Historically, the first documented observation of a solitary water wave was made by a Scottish engineer, John Scott Russell, in August 1834, when he saw a rounded smooth well-defined heap of water detach itself from the prow of a barge brought to rest and proceed without change of shape or diminution of speed for over two miles along the Union Canal linking Edinburgh with Glasgow. He described his observations in the following delightful terms:

> *I was observing the motion of a boat which was rapidly drawn along a narrow channel by a pair of horses, when the boat suddenly stopped-not so the mass of water in the channel which it had put in motion; it accumulated round the prow of the vessel in a state of violent agitation, then suddenly leaving it behind rolled forward with great velocity, assuming the form of a large solitary elevation, a rounded, smooth and well-defined heap of water, which continued its course along the channel apparently without change of form or diminution of speed. I followed it on horseback, and overtook it still rolling on at a rate of some eight or nine miles an hour, preserving its original figure some thirty feet long and a foot to a foot and a half in height. Its height gradually diminished, and after a chase of one or two miles I lost it in the windings of the channel. Such, in the month of August 1834, was my first chance interview with that singular and beautiful phenomenon which I have called the Wave of Translation, ...*
>
> John Scott Russell, Report on Waves (1844)

These observations were followed by extensive wave-tank experiments which established the following important properties of *solitary water waves*.
(i) These localized waves are bell-shaped and travel with permanent form and velocity.

3

(ii) In water of undisturbed depth h a wave of elevation a_m, towards which the crest points, propagates with velocity

$$v = \sqrt{g(h+a_m)}, \tag{1.4}$$

where g is the acceleration due to gravity.

(iii) An initial elevation of water might, depending on the relation between its height and length, evolve into a pure solitary wave, a single solitary wave plus a residual wave train, or two or more solitary waves with or without a residual wave train, as represented in Fig.1.5.

(iv) Solitary waves can cross each other without change of any kind.

(v) Solitary waves of depression are not observed: an initial depression is transformed into an oscillatory wave train of gradually increasing length and decreasing amplitude, as shown in Fig.1.6.

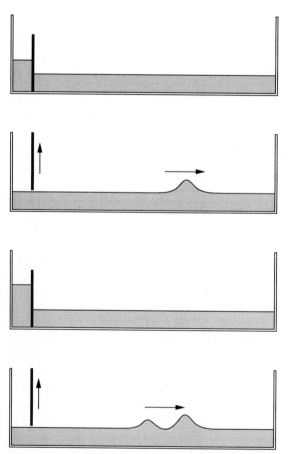

Fig.1.5. (a) Evolution of an initial elevation of water into **(b)** a pure solitary wave without a residual wave train **(c)** Evolution of an initial elevation of water into: **(d)** two solitary waves.

4

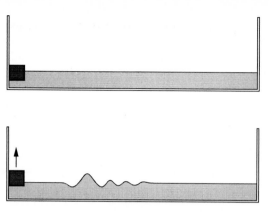

Fig.1.6. Evolution of an initial depression which is transformed into an oscillatory wave train of gradually increasing length and decreasing amplitude

At the time of their publication, the observations of J. Scott Russell appeared to contradict the nonlinear shallow-water wave theory of Airy (1845), which predicts that a wave with elevation of finite amplitude cannot propagate without change of form: it steepens and eventually breaks. On the other hand, Stokes (1847) showed that waves of finite amplitude and permanent form are possible in deep water, but they are periodic. He showed, in terms of a power series in the amplitude a_m, that the surface elevation of the waves is a sum of a basic sinusoidal term and harmonics which propagate with velocity v. This velocity depends both on wave number $k=2\pi/\lambda$, where λ is the wavelength, and the amplitude squared:

$$v = \sqrt{\frac{g}{k}(1 + k^2 a_m^2)} \qquad (1.5)$$

In this relation the absolute magnitude of the nonlinear correction is small: it does not exceed about 10% for the steepest possible waves.

In a little-noticed paper, H. Bazin (1865) reported experiments, performed in the long branch of the canal de Bourgogne close to Dijon (France), which confirmed Russell's observations.

The controversy over Russell's observations arises because the nonlinear shallow-water theory neglects dispersion, which generally tends to prevent wave steepening. It was resolved by Joseph Boussinesq (1871) and independently by Lord Rayleigh (1876), who made important contributions to the topic of solitary waves by showing that, if one ignores dissipation, *the increase in local wave velocity associated with finite amplitude is balanced by the decrease associated with dispersion, leading to a wave of permanent form.* In 1895 Korteweg and de Vries in fact derived a model equation (incorporating the effects of surface tension) which describes the unidirectional propagation of long waves in water of relatively shallow depth. This equation has become much celebrated and it is now known as the Korteweg–de Vries equation or KdV equation for short. Korteweg and de Vries showed that periodic solutions, which they named cnoidal waves, could be found in closed form and without further approximations. Moreover, they found a localized solution which represents a single hump of positive elevation and is also found in the limit of infinite wavelength or spatial period of the cnoidal wave as represented in Fig.1.7. This hump is the solitary wave discovered experimentally by Scott Russell.

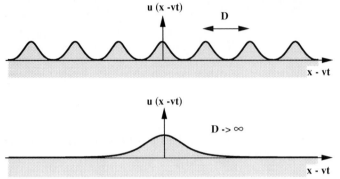

Fig.1.7. Representation of (**a**) a cnoidal wave and (**b**) a solitary wave.

A remarkable member of the class of the solitary waves which can be formed in the estuary of a river is the *tidal bore*, known as the *mascaret* in France. A tidal bore is a moving wall of water that carries the tide up some rivers that empty into the sea (for an introduction to the phenomenon, see: Lynch 1982). For a river to have a tidal bore two conditions must be met. First, the tides in the adjoining tidal body must be exceptionally high: a difference of more than six meters between high and low water is generally needed. Second, the river must be shallow, with a gently slopping bottom and a broad funnel-shaped estuary. As the incoming tide is forced into the narrowing river mouth, it builds in height until a single wave forms and moves upstream. The bores are distributed widely throughout the world, they can be observed (see Fig. 1.7(c)) in various rivers of different countries.

Fig. 1.7. (c) A tidal bore on the river Dee near Chester England (Photo by Eric Jones, Proudman Oceanographic Laboratory; reproduced by kind permission)

1.3 Discovery of the soliton

For a long time, the solitary wave was considered a rather unimportant curiosity in the mathematical structure of nonlinear wave theory. However, a remarkable discovery, which at first sight had nothing to do with solitary waves, was made by E. Fermi, J. Pasta, and S. Ulam as they studied the heat-transfer problem, that is, the flow of incoherent energy in a solid modeled by a one dimensional lattice consisting of equal masses connected by nonlinear springs (Fig.1.8). In fact, the one-dimensional monoatomic lattice is possibly the simplest discrete structure with which to study lattice dynamics. It provides a model of quasi-one-dimensional crystals within which a clear understanding of the dynamics is more readily attainable. When the interactions between particles are harmonic, that is, when particle displacements are very small deviations from equilibrium positions, the equations of motion of the particles can be decoupled and the dynamics of the lattice can be described by superposition of normal modes, represented by sinusoidal waves, which are mutually independent.

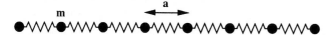

Fig.1.8. The monoatomic chain model.

In a crystal lattice, in contrast to sound waves, which are connected to coherent energy transport, heat conduction is connected to incoherent or disordered energy transport. If, for example, a normal mode is excited, its energy is not transferred to the other normal modes, which are independent. Consequently, a lattice with harmonic oscillations never reaches the state of thermal equilibrium; it is said to be *nonergodic*. In 1914, Debye suggested that if nonlinearity exists, the normal modes will interact and this will inhibit the propagation of thermal energy leading to a finite thermal conductivity. This problem was also examined by Peierls (1961) and it was assumed that nonlinear interactions would cause the energy to flow into each mode, resulting in energy equipartition or thermal equilibrium.

In 1955, E. Fermi, J. Pasta, and S. Ulam (FPU) tried to verify this assumption by computer simulations on a one-dimensional lattice. They examined the dynamical behavior of a chain (Fig.1.8) with nonlinear interactions (see Section 9.2) between the atoms (mass points), expecting that the initial energy would eventually be shared among all the degrees of freedom of the lattice. Much to their surprise, the system did not approach energy equipartition, that is, the energy did not spread throughout all the normal modes, but returned almost periodically to the originally excited mode and a few nearby modes as sketched in Fig.1.9. This remarkable near recurrence phenomenon, known nowadays as the FPU problem, was further examined by Ford (1961) and Jackson (1963), and it was confirmed that the nonlinear terms did not guarantee the approach of the system to thermal equilibrium.

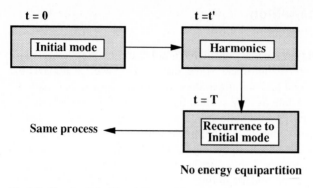

No energy equipartition

Fig.1.9. Sketch of the Fermi–Pasta–Ulam recurrence.

The unexpected results of FPU motivated N. Zabusky and M. Kruskal to reinvestigate this problem. The fact that only the lowest-order modes, with long wavelength, of the discrete lattice contributed to the phenomenon led them by a continuum approximation to the study of the KdV equation. From numerical simulations they found that robust pulse-like waves can propagate in a system modeled by such an equation. These solitary waves, which can pass through each other and preserve their shapes and speed after the collision, were called *solitons* to indicate these remarkable quasi-particle properties. This soliton concept allowed Zabusky and Kruskal to explain why almost recurrence occured. Specifically, from numerical solutions of the KdV equation used to model in the continuum approximation a nonlinear atomic lattice with periodic boundary conditions, they made the following observations. A sinusoidal initial condition (see Fig.1.10, where the evolution of only one sinusoidal period is sketched) with long wavelength, launched on the ring of atoms, evolves into a number of solitons which travel around the lattice with different velocities. These solitons collide, but preserve their individual shapes and velocities with only a small change in their phases. As time increases, there is an instant when the solitons collide at the same point, and the initial state comes close to recurrence. In fact, the recurrence does not quite occur, owing to the small phase shifts.

It is interesting to note that, prior to the investigations of Zabusky and Kruskal, analytical expressions describing collision events between solitary wave solutions of an equation, now called the Sine–Gordon equation, were found by Seeger, Donth, and Kochendörfer (1953) when studying dislocations in solids. Furthermore, Perring and Skyrme (1962) were interested in the kink-solitary-wave solutions (see Fig.1.11) of the Sine-Gordon equation as a simple model of elementary particles. Their computer experiments and analytic solutions showed that these solitary waves preserved their kink shapes and velocities after having collided.

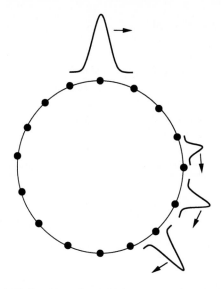

Fig.1.10. Breaking of an initial state (one sinusoidal period) into solitons; the recurrence to the initial state will occur when the solitons collide at the same point.

In 1967 Gardner et al. made an important contribution to the development of the theory. They showed that, if the initial shape of the wave is sufficiently localized the analytical solution of the KdV equation can be obtained. That is, the results obtained through the so-called *inverse-scattering method* (see Chap.10) show that at sufficiently long times the initial wave pulse evolves into one or more solitons and a dispersive small-amplitude tail. The total number of solitons depends on the initial shape. These theoretical results were in remarkable agreement with many of the experimental results obtained by John Scott Russell more than one hundred and fifty years ago.

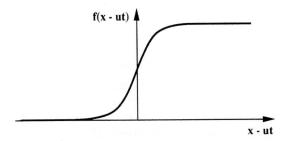

Fig.1.11. A kink soliton represented in its reference frame, which travels with velocity u.

At approximatively the same time these advances were being made, Lighthill (1965), using a theory developed by Whitham (1965), discovered theoretically that Stokes waves are unstable to infinitesimal modulational perturbations. This surprising result was then theoretically and experimentally confirmed by Benjamin and Feir (1967). They explained this instability in terms of resonant wave interaction theory, pioneered by Phillips (1960), which has a general applicability to a variety of phenomena (Benney 1962).

These findings generated great interest in the problem of unsteady time evolution of deep-water wave trains and wave packets. Then, it was shown by Zakharov (1968) that the time evolution of the envelope of a weakly nonlinear deep-water wave train is described by an equation called the nonlinear Schrödinger equation (NLS equation for short). Furthermore, this equation was solved exactly by Zakharov and Shabat (1971) by using the newly discovered inverse-scattering method. They showed that the exact solutions are deep-water wave-*envelope solitons* and that an initial wave packet eventually evolves into a number of envelope solitons and a dispersive tail. These solutions were verified experimentally by Yuen and Lake (1975). In the case of deep water, an envelope soliton consists of a sech-shaped hyperbolic secant envelope which modulates a periodic (cosine) wave, a dozen oscillations lying in the sech envelope, as represented in Fig.1.12. This may be the origin of the fisherman's or surfer's prediction that the seventh wave will be the largest!!!

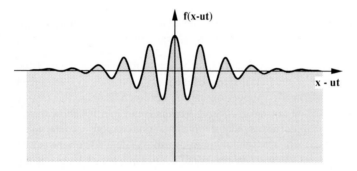

Fig.1.12. An envelope soliton with a dozen oscillations in the envelope.

Parallel to these investigations, the instability of electromagnetic waves propagating in nonlinear dispersive media was predicted (Ostrovskii 1963, Bespalov and Talanov 1966, Karpman 1967). Hasegawa and Tappert (1973) showed theoretically that the envelope of a light wave propagating in an optical fiber can also be described by the NLS equation. As a result, the existence of optical-envelope solitons, now currently called *bright solitons*, was predicted. This prediction was verified in 1980 by Mollenauer et al who observed bright-soliton propagation in a single fiber. Physically, these optical solitons originate from a simple kind of nonlinearity manifested through the intensity dependence of the refractive index. Nowadays, from the standpoint of technological applications, such optical-fiber solitons are a very important type of soliton.

In 1983, envelope solitons of the NLS type were also predicted theoretically by Zvezdin and Popkov for magnetostatic waves in magnetic films and soon observed experimentally, in ferrite films by Kalinikos and coworkers (1983). Since then, it appears that magnetic envelope solitons are an important subject area for the study of nonlinear high frequency waves in ferrite thin films and should stimulate interesting applications in the near future.

10

1.4 The soliton concept in physics

Before proceeding, we should explain the sense in which we use the term soliton. First, from the above results we present the following working definitions.

> *A solitary wave, as discovered experimentally by Scott Russell, is a localized wave that propagates along one space direction only, with undeformed shape. A soliton, as discovered numerically by Zabusky and Kruskal, is a large-amplitude coherent pulse or very stable solitary wave, the exact solution of a wave equation, whose shape and speed are not altered by a collision with other solitary waves.*

Thus, to the mathematician the word soliton has a quite specific and ideal connotation in the context of systems with exact solutions. Whilst these exact systems, called *integrable systems*, are valuable models, the physicist, who studies the real world, is rarely if ever concerned with the ideal soliton form. Rather, he is presented with real versions, where nonsolitonic or perturbational effects such as frictional loss mechanisms, external driving forces, defects, and so forth, are inevitable; thus the solitary waves or solitons, though long lived, are only metastable. Certainly, it is the existence of such quasi-soliton properties and not strict soliton properties that is the greater physical concern. Bearing this in mind, throughout this book we will use the keyword soliton with this weaker meaning which has been most widely adopted by physicists. For them, the soliton concept (Scott et al, 1973) has highlighted the importance of a particular type of energy propagation that takes the form of:

> *localized, finite energy states which are fundamentally nonlinear objects and so cannot be reached by perturbation theory from any linear state.*

Qualitatively, the solitary wave or soliton can be understood as representing a balance between the effect of dispersion and that of nonlinearity. Solitary waves can also be found resulting from a dynamic balance between dissipation and nonlinearity.

In this book, from the beginning to chapter 10 we concentrate on reversible nonlinear systems where dispersion dominates and dissipative effects are small enough to be neglected or in some cases to be considered weak perturbations. Chapter 11 of this new edition, is devoted to irreversible systems where dispersion can be ignored but, where dissipation can balance the effects of nonlinearity leading to a diffusive solitary wave or diffusive soliton. Striclty speaking, such a localized entity is not a soliton, however, we still use the soliton terminology in order to describe such a robust nonlinear wave which can play an important role in the dynamics of real stongly dissipative systems.

2 Linear Waves in Electrical Transmission Lines

Nowadays, linear transmission lines provide vital links in virtually all communications and computer systems, and the parallel-wire line is still widely used today in open-wire form, in coaxial cables and microstrips. The standard two-conductor transmission line is an important familiar system, that is able to support the propagation of transverse electromagnetic modes and is of great interest in many practical situations. We have all often studied this electrical circuit or variation of it in elementary electronics, physics, or mathematics courses (Ramo et al. 1965, Davidson 1978, Badlock and Bridgeman 1981). In fact, the study of linear transmission lines is an old problem: in their simplest forms they date from the early days of the electric telegraph and the telephone and their defining linear equations were sometimes called the telegraphist equations.

In this chapter, we review and illustrate the main properties of linear waves (Crawford 1965; Elmore and Heald 1985; Hirose and Lonngren 1985; Main 1993) propagating in one space dimension by considering simple linear transmission lines. We first consider the one-dimensional propagation of linear nondispersive waves on the basic linear transmission line. Second, we examine the specific effects on wave propagation caused by dispersion.

2.1 Linear nondispersive waves

We start with the simple example of a continuous two-wire transmission line (Fig. 2.1a), where a shunt voltage v(x, t) and a series current i (x, t) can be defined which depend on the distance x along the conductors and the time t. Thus, the lossless grounded transmission line can be represented by an equivalent electrical circuit, as shown in Fig. 2.1 b, where ℓ represents the inductance per unit length and c the capacitance per unit length. This representation of a continuous transmisssion line by distributed ℓc unit sections of length dx is useful, and aids the imagination. The reader should note that this line is the analog of the distributed mass–spring system or vibrating string, where the mass m per unit length and force constant K per unit length are analogous to the inductance ℓ and capacitance per unit length.

We consider a lossless transmission line where the significant parameters are the series inductance and the shunt capacitance. In fact, in a real transmission line there are always some losses, characterized by the resistance of the line conductors and the conductance between them, that cause deviations from the ideal system. However, in a first approximation these perturbations can be neglected for

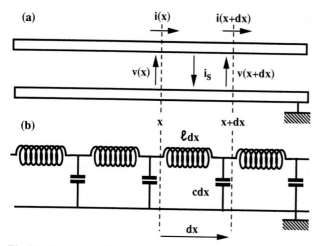

(a)

i(x)   i(x+dx)

v(x) i_s v(x+dx)

(b)   x   x+dx

ℓdx

cdx

dx

Fig.2.1. (**a**) A continuous two-wire transmission line which is assumed lossless and is grounded. (**b**) Its equivalent electrical circuit.

transmission lines with weakly dissipative elements, as will be the case in this chapter. From Kirchhoff's law, the difference between the shunt voltage at x and x+dx is given by

$$v(x) - v(x+dx) = dx \frac{\partial \Phi}{\partial t} , \qquad (2.1a)$$

where Φ is the total magnetic flux per unit length between the two conductors given by

$$\Phi = \ell i. \qquad (2.1b)$$

In the linear case, ℓ is independant of i and (2.1) becomes

$$v(x) - v(x+dx) = dx \, \ell \, \frac{\partial i(x)}{\partial t} . \qquad (2.2)$$

Note that the right-hand side of (2.2) should have been written

$$dx \, \ell \, \frac{\partial i(x+dx)}{\partial t} .$$

In fact, we have assumed that dx is very small and that we may Taylor-expand the current as

$$i(x+dx) \cong i(x) + dx \frac{\partial i}{\partial x} + \dots .$$

13

Thus, $\partial/\partial t \, [i(x+dx)]$ can be approximated by $\partial/\partial t \, [i(x)]$. Next, the difference between the series current at x and at x+dx is equal to the shunt current $i_s = \partial q/\partial t$, where q represents the charge per unit length:

$$i(x) - i(x+dx) = dx\frac{\partial q}{\partial t} \, . \tag{2.3a}$$

When the charge q is a linear function of the voltage, the capacitance c is independant of v and one has

$$q = cv, \tag{2.3b}$$

and (2.3a) becomes

$$i(x) - i(x+dx) = dxc \, \frac{\partial}{\partial t} \, v(x). \tag{2.4}$$

Here, the voltage at (x+dx) was also Taylor expanded. When dx is very small, (2.2) and (2.4) can be written as two simultaneous differential equations for the voltage v(x,t) and the curent i (x,t):

$$\frac{\partial v}{\partial x} = - \ell \, \frac{\partial i}{\partial t} \, , \qquad \frac{\partial i}{\partial x} = - c\frac{\partial v}{\partial t} \, . \tag{2.5}$$

We differentiate the first equation of (2.5) with respect to x and the second one with respect to t to obtain

$$\frac{\partial^2 v}{\partial x^2} = - \ell \, \frac{\partial^2 i}{\partial x \partial t} \, , \qquad \frac{\partial^2 i}{\partial t \partial x} = - c\frac{\partial^2 v}{\partial t^2} \, .$$

Since $\partial^2 i/\partial x \partial t = \partial^2 i/\partial t \partial x$, the current i can be eliminated to give the following linear wave equation:

$$\frac{\partial^2 v}{\partial t^2} - v_0{}^2 \, \frac{\partial^2 v}{\partial x^2} = 0, \tag{2.6a}$$

sometimes called "*the telegraphist's equation*", which dates from the 19th century. A similar equation for the current can be obtained by differentiating the second equation of (2.5) with respect to t and the first with respect to x:

$$\frac{\partial^2 i}{\partial t^2} - v_0{}^2 \, \frac{\partial^2 i}{\partial x^2} = 0. \tag{2.6b}$$

Let us focus on the wave equation (2.6a) for v. (Obviously, all the results discussed in the following are also valid for the current i.) In the linear wave equation (2.6a), the quantity

$$v_0 = (\ell c)^{-1/2}, \tag{2.7}$$

is the propagation velocity, which depends only on the inductance per unit length ℓ and the capacitance per unit length c. Equation (2.6a) can be rewritten in terms of new variables

$$X = x - v_0 t, \quad T = x + v_0 t, \tag{2.8}$$

to give

$$\partial^2 v / \partial X \partial T = 0.$$

This is immediately integrated to show that the general solution is the sum of arbitrary functions f and g

$$v = f(X) + g(T) = f(x - v_0 t) + g(x + v_0 t), \tag{2.9}$$

which correspond to waves propagating along the positive x direction, and along the negative x direction, respectively. This corresponds to the fact that (2.6a) is invariant under the transformation $x \rightarrow -x$, $t \rightarrow -t$. In other words, it is time reversible and admits bidirectional wave propagation. By contrast, in Chapter 3 we shall discuss unidirectional equation of evolution.

According to Fourier's theorem, any arbitrary disturbance can be decomposed into sinusoidal wave components (or harmonics) of different wavelengths and amplitudes. Consequently, the analysis of sinusoidal wave propagation is very important: it can serve as the basis for discussing the propagation of other shapes of waves. Namely, the functions f and g can take any form: sinusoidal, pulse, square, triangle, Gaussian and so on; no matter what waves are created, they all propagate with constant profile and with the same speed v_0 determined by the parameters of the *linear and dispersionless transmission line*.

2.2 Sinusoidal-wave characteristics

Let us consider a wave propagating along the transmission line in the positive x direction. In this case $g = 0$ and (2.9) reduces to

$$v(x, t) = f(x - v_0 t). \tag{2.10}$$

Moreover we assume that this wave is a sinusoidal wave train and we examine the properties of this simple periodic wave. When we examine the shape of the wave at

15

a given point, say x = 0, of the transmission line by using an electrical probe connected to an oscilloscope, we observe, as sketched in Fig.2.2a, a periodic voltage which is a sinusoidal function of time

$$v(t) = V_0\cos \omega t. \tag{2.11}$$

Here, ω is the angular frequency of the periodic signal and V_0 is the amplitude. The angular frequency is related to the period T by

$$\omega = \frac{2\pi}{T}.$$

This relation means that the phase of the wave changes from 0 to 2π when t changes from 0 to T. We now connect different probes at equidistant points along the transmission line, and at time t, say t = 0, we plot the measured voltage at each point versus the x coordinate (see Fig.2.2b). We obtain

$$v(x) = V_0\cos kx. \tag{2.12}$$

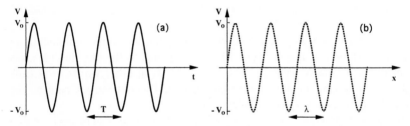

Fig.2.2. (a) Sinusoidal variation of the voltage as a function of time t which can be observed at a given point x of the transmission line. (b)Spatial sinusoidal variation of the voltage along the transmission line observed at a given time t.

Here, the quantity k represents the periodicity in the x coordinate, it is called *the wave number*. It can be regarded as the spatial frequency of the wave and is related to *the wavelength* λ by

$$k = \frac{2\pi}{\lambda}.$$

This relation means that the phase of the wave changes from 0 to 2π when x changes from zero to λ. Comparing (2.10), (2.11), and (2.12) we can write

$$v(x,t) = V_0\cos k(x-v_0 t) = V_0\cos (kx-\omega t).$$

The argument $\theta(x,t) = k(x-v_0t) = (kx - \omega t)$ is called *the phase of the wave*, and points of constant phase are those where the waveform has the same value, say a maximum (crest) or a minimum (trough). Then, from the above relation we get

$$v_0 = \frac{\omega}{k} = \text{constant.} \qquad (2.13)$$

The quantity v_0 is called *the phase velocity*, since it is the rate at which the phase of the wave (minima and maxima) propagates. Relation (2.13), is called *the dispersion relation*. It shows again that waves described by equations of the form (2.6) all have constant propagation velocities irrespective of wave frequencies or wavelengths. If we plot ω as a function of k, the graph is simply a straight line (Fig.2.3) and its slope gives the propagation velocity or constant *phase velocity* v_0.

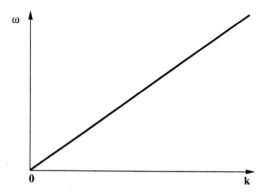

Fig.2.3. Frequency ω versus wave vector k for a nondispersive wave.

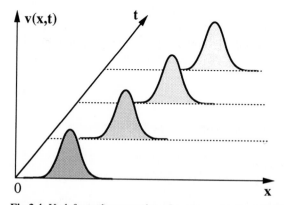

Fig.2.4. Undeformed propagation of a wave pulse in a nondispersive transmission line.

17

Waves, characterized by a linear dispersion relation with a constant propagation velocity are called nondispersive. Linear nondispersive waves are described by (2.6). As represented in Fig.2.4 a pulse or a wave pattern initially created is simply translated without deforming its shape, as time goes on.

2.2.1 Wave energy density and power

Let us now consider the energy of the wave. We multiply (2.5a) by i and (2.5b) by v and add the equations to get

$$\ell\, i \frac{\partial i}{\partial t} + cv \frac{\partial v}{\partial t} = - (i \frac{\partial v}{\partial x} + v \frac{\partial i}{\partial x}).$$ (2.14)

This equation can be rewritten in the form

$$\frac{\partial}{\partial t} (\frac{1}{2} cv^2 + \frac{1}{2} \ell i^2) = - \frac{\partial}{\partial x} (vi).$$ (2.15)

Here, the terms $1/2(cv^2)$ and $(1/2(\ell i^2)$ represent the electrical energy and magnetic energy per unit length, respectively. We now introduce the instantaneous wave energy density $W(x, t)$ and the wave power $P(x, t)$ as

$$W(x, t) = \frac{1}{2} cv^2 + \frac{1}{2} \ell i^2, \qquad P(x, t) = vi,$$ (2.16)

and express (2.15) as a conservation equation

$$\frac{\partial}{\partial t} W(x,t) + \frac{\partial}{\partial x} P(x,t) = 0.$$ (2.17)

This equation tells us that the time rate of increase (decrease) of energy density $W(x, t)$ is balanced by a decrease (increase) in the energy flow or instantaneous power $P(x, t)$. For example, let us apply this conservation equation to the case of a simple sinusoidal wave propagating in the positive x direction,

$$v = V_o \cos (kx - \omega t).$$ (2.18)

From (2.5b) we calculate $i = (cv\omega/k)$ and using (2.7) we find that

$$\frac{1}{2} \ell i^2 = \frac{1}{2} cv^2 = \frac{1}{2} cV_o^2 \cos^2(kx - \omega t).$$ (2.19)

Thus, *the energy density is proportional to the amplitude squared.* The energy conservation equation yields

$$\frac{\partial}{\partial t} W(x,t) + \frac{\partial}{\partial x} P(x,t) = \frac{\partial}{\partial t}(cv^2) + \frac{\partial}{\partial x}(v_0 cv^2), \qquad (2.20)$$

that is, in this simple case we have

$$P = v_0 W. \qquad (2.21)$$

2.3 The group-velocity concept

Let us consider the superposition of two sinusoidal waves of equal amplitude, both propagating in the positive x direction, which differ slightly in their frequencies and wave numbers:

$$V_1 = V_0 \cos(kx - \omega t), \qquad V_2 = V_0 \cos[(k + \delta k)x - (\omega + \delta\omega)t]. \qquad (2.22)$$

Then the combination has a wave form

$$V = V_1 + V_2 = 2V_0 \cos[\frac{1}{2}(\delta k\, x - \delta\omega\, t)]\cos[(k + \frac{\delta k}{2})x - (\omega + \frac{\delta\omega}{2})t]. \qquad (2.23)$$

This is the familiar expression for beats. The new waveform V(x, t) depicted in Fig. 2.5 is a wave with modulated amplitude. It consists of a progressive carrier wave with high frequency $(\omega + \delta\omega/2)$ and large wavenumber $(k + \delta k/2)$ and a modulated amplitude or envelope wave

$$\mathcal{V} = 2V_0 \cos[\frac{1}{2}(\delta k\, x - \delta\omega\, t)].$$

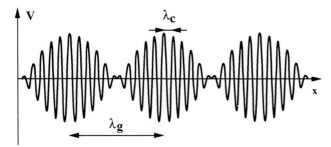

Fig.2.5. Linear superposition of two sinusoidal waves resulting in a wave with modulated amplitude.

Since δk and $\delta\omega$ are small, the period and wavelength of the envelope wave are both large: it varies slowly with period $4\pi/\delta\omega$ and wavelength $4\pi/\delta k$. Thus the function V(x, t) that is the product of two sinusoidal functions appears as a series of periodically repeating groups as shown in Fig.2.5. Each group is formed by several short waves. From (2.23) we see that the carrier wave with short wavelength propagates with the phase velocity

$$v = \frac{(2\omega + \delta\omega)}{(2k + \delta k)} \, ,$$

and the envelope wave (which contains the groups) with large wavelength propagates with the velocity

$$\frac{(\delta\omega/2)}{(\delta k/2)} = \frac{\delta\omega}{\delta k} \, .$$

When $\delta\omega$ and δk become infinitesimally small, the term $\delta\omega/\delta k$ approaches the value

$$v_g = \frac{d\omega}{dk} \, . \qquad (2.24)$$

where v_g is defined as *the group velocity*. Thus, the groups formed by several short waves propagate with the group velocity. Using the dispersion relation $\omega = v_o k$ obtained for the transmission line considered before, and relations (2.13) and (2.24) we find that $v = v_o = v_g$. In this simple nondispersive transmission line the group velocity is equal to the phase velocity.

Remarks

(i) By using (2.13) we can rewrite relation (2.24) as

$$v_g = \frac{d(kv_o)}{dk} = v_o + k\frac{dv_o}{dk} \, .$$

Since $k = 2\pi/\lambda$, we have $dk = -2\pi d\lambda/\lambda^2$ and can obtain the group velocity in terms of the phase velocity v_o and the wavelength λ:

$$v_g = v_o - \lambda\frac{dv_o}{d\lambda} \, .$$

(ii) Instead of considering the superposition of only two waves one can consider the superposition of an infinite number of waves with amplitudes and wave numbers which vary in over a small range $(k_o-\delta k, k_o+\delta k)$. Under these assumptions by means of an integral we may write

$$V = \int_{k_o-\delta k}^{k_o+\delta k} V(k)\exp[i(kx-\omega t)]dk.$$

20

The argument of the exponential can be put in the form $k_0x - \omega_0t + (k-k_0)x - (\omega-\omega_0)t$

Thus, one can write

$$V = \exp[i(k_0x-\omega_0t)] \int_{k_0-\delta k}^{k_0+\delta k} V(k)\exp[i(k-k_0)x-(\omega-\omega_0)t]dk,$$

which represents a group of waves. From this expression we see that the carrier wave propagates with phase velocity $v = \omega_0/k_0$ and if $(k-k_0) = \delta k$ and $(\omega-\omega_0) = \delta\omega$ are small enough the amplitude or envelope wave propagates with the group velocity given by (2.24).

2.4 Linear dispersive waves

Interesting dispersive phenomena take place when the phase velocity and the group velocity of a wave depend on its wave number (or wavelength). Specifically, the different wave components which propagate at different velocities separate or disperse from each other, as we shall see in this section .

2.4.1 Dispersive transmission lines

Let us consider a continuous transmission line of the previous type which has now an additional inductance ℓ_2 per unit length in parallel with c as shown in Fig.2.6. The set (2.5) of fundamental first-order equations is now replaced by

$$\frac{\partial V}{\partial x} = -\ell\frac{\partial i}{\partial t}, \qquad \frac{\partial i}{\partial x} = -(j_1+j_2), \qquad j_1 = \frac{\partial q}{\partial t}\,dx, \qquad \frac{\ell_2}{dx}\frac{\partial j_2}{\partial t} = V. \qquad (2.25)$$

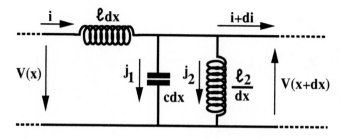

Fig.2.6. Dispersive transmission line structure with an inductance ℓ_2 in parallel with the capacitance c.

Combining the equations of (2.25) yields

$$\frac{\partial^2 V}{\partial t^2} - v_0^2 \frac{\partial^2 V}{\partial x^2} + \omega_0^2 V = 0, \tag{2.26}$$

where v_0 is given by (2.7) and where we have set $\omega_0^2 = 1/\ell_2 c$. This second order linear equation (called the Klein–Gordon equation) is dispersive. Indeed, assuming as in the previous section a sinusoidal voltage waveform

$$V(x,t) = V_0 \cos(kx - \omega t) = V_0 \operatorname{Re} [\, e^{i(kx - \omega t)}],$$

where Re denotes the real part, we derive the dispersion relation

$$\omega = \sqrt{\omega_0^2 + v_0^2 k^2} \, . \tag{2.27}$$

The phase velocity v and the group velocity v_g both depend on wavenumber k and are given by

$$v(k) = \frac{\omega}{k} = \frac{1}{k}\sqrt{\omega_0^2 + v_0^2 k^2} \, , \tag{2.28a}$$

and

$$v_g(k) = \frac{d\omega}{dk} = \frac{v_0^2 k}{\sqrt{\omega_0^2 + v_0^2 k^2}} \, . \tag{2.28b}$$

For k >0, the diagrams corresponding to (2.27), (2.28a), and (2.28b) are represented in Fig.2.7 and 2.8a,b. If we compare to Fig.2.3, we see that the dispersion arises from the term ω_0. Indeed, for $\omega_0 = 0$ the above expression for ω reduces to (2.13), which characterizes the nondispersive transmission line. From (2.27) we remark that the wave number k is imaginary for $\omega < \omega_0$, that is, ω_0 is a cutoff frequency and waves with a frequency below ω_0 are evanescent: they cannot propagate.

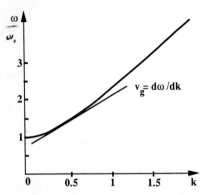

Fig.2.7. Representation of the dispersion relation (2.27).

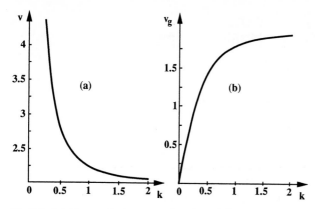

Fig.2.8. (**a**) Phase velocity v as a function of the wave number k. (**b**) Group velocity v_g as a function of k.

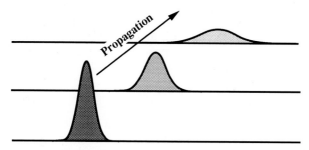

Fig.2.9. A newly created voltage-pulse spreads out and disperse, as it propagates along a dispersive electrical transmission line.

Qualitatively, a pulse initially well defined with a given narrow width deforms its shape or disperses as it propagates: it becomes widely spread as sketched in Fig.2.9. In fact, the pulse can be regarded as the superposition of sinusoidal wave trains or harmonics with different wave numbers, each traveling with its own phase speed v(k) given by (2.28a). As time evolves, these different component waves *disperse* with the result that a single concentrated pulse spreads out. This dispersive effect will be examined in more detail in Section 2.6.

2.4.2 Electrical network

Let us now consider a discrete electrical transmission line, i.e, an electrical network which is realized with identical components: linear inductors L and linear capacitors C, as represented in Fig. 2.10. This electrical network, with a large number N of identical unit sections, is the analog of the well-known one-dimensional atomic lattice.

23

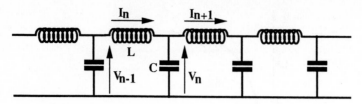

Fig. 2.10. Equivalent circuit of an electrical network.

In a first approximation, we neglect the small dissipation and inhomogeneities introduced by the components. $V_n(t)$ is the voltage across the nth capacitor and $I_n(t)$ is the current across the nth inductance. Considering a unit LC section, from Kirchhoff's law one obtains

$$V_{n-1} - V_n = \frac{d\Phi_n}{dt}, \qquad I_n - I_{n+1} = \frac{dQ_n}{dt}, \tag{2.29}$$

where the magnetic flux Φ_n and the electric charge are related to the current I_n and the voltage V_n by the following relations: $\Phi_n = LI_n$ and $Q_n = CV_n$. Combining these relations with (2.29) and rearranging one gets the system of differential linear equations for the discrete voltage $V_n(t)$ or the discrete current $I_n(t)$:

$$\frac{d^2V_n}{dt^2} = \frac{1}{LC}(V_{n+1} + V_{n-1} - 2V_n), \qquad n = 1, 2, ..., \tag{2.30a}$$

$$\frac{d^2I_n}{dt^2} = \frac{1}{LC}(I_{n+1} + I_{n-1} - 2I_n), \qquad n = 1, 2, ... \tag{2.30b}$$

Let us consider a particular harmonic solution of (2.30a),

$$V_n(t) = V_0 \, \text{Re} \, [\, e^{i(\omega t - \kappa n)}], \tag{2.31}$$

with constant amplitude V_0. We assume similar expressions for $V_{n+1}(t)$ and $V_{n-1}(t)$, but with $\kappa(n+1)$ and $\kappa(n-1)$ instead of κn in the exponential. Note that n is an index instead of a length, consequently in order to ensure that κ has the correct units it can be assumed that $n = x/\delta$, that is, n is expressed in terms of a small quantity $\delta \ll 1$, which represents a hypothetical unit cell length. One has $(\kappa/\delta) = 2\pi/\lambda = k$. Under these conditions the wavelength λ is expressed in unit cells and the velocities will be expressed in cell/second. Inserting (2.31) into (2.30a) yields

$$-\omega^2 V_n = \frac{1}{LC}(e^{i\kappa} + e^{-i\kappa} - 2)V_n.$$

Putting $\omega_c = 2(LC)^{-1/2}$ one obtains *the dispersion relation* for the dispersive modes or waves that can propagate through the electrical network:

$$\omega^2 = \omega_c^2 \sin^2 \frac{\kappa}{2}. \tag{2.32a}$$

24

Since the phase velocity of the wave is equal to ω/κ, we can get both positively and negatively traveling waves by keeping the frequency ω positive and letting the wave number κ assume both positive and negative values. With this convention the dispersion relation becomes

$$\omega = \omega_c \left| \sin\frac{\kappa}{2} \right|. \tag{2.32b}$$

In this case, for $0 \le \kappa \le \pi$, the $\omega = f(\kappa)$ diagram corresponding to (2.32b) is depicted in Fig.2.11, for $-\pi \le \kappa \le 0$ it is symmetric to the ω axis (not represented here). The phase velocity depends on the mode considered, and is given by

$$v(\kappa) = \frac{\omega_c}{\kappa} \left| \sin\frac{\kappa}{2} \right|. \tag{2.33}$$

The group velocity v_g is different from v, and is given by

$$v_g(\kappa) = \frac{\omega_c}{2} \left| \cos\frac{\kappa}{2} \right|. \tag{2.34}$$

We note that $v_g = 0$ at the edges of the zone $\kappa = \pm\pi$ (called the Brillouin zone in the case of atomic lattices). One has $\omega = \omega_c$ which is the cutoff frequency: no mode with $\omega > \omega_c$ can propagate along the electrical network. In electronics such a network is called a low-pass filter.

Here, in contrast to the previous transmission line which was continuous, the dispersion is spatial: it arises from the discrete structure of the electrical network with periodic LC sections. As in the previous case (see Fig.2.11) a pulse will broaden when it propagates along the network.

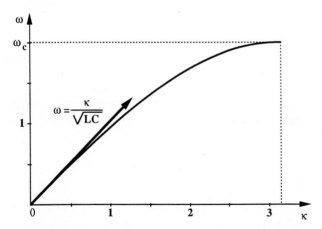

Fig. 2.11. $\omega = f(\kappa)$ diagram representing the dispersion relation of dispersive waves on an LC transmission line or low-pass filter. The straight line of slope $1/\sqrt{LC}$ represents the nondispersive case $\kappa \approx 0$.

2.4.3 The weakly dispersive limit

Let us now consider the weakly dispersive limit of the discrete system (2.30a) of linear equations. Specifically, if the voltage (current) varies slowly from one unit section to the other, the discrete index n can be assumed to be a continuous variable x and we can use the Taylor expansion for the right-hand side (rhs) of (2.30a). We have

$$V_n(t) \rightarrow V(x, t), \tag{2.35a}$$

and

$$V_{n\pm1} = V \pm \frac{\partial V}{\partial n} + \frac{1}{2}\frac{\partial^2 V}{\partial n^2} \pm \frac{1}{3!}\frac{\partial^3 V}{\partial n^3} + \frac{1}{4!}\frac{\partial^4 V}{\partial n^4} + ... \tag{2.35b}$$

Using $x = n\delta$ and substituting (2.35) in (2.30a) yields the following wave equation

$$\frac{\partial^2 V}{\partial t^2} - \frac{\delta^2}{LC}\frac{\partial^2 V}{\partial x^2} = \frac{\delta^4}{12LC}\frac{\partial^4 V}{\partial x^4} + ... \tag{2.36}$$

This wave equation (2.36) is linear and dispersive, its leading dispersive term is of order δ^4, resulting from the Taylor expansion of (2.30a). The corresponding dispersion relation is obtained by inserting into (2.36) a sinusoidal solution of the form (2.31) with $\kappa = k\delta$ or by expanding the dispersion relation (2.32b). We obtain

$$\omega(k) = v_o k \sqrt{1 - \frac{\delta^2 k^2}{12}}, \tag{2.37a}$$

where

$$v_o = \frac{\delta}{\sqrt{LC}}. \tag{2.37b}$$

In Fig.2.11 this weak dispersion corresponds to the small-k ($=\kappa/\delta$) values (long-wavelength limit). The phase velocity also depends on the wave number, and is given by

$$v(k) = v_o \sqrt{1 - \frac{\delta^2 k^2}{12}}. \tag{2.38}$$

Remarks

(i) Equations (2.36), (2.37), and (2.38) are derived with long wave phenomena in mind, that is, the wave number k must be small, otherwise, from a mathematical point of view, ω would become imaginary for large k.

(ii) To order δ^2, (2.36) reduces to the dispersionless wave equation (1.6a) for the continuous transmission line. This case corresponds to the very-long-wavelength limit, i. e, the limit where $k = 2\pi/\lambda \ll 1$. The dispersion relation (2.37a) reduces to a linear relation similar to the dispersion relation (2.13) of the continuous transmission line, $\omega \approx v_0 k = \kappa/\sqrt{LC}$, which is represented by a straight line of slope $v_0/\delta = 1/\sqrt{LC}$ in Fig.2.11 (the symmetric line with slope $-v_0/\delta$ is not represented). Here v_0 is the propagation velocity of the waves with wavelengths that are very large compared to the discreteness of the electrical network. In this limit, the length of the wave is so long that *the wave does not feel the network*, which behaves like a continuous transmission line.

(iii) One can combine both kinds of dispersion encountered in the two previous examples by considering a network with unit sections similar to the one described in Fig. 2.6, but with discrete elements. In this case, instead of (2.25) we have

$$V_{n-1} - V_n = L \frac{dI_n}{dt}, \qquad I_n - I_{n+1} = \frac{dQ_n}{dt} + I_2, \qquad L_2 \frac{dI_2}{dt} = V_n,$$

where I_2 is the current in inductance L_2, which is in parallel with C. Combining these equations we get

$$\frac{d^2V_n}{dt^2} = \frac{1}{LC}(V_{n+1} + V_{n-1} - 2V_n) - \frac{V_n}{L_2C}, \qquad n = 1, 2, \dots.$$

The corresponding dispersion relation is

$$\omega^2 = \omega_0^2 + 4u_0^2 \sin^2 \frac{\kappa}{2},$$

where $u_0 = \sqrt{1/LC}$, $\omega_0 = \sqrt{1/L_2C}$ and $\omega_{max} = \sqrt{\omega_0^2 + 4u_0^2}$ are the low and high cut-off frequencies, respectively. This relation has combined features of dispersion relations (2.27) and (2.32b). The electrical network behaves as a pass-band filter.

2.5 Evolution of a wavepacket envelope

In Section 2.3 we introduced the group velocity by considering the propagation of a pulse. On the other hand, it is instructive to introduce the concept of group velocity dispersion by considering a wave packet that is a wave which consists of a sinusoidal carrier wave of constant frequency ω_0 and wave number k_0, modulated by a waveform ψ. We assume that this wave envelope ψ varies slowly in time and space compared to the variations of the carrier wave:

$$V = \text{Re} \, [\psi \, (X, T) \, e^{i \, (\omega_0 t - k_0 x)} \,].$$ (2.39)

Here, $X = \varepsilon x$ and $T = \varepsilon t$, where ε is a small parameter : $\varepsilon \ll 1$, represent the slow space and time variables ; Re means the real part. On figures 2.12a,b the modulated wave and its Fourier spectrum are represented. As the envelope of the wavepacket is slowly varying, it contains a large number of crests of the carrier wave and the amplitude distribution is concentrated in wave numbers close to the value $k = k_0$.

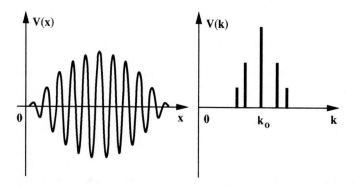

Fig. 2.12. (a) Wave packet with a slowly varying envelope compared to the rapid variations of the carrier wave. (b) Fourier spectrum of this wavepacket.

Under these conditions, if the transmission line is dispersive (see Sect. 2.4), the angular frequency is a function of the wave number, and

$$\omega = \omega \, (k),$$ (2.40)

and we can expand ω as a function of k in a Taylor series about the wave number ko of the sinusoidal carrier wave:

$$\omega - \omega_0 = (\frac{\partial \omega}{\partial k})(k-k_0) + \frac{1}{2} (\frac{\partial^2 \omega}{\partial k^2}) \, (k-k_0)^2 + ...$$ (2.41)

Here, the derivatives are evaluated at the wave number $k = k_0$, that is, at frequency ω_0. In relation (2.41) we recognize the coefficient

$$(\frac{\partial \omega}{\partial k}) = v_g , \quad \text{at } k = k_0,$$ (2.42)

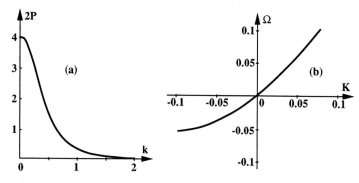

Fig. 2.13. (a) Group velocity dispersion 2P, corresponding to v_g as given by (2.28b), as a function of k. **(b)** Representation of the dispersion relation of the envelope wave.

which is the linear group velocity v_g evaluated at the wave number k_0. The second coefficient

$$(\frac{\partial^2\omega}{\partial k^2}) = (\frac{\partial v_g}{\partial k}) = 2P, \text{ at } k = k_0, \tag{2.43}$$

represents *the group velocity dispersion*. Both quantities v_g and 2P can be calculated from the dispersion relation (2.40) if it is known explicitly. For example, for the dispersive transmission line we considered previously: to v_g, represented in Fig.2.8b, corresponds 2P, represented in Fig. 2.13a. Putting

$$\omega - \omega_0 = \Omega, \qquad k - k_0 = K, \tag{2.44}$$

we can rewrite (2.41) as

$$\Omega = v_g K + PK^2. \tag{2.45}$$

Relation (2.45) is the dispersion relation of the envelope wave, it is valid in the vicinity of a particular frequency ω_0 and related wave number k_0 (Fig.2.13b). Now, using the Fourier variables Ω and K, which correspond to the slow space and time variables X and T, we can write the Fourier transform $\mathcal{F}$ of the envelope function

$$\psi (K, \Omega) = \mathcal{F}[\psi(X;T)] = \int_{-\infty}^{\infty} \psi(X,T) \, e^{i(\Omega T - KX)} \, dXdT,$$

and the inverse transform

$$\psi(X;T) = \mathcal{F}^{-1}[\psi (K, \Omega)] = \frac{1}{(2\pi)^2} \int_{-\infty}^{\infty} \psi(K, \Omega) \, e^{-i(\Omega T - KX)} \, dKd\Omega.$$

29

At this point we remark that

$$\frac{\partial \psi}{\partial X} = iK\mathcal{F}^{-1}[\psi(K, \Omega)], \qquad \frac{\partial \psi}{\partial T} = -i\Omega\mathcal{F}^{-1}[\psi(K, \Omega)].$$

In the dispersion relation (2.45) of the envelope wave the terms Ω and K are small, say of order ε with $\varepsilon \ll 1$. They can be replaced by the operators $i\varepsilon \partial/\partial T$ and $-i\varepsilon \partial/\partial X$, and then let the resulting expression operate on the envelope function $\psi(X,T)$. We get

$$i\left[\varepsilon\frac{\partial \psi}{\partial T} + \varepsilon v_g \frac{\partial \psi}{\partial X}\right] + \varepsilon^2 P \frac{\partial^2 \psi}{\partial X^2} = 0. \tag{2.46}$$

The first two terms of (2.46) represent the undistorted propagation of the wave envelope at group velocity v_g, and the third term represents its linear distorsion. If one sets $P = 0$, the solution of (2.46) can be expressed in terms of an arbitrary function of $(X - v_g T)$. Consequently, choosing a frame of reference which moves at the group velocity, we find that the substitution

$$\xi = X - v_g T, \qquad \tau = \varepsilon T, \tag{2.47a}$$

yields

$$\frac{\partial}{\partial X} = \frac{\partial}{\partial \xi}, \qquad \frac{\partial}{\partial T} = \varepsilon \frac{\partial}{\partial \tau} - v_g \frac{\partial}{\partial \xi}. \tag{2.47b}$$

This allows us to eliminate the term in $\partial \psi / \partial X$ and to transform (2.46) into

$$i\frac{\partial \psi}{\partial \tau} + P \frac{\partial^2 \psi}{\partial \xi^2} = 0. \tag{2.48}$$

This linear equation governs the evolution of the wave envelope (the term ε^2 which multiplies both terms has been dropped out). We first note that if one replaces i by $i\hbar$ and P by $\hbar^2/2m$, with $\hbar = h/2\pi$, where h is the Planck constant, the linear equation (2.48) which describes the envelope evolution is formally similar to the linear Schrödinger equation in quantum mechanics for a free particle of mass m. Equation (2.48) can be solved (see Appendix 2A) to give

$$\psi(\xi,\tau) = \frac{1}{2\pi} \int_{-\infty}^{\infty} \psi(\kappa,0) \exp[-i(P\kappa^2\tau - \kappa\xi)] \, d\kappa, \tag{2.49}$$

where $\psi(\kappa,0)$ is the Fourier transform of the wave packet $\psi(\xi,0)$ at time $\tau = 0$, and κ is a wave number associated with ξ. This solution shows that the group-velocity dispersion 2P modifies the phase of each Fourier component of the wave packet envelope by an amount that depends on the square of the wavenumber κ and the time τ. As a result, the wave packet deforms as it propagates.

2.6 Dispersion-induced wavepacket broadening

As a simple familiar example (which is discussed in many textbooks because the calculations are tractable), let us suppose that the wave packet consists of a Gaussian pulse, illustrated in Fig.2.14. This bell-shaped envelope function, at $\tau = 0$, may be expressed by

$$\psi(\xi,0) = A \exp(-\xi^2 / 2L_o^2). \tag{2.50}$$

It is symmetrical about the origin, where it has the amplitude A; L_o is the half width (at an amplitude of $1/\sqrt{e}$ amplitude). In practical cases one uses the full width at half maximum (or half height) L_{fw}. For a Gaussian pulse one has $L_{fw} = 2\sqrt{Ln2}\, L_o$. As shown in Appendix 2B, one can calculate the envelope function ψ in terms of the variables X and T, related to ξ and τ by (2.47a), with the result that

$$\psi(X,T) = \frac{A}{\sqrt{1 + \dfrac{2iP\varepsilon T}{L_o^2}}} \exp\left[- \frac{(X - v_gT)^2}{2L_o^2(1 + \dfrac{2iP\varepsilon T}{L_o^2})}\right]. \tag{2.51}$$

The real part of this expression represents the physical part of the envelope of the wave packet. Multiplying (2.51) by its complex conjugate $\psi^*(X,T)$ yields the modulus of the wave packet envelope squared:

$$|\psi(X,T)|^2 = \frac{A^2}{\sqrt{1 + \left(\dfrac{2P\varepsilon T}{L_o^2}\right)^2}} \exp\left[- \frac{(X- v_gT)^2}{L_o^2(1 + \left(\dfrac{2P\varepsilon T}{L_o^2}\right)^2)}\right]. \tag{2.52}$$

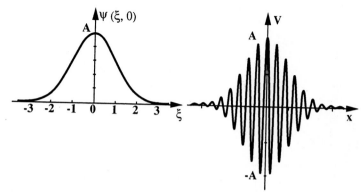

Fig. 2.14. Gaussian envelope function at initial time $\tau = 0$ and corresponding voltage wavepacket in laboratory coordinates as given by (2.39) and (2.50).

This quantity is proportional to the energy content of the wave packet. We note that, as expected, owing to the presence of the group velocity dispersion 2P, the wave packet disperses as it propagates. That is, at time T =0, one recovers, as one should, the square $|\psi(\xi,0)|^2$ of the absolute value of the initial wave packet (2.50). The width which was initially L_o at time T=0 becomes L at time T:

$$L = L_o \sqrt{1 + \left(\frac{T}{T_d}\right)^2}.$$ (2.53)

The extent of broadening is governed by the dispersion time $T_d = L_o^2/2|P|\epsilon$. The amplitude simultaneoulsy decreases with time as the wave packet travels along with velocity v_g. One can define a dispersion length $L_d = v_g T_d$ that provides a length scale over which the dispersive effects become important. However, the wave packet continues to be Gaussian in form in such a manner (see Fig.2.15) that the area under the curve $|\psi(X,T)|^2$ remains constant. Obviously, all the energy is where the packet is and this constancy of area expresses the conservation of the pulse energy. The pulse moves as whole with the group velocity v_g, and therefore so does the energy. We note that the broadening of the wave packet is independent of the sign of the group-velocity dispersion.

The wave packet envelope (2.51) can be written in the form

$$\psi(X,T) = |\psi(X,T)| \exp [- i\phi(X, T)],$$ (2.54)

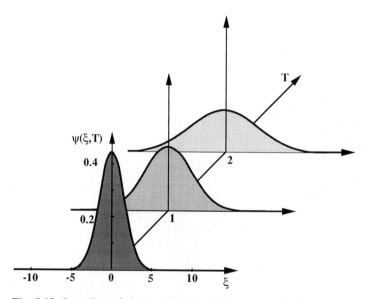

Fig. 2.15. Spreading of the envelope ψ of the wave packet in the frame moving at group velocity v_g: here $\xi = X-v_g T$, as given by (2.47a).

32

where the modulus is given by the square root of (2.52) and the phase $\phi(X, T)$ is calculated to be

$$\phi(X, T) = \arctan\left(\frac{2PT}{L_0^2}\right) + \frac{(X - v_g T)^2}{2L_0^2} \cdot \frac{\frac{2PT}{L_0^2}}{[1 + (\frac{2PT}{L_0^2})^2]} . \qquad (2.55)$$

Then, from (2.39) and (2.54) one can express the voltage in the form

$$V(x, t) = |\psi(\varepsilon x, \varepsilon t)| \cos[k_0 x - \omega_0 t + \phi(\varepsilon x, \varepsilon t)], \qquad (2.56)$$

where the parameter ε reminds us that the modulus and the phase of the envelope wave are changing slowly in space and time. The total phase of the wave packet is thus

$$\Phi = k_0 x - \omega_0 t + \phi(\varepsilon x, \varepsilon t). \qquad (2.57)$$

Remark

Relation (2.57) suggests that we can define a local wave number $k(x,t)$ and a local frequency $\omega(x, t)$ by

$$k(x,t) = \frac{\partial \Phi}{\partial x}, \qquad \omega(x, t) = -\frac{\partial \Phi}{\partial t} .$$

In the present example of a Gaussian wave packet, using (2.55) we find that

$$k(x,t) = k_0 + \frac{\varepsilon x - v_g \varepsilon t}{L_0^2} \cdot \frac{\frac{2P\varepsilon t}{L_0^2}}{[1 + (\frac{2P\varepsilon t}{L_0^2})^2]}$$

and a more complicated relation for $\omega(x, t)$ which is not written here. The above relation shows that the local wave number consists of the original wave number k_0 of the carrier wave and a term which depends linearly on the distance x and is proportional to the group-velocity dispersion. Next, cross derivation of $k(x, t)$ and $\omega(x, t)$ gives

$$\frac{\partial k}{\partial t} + \frac{\partial \omega}{\partial x} = 0.$$

From the definition of the wave number at an instant t a given unit length will contain k waves or k wave crests, so that $\partial k/\partial t$ gives the rate of change of wave crests per unit time in the unit length. On the other hand, as the frequency represents the number of waves or wave crests passing through a fixed point per unit time, the term $\partial\omega/\partial x$ gives the net flux of waves across the unit length. Thus, this relation is a conservation relation which expresses that the wave crests are neither created nor annihilated. It may be written as

$$\frac{\partial k}{\partial t} + \frac{d\omega}{dk}\frac{\partial k}{\partial x} = 0,$$

where $d\omega/dk = v_g$ is the group velocity as defined previously. Then, the local wave number propagates at group velocity.

Appendix 2A. General solution for the envelope evolution

Equation (2.48) is readily solved by using the Fourier transform method. Namely, if $\psi(\kappa,\tau)$ is the spatial Fourier transform of $\psi(\xi,\tau)$, then

$$\psi(\xi,\tau) = \frac{1}{2\pi}\int_{-\infty}^{\infty}\psi(\kappa,\tau)e^{i\kappa\xi}\,d\kappa, \qquad (2A.1)$$

where k is a wavenumber. Substituting (2A.1) in (2.48) yields the differential equation

$$i\frac{\partial\psi(\kappa,\tau)}{\partial\tau} = P\kappa^2\,\psi(\kappa,\tau), \qquad (2A.2)$$

whose solution is given by

$$\psi(\kappa,\tau) = \psi(\kappa,0)\exp(-iP\kappa^2\,\tau). \qquad (2A.3)$$

By substituting (2A.3) in (2A.1) we get the general solution of (2.48):

$$\psi(\xi,\tau) = \frac{1}{2\pi}\int_{-\infty}^{\infty}\psi(\kappa,0)\exp[-i(P\kappa^2\tau-\kappa\xi)]\,d\kappa, \qquad (2A.4)$$

where $\psi(\kappa,0)$ is the Fourier transform of the wave packet $\psi(\xi,0)$ at time $\tau = 0$.

34

Appendix 2B. Evolution of the envelope of a Gaussian wavepacket

The Fourier transform of the Gaussian wave packet (2.50) is

$$\psi(\kappa,0) = A \int_{-\infty}^{\infty} \exp(-\xi^2/2L_0^2) \exp(-i\kappa\xi)d\xi. \qquad (2B.1)$$

In this expression the exponent can be written as

$$\frac{\xi^2}{2L_0^2} + i\kappa\xi = (\frac{\xi}{L_0\sqrt{2}} + i L_0 \frac{\kappa}{\sqrt{2}})^2 + \frac{1}{2} L_0^2\kappa^2. \qquad (2B.2)$$

Setting

$$(\frac{\xi}{L_0\sqrt{2}} + i L_0 \frac{\kappa}{\sqrt{2}}) = u,$$

and using the result

$$\int_{-\infty}^{\infty} \exp(-u^2) du = \sqrt{\pi},$$

we find that

$$\psi(\kappa,0) = AL_0\sqrt{2\pi} \exp(-\frac{1}{2} L_0^2\kappa^2). \qquad (2B.3)$$

Substituting (2B.3) in (2A.4) we get the expression for the packet at a later time t:

$$\psi(\xi,\tau) = \frac{AL_0}{\sqrt{2\pi}} \int_{-\infty}^{\infty} \exp(-\frac{1}{2} L_0^2\kappa^2) \exp[-i(P\kappa^2\tau - \kappa\xi)] d\kappa. \qquad (2B.4)$$

We must therefore evaluate this integral. Using the same procedure as previously, the exponent is written as

$$-\frac{1}{2} L_0^2\kappa^2 - i P\kappa^2\tau + i\kappa\xi = -(i \frac{\xi}{2\sqrt{L_0^2/2 + iP\tau}} - \kappa\sqrt{L_0^2/2 + iP\tau})^2$$
$$-\frac{\xi^2}{4(L_0^2/2 + iP\tau)}.$$

Then, setting

$$U = i \, \frac{\xi}{2\sqrt{L_o^2/2 + iP\tau}} - \kappa\sqrt{L_o^2/2 + iP\tau},$$

the integral (2B.4) is evaluated, in terms of the variables X and T, with the result that

$$\psi(X,T) = \frac{A}{\sqrt{1 + \dfrac{2iP\varepsilon T}{L_o^2}}} \exp\left[-\frac{(X - v_g T)^2}{2L_o^2\left(1 + \dfrac{2iP\varepsilon T}{L_o^2}\right)}\right]. \qquad (2B.5)$$

3 Solitons in Nonlinear Transmission Lines

Although solitary waves and solitons were originally discovered in the context of water waves and lattice dynamics, consideration of these physical systems (which will be considered in Chaps.5 and 8) leads to calculations far too involved for pedagogical purposes. Thus, for an introduction to the soliton concept, we therefore consider simple wave propagation in electrical nonlinear transmission lines and electrical networks.

   The natural extension of linear to nonlinear transmission lines where robust localized waves with permanent profile known as *solitary waves or solitons*, can propagate, was realized only in the last two decades. Nonlinear electrical transmission lines (Scott 1970, Lonngren and Scott 1978, Ostrowski 1976 Peterson 1984, Remoissenet and Michaux 1990) are perhaps the most simple experimental devices to observe and study quantitavely the propagation and properties of nonlinear waves. Moreover, they are relatively inexpensive and easy to construct, allowing one to become quickly familiar with the essential soliton properties. Soliton propagation on nonlinear transmission lines is now used for picosecond impulse generation and broadband millimeter-wave frequency multiplication (Jäger 1985, Rodwell et al. 1994).

   In Chap. 2 we have encountered dispersion, which leads to a progressive broadening of a pulse as it propagates along a linear dispersive transmission line. In this chapter we first examine the effect of nonlinearity on the shape of the wave propagating along a nonlinear dispersionless transmission line. Then, we discover the soliton when examining the remarkable case where dispersion and nonlinearity can balance to produce a pulse-like wave with a permanent profile. Finally, we describe experiments on pulse solitons.

3.1 Nonlinear and dispersionless transmission lines

If in the transmission line represented in Fig.3.1 the amplitude of the voltage v and the current i increases, the nonlinear effects cannot be ignored. In other words, the capacitance per unit length c and the inductance per unit length ℓ are no longer constant but are functions of the voltage ($c = c(v)$) and current ($\ell = \ell(i)$), respectively. However, for standard components even with very large voltage and current amplitudes the nonlinear effects are small. To observe them it is preferable to use specific components with appreciable nonlinearities. With such components, instead of linear partial differential equations (2.5) we have a system of first-order

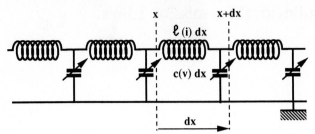

Fig.3.1. Equivalent electrical circuit of a transmission line with nonlinear inductance ℓ (i) and nonlinear capacitance c(v) per unit length dx.

nonlinear differential equations

$$\frac{\partial v}{\partial x} = -\ell \text{ (i)} \frac{\partial i}{\partial t} , \qquad (3.1a)$$

$$\frac{\partial i}{\partial x} = -c(v) \frac{\partial v}{\partial t} . \qquad (3.1b)$$

For the sake of simplicity let us restrict ourselves to the simple practical case where ℓ is independent of the current i but where the capacitance is voltage dependent. For example, if the nonlinear capacitance is that of a reverse-biased pn diode junction the capacitance–voltage relation can be approximated by a polynomial expansion

$$c(v) = C_o (1 + a_1 v + a_2 v^2 + ...). \qquad (3.2a)$$

For a small enough voltage, one can keep the first two terms in expansion (3.2a) and write

$$c(v) \approx C_o (1 - 2bv), \qquad (3.2b)$$

where the nonlinear coefficient $b = -a_1/2$ is small. We first note (see Chap.1) that the first effect of nonlinearity is to create harmonics. Namely, if we assume that initially

$$v = A \cos(\omega t - kx),$$

we then obtain second harmonic terms on the right-hand side of (3.1b):

$$\frac{\partial i}{\partial x} = AC_o\omega \sin(\omega t - kx) - A^2 bC_o\omega \sin 2(\omega t - kx).$$

Now, in analogy with the solution of the linear nondispersive line (see Sect.2.1),

38

we assume that (3.1) has a solution of the form

$$v = f \{x \pm [\ell c(v)]^{-1/2}t \} \tag{3.3}$$

and that the velocity of propagation

$$[\ell c(v)]^{-1/2} = \frac{1}{\sqrt{\ell C_0 (1-2bv)}} \approx \frac{1}{\sqrt{\ell C_0}} (1 + bv), \tag{3.4}$$

of a waveform is voltage dependent. The solution (3.3) for the voltage together with the following solution for the current:

$$i = \int_0^v [c(v)/\ell]^{1/2} dv, \tag{3.5}$$

can be shown to be correct as verified in Appendix 3A.

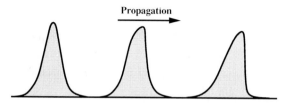

Fig.3.2. Sketch of the formation of a shock wave in a nonlinear nondispersive transmission line.

If as in (3.2) the capacitance c(v) decreases with increasing voltage v, we expect that high-voltage parts of the waveform will progagate faster than the more slowly moving low-voltage parts. Qualitatively, as time evolves, the peak of a voltage pulse can overtake the bottom, and a wave with a steepening front, leading progressively to a jump discontinuity, can develop as represented in Fig.3.2. Such a wave is called *a shock wave*. Shock waves are known in fluid dynamics; for example, they are created by a sound source moving faster than the sound velocity.

Analytically, one can show that owing only to nonlinearity, a physically meaningful solution is the one that contains a propagating jump discontinuity. The calculations are tractable if one considers a pulse which at initial time t = 0 has a parabolic shape (Bathnagar 1979). According to the analytical results presented in Appendix 3A, we have sketched the evolution of the pulse in Fig.3.3a. As time increases, the initial parabolic waveform is progressively deformed. It is instructive to compare this result to the evolution of the parabolic pulse in the linearized case i.e, when b = 0 in (3.11). In this case (see Sect.1.1), one has a pulse moving with constant velocity $(\ell C_0)^{-1/2}$ without change of form, as shown in Fig.3.3b.

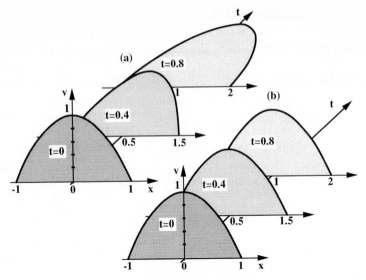

Fig.3.3. (a) Progressive deformation of the initial parabolic profile, at t=0, of the pulse and its evolution into a jump discontinuity; note that for t =0.8 the wave breaking is not observed on a real transmission line because the wave of voltage cannot be multivalued (b) In the linearized case b=0, the pulse moves with constant velocity and without change of form.

Next, by differentiating (3.1a) with respect to x and Eq. (3.1b) with respect to t, and eliminating the current, one obtains the wave equation

$$\frac{\partial^2 v}{\partial x^2} = \ell \frac{\partial}{\partial t}[c(v)\frac{\partial v}{\partial t}].$$

$$= L \frac{\partial c(v)}{\partial t}\frac{\partial v}{\partial t} + L\, c(v) \frac{\partial^2 v}{\partial t^2} = L\frac{\partial c}{\partial v}\left(\frac{\partial v}{\partial t}\right)^2 + L\, C_o(1-2bv)\frac{\partial^2 v}{\partial t^2}$$

$$= -2bC_oL\left(\frac{\partial v}{\partial t}\right)^2 + L\,c_o\frac{\partial^2 v}{\partial t^2} - 2bLc_o \quad (3.6)$$

By using relation (3.2b) this second-order *nonlinear* partial differential equation is transformed into

$$\frac{\partial^2 v^2}{\partial t^2} = \frac{\partial}{\partial t}\left(\frac{\partial v^2}{\partial t}\right) = \frac{\partial}{\partial t}\left(2v\frac{\partial v}{\partial t}\right) = 2\left(\frac{\partial v}{\partial t}\right)^2 + 2v\frac{\partial^2 v}{\partial t^2}$$

$$\frac{\partial^2 v}{\partial x^2} - \ell C_o \frac{\partial^2 v}{\partial t^2} = -\ell C_o b \frac{\partial^2 v^2}{\partial t^2}. \qquad = -2Lc_o b\left(\frac{\partial v}{\partial t}\right)^2 - 2Lc_o bv\frac{\partial^2 v}{\partial t^2} \qquad (3.7)$$

The left-hand side of (3.7) contains the linear terms . The right-hand side accounts for the *nonlinearity*. Since (3.7) is a combination of (3.1a,b), its solutions are *shock wave solutions* of the form (3.3).

Remark

Note that in the above transmission line, instead of using a nonlinear capacitance c(v) and a linear inductance ℓ, we can use a nonlinear inductance ℓ(i) with a linear capacitance c. In practice, in the past the use of nonlinear capacitance rather than

40

nonlinear inductance in transmission lines was mainly attributable to the low-loss behavior of varicap (varactor) diodes, compared to the lossy behavior of multidomain ferrimagnetic materials. The situation has now changed with the developement of thin single-domain ferromagnetic films, which can be used in distributed structures at ultra-high frequencies.

3.2 Combined effects of dispersion and nonlinearity

Until now, we have encountered dispersion and nonlinearity separately. We have seen that for a linear dispersive transmission line a newly created pulse spreads out and disperses as it propagates. On the contrary, for a nonlinear nondispersive transmission line, the profile of an initial pulse deforms and its wavefront tends to become abrupt (with a jump discontinuity). In the following, we shall see that in a nonlinear dispersive transmission line the dispersion can balance the effects of nonlinearity, leading to a pulse-like wave that is a *solitary wave* or a *soliton* which can propagate with constant velocity and profile.

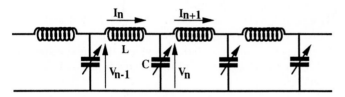

Fig.3.4. Equivalent circuit of an electrical network with linear inductance L and nonlinear capacitance $C(V_n)$ in each unit section.

To show this, we consider the elementary LC network depicted in Fig.3.4, with linear inductors L and with *nonlinear* capacitors C. As previously, the differential capacitance $C(V_n)$ is supposed to depend nonlinearly on the voltage V_n:

$$C(V_n) = \frac{dQ_n(V_n)}{dV_n}, \tag{3.8}$$

where $Q_n(V_n)$ denotes the charge stored in the nth capacitor. The inductance is assumed to be current independent: $d\Phi_n = LdI_n$. From Sect.2.4 we know that the spatial dispersion will arise from the discrete structure of the components. From equations similar to (2.29) and using (3.8) we derive the following equation:

$$\frac{d^2Q_n}{dt^2} = \frac{d}{dt}[C(V_n)\frac{dV_n}{dt}] = \frac{1}{L}(V_{n+1} + V_{n-1} - 2V_n), \qquad n = 1, 2, ... \tag{3.9}$$

As in Sect.3.1 we assume that for a small enough voltage the capacitance–voltage relation can be approximated by

$$C(V_n) = C_o(1 - 2bV_n). \tag{3.10}$$

41

Substituting (3.10) in (3.9) we find that

$$LC_o \frac{d^2V_n}{dt^2} - LC_o\, b\, \frac{d^2V_n{}^2}{dt^2} = (\, V_{n+1} + V_{n-1} - 2V_n\,), \qquad n = 1, 2, \ldots \quad (3.11)$$

Comparing (3.11) to (3.9) we note in passing that the charge–voltage relation is

$$Q_n = C_o (\, V_n - bV_n{}^2). \qquad (3.12)$$

The system (3.11) of nonlinear discrete equations cannot be solved analytically, but only by using numerical methods (Eilbeck 1991). Thus, to get approximate solutions we employ the continuum limit (see (2.35b)) and set x= nδ (as in Sect. 2.4) to get

$$V_{tt} - (\delta^2/LC_o)V_{xx} = (\delta^4/\, 12LC_o)V_{xxxx} + b\, V^2{}_{tt}. \qquad (3.13)$$

Here, we have used *the convention of writing a derivative $\partial V/\partial x$ as V_x, $\partial^2 V/\partial x^2$ as V_{xx}, and so forth.* It will be employed in the text whenever it is convenient.

Equation (3.13) describes waves that can travel both to the left and to the right; it is *a weakly dispersive and nonlinear wave equation.*. Its left hand side is none other than the linear wave equation (2.6a). It contains the same dispersive term than (2.36), which is due to discreteness of the electrical network, and the same nonlinear term than (3.7).

3.3 Electrical solitary waves and pulse solitons

In (3.13) the nonlinear term can balance the dispersive term if it is of the same order, say $b \sim O(\delta^4)$. Under this condition we look for a solution with a permanent profile, i.e, a localized wave solution of (3.13) that does not change its shape as it propagates with constant velocity v (contrary to Sect.3.1, v now denotes a velocity rather than a voltage). Setting $v_o = \delta/\sqrt{LC}$, this *solitary-wave* solution (see Appendix 3B) is

$$V = \frac{3}{2b} \frac{v^2 - v_o{}^2}{v^2} \operatorname{sech}^2 \left(\frac{\sqrt{3(v^2 - v_o{}^2)}}{v_o} \left(n - \frac{v}{\delta}\, t\, \right) \right). \qquad (3.14)$$

Note that in (3.14) we have $[n - (v/\delta)t] = (x - vt)/\delta$. Thus, for practical cases the distance is expressed in sections or cells, and the velocities are expressed in sections or cells per second. Equation (3.14) represents a pulse-like wave with amplitude

$$V_m = \frac{3}{2b} \frac{v^2 - v_o{}^2}{v^2}, \qquad (3.15)$$

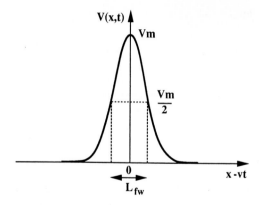

Fig.3.5. Representation of the solitary wave solution V(x,t) given by (3.14) traveling with velocity v.

which depends on the velocity v. Let us calculate its width L_{fw}, which is constant for a given amplitude V_m (or velocity v). This is defined (see Sect. 2.6) at half height $V_m/2$ in Fig.3.5. Assuming that $t = 0$ and $x/\delta = n = L_{fw}/2$, one has

$$V\left(\frac{L_{fw}}{2}, 0\right) = V_m \operatorname{sech}^2\left(\frac{\sqrt{3(v^2 - v_0^2)}}{v_0}\left(\frac{L_{fw}}{2}\right)\right) = \frac{V_m}{2}. \qquad (3.16)$$

Simplifying by substituting for V_m and evaluating $\operatorname{sech}^{-1}(1/\sqrt{2}) = 0.88$ yields

$$L_{fw} = 1.76 \frac{v_0}{\sqrt{3(v^2 - v_0^2)}} \approx \frac{v_0}{\sqrt{(v^2 - v_0^2)}}. \qquad (3.17)$$

The amplitude and width at half height of the solitary wave are useful quantities when one compares the experimental and theoretical waveforms.

For waves traveling in one direction only, and in the weak-amplitude limit one can reduce (3.13) to the well-known standard nonlinear equation, known as the *Korteweg-de Vries* (KdV) equation, which was first used to model water-wave propagation (see Chaps. 1 and 5). It admits *particular solitary-wave solutions, called solitons*. This equation can be derived (see Appendix 3C) by using a technique known as the reductive perturbation technique (Taniuti and Wei 1968). Specifically, we introduce a small parameter $\varepsilon \ll 1$ and perform the following change of variables:

$$s = \varepsilon^{1/2} (n - c_0 t), \qquad T = \varepsilon^{3/2} t, \qquad (3.18)$$

where

$$c_0 = 1/\sqrt{LC_0} = v_0/\delta, \qquad (3.19)$$

and express the voltage in a perturbation series

$$V = \varepsilon V_1 + \varepsilon^2 V_2 + \ ....$$

(3.20)

Here the parameter ε indicates the magnitude of the rate of change, the coefficients $\varepsilon^{1/2}$ and $\varepsilon^{3/2}$ are chosen in order to balance the nonlinear term, and the 4th derivative (dispersive term) and the time derivative are of the same order in ε. In terms of the long space and time variables s and T, we obtain

$$\frac{\partial V}{\partial T} + bc_o V \frac{\partial V}{\partial s} + \frac{c_o}{24} \frac{\partial^3 V}{\partial s^3} = 0.$$

(3.21)

Equation (3.21) is one of the possible forms of the KdV equation, known since 1895 (see Chaps. 1 and 5) *when dissipation is ignored*. The KdV equation is a useful approximation in the modeling of many physical systems when one wishes to combine a simple weak nonlinearity and a simple weak dispersion, as in the present case for the transmission line. In terms of the original laboratory coordinates (x,t) the solution is (see Appendix 3C)

$$V = \varepsilon V_1 = \frac{\varepsilon u}{8b} \ \text{sech}^2 \ [\ \frac{\sqrt{\varepsilon u}}{2} \ (n - \frac{v_s}{\delta} t\ )\ ].$$

(3.22)

Here, one has

$$v_s = (1 + \varepsilon u/24)v_o,$$

(3.23)

where u is a parameter (see Appendix 3C). This localized or pulse solution of the KdV equation is the so-called *solitary-wave*, first described by John Scott Russell in his "Report on Waves" in 1844 (see Chaps. 1 and 5). Its shape and speed are not altered by a collision with other solitary waves, for this reason it is called a *soliton*.

In (3.22) and (3.23), the term ε, which can be dropped, reminds us that the amplitude of the KdV soliton is weak. The amplitude V_m and the width L_s (at half height) of the pulse are useful quantities when one compares the theoretical and experimental waveforms. The amplitude of the soliton is

$$V_{ms} = \frac{u}{8b} = \frac{3}{b} (\frac{v_s}{v_o} - 1),$$

(3.24)

and depends on the velocity v_s: u >0 implies that $v_s/v_o > 1$, when the soliton velocity is higher than the velocity v_o of the linear waves. In the same way as for the solitary wave, the width L_s of the KdV soliton, defined at half height $V_m/2$, can be calculated. We find that

$$L_s = \frac{3.52}{\sqrt{u}} = \frac{1.76}{\sqrt{6(v/v_o-1)}} \ .$$

(3.25)

The width L_s of the KdV soliton is velocity (or amplitude) dependent and we note that the product

$$V_{ms}.L_s^2 = (3.52)^2/8b = C^{te}. \tag{3.26}$$

It is instructive to compare the above KdV soliton parameters with those of the solitary wave (3.14) when its velocity is close to v_0. We set $v = v_0 + (dv_0)$, where (dv_0) is a small quantity. Under this assumption one finds that

$$V_m \approx \frac{3}{b}\left(\frac{v}{v_0} - 1\right) = V_{ms}, \qquad L \approx \frac{1.76}{\sqrt{6(v/v_0-1)}} = L_s. \tag{3.27}$$

Consequently, for $v \approx v_0$ the waveforms of the solitary wave and the KdV soliton are similar and in practice one cannot tell the difference.

The basic properties of the *KdV soliton* (3.22) can be summarized.
(i) *Its amplitude increases with its velocity.*
(ii) *Its width is inversely proportional to the square root of its velocity.*
(iii) *It is a unidirectional wave pulse, i.e, its velocity cannot be negative for solutions of the KdV equation, since in the integration $\sqrt{u}$ must be real.*
(iv) *The sign of the soliton solution depends on the sign of the nonlinear coefficient in the KdV equation.*

Remarks

(i) The continuum limit used to derive (3.13) has some limitations for the following reasons. As previously mentioned (see Chap.2) the dispersion relation (2.37a) which corresponds to a linearized version of (3.13) supposes that k must remain very small. Otherwise, when nonlinear effects exist, short wavelengths are present (generated by harmonic generation, for example) and (2.37a) may break down. As pointed out by Rosenau (1987), the problem is mathematically ill posed, and this will cause, while ω is imaginary for $k>\sqrt{12}/\delta$ in (2.37a), the solution to equation (3.13) to blow up in a finite time. In order to prevent the boundness of (2.37a) he proposed rewriting it as

$$\omega^2 = v_0^2 k^2/(1+\delta^2 k^2/12).$$

Consequently, one has $\omega^2 = v_0^2 k^2(1-\delta^2 k^2/12)$ for $k\ll1$ and $\omega^2=12v_0^2/\delta^2$ for $k\gg1$. According to the modified dispersion relation, the equation of propagation (3.13) is replaced by a regularized equation

$$\frac{\partial^2 V}{\partial t^2} - \frac{\delta^2}{LC_o}\frac{\partial^2 V}{\partial x^2} = \frac{\delta^4}{12LC_o}\frac{\partial^4 V}{\partial x^2 \partial t^2} + b\frac{\partial^2 V^2}{\partial t^2}.$$

Nevertheless, in the previous analysis and experiments we consider weakly nonlinear and dispersive waves (k<<1) and keep (3.13) as an approximate model equation.

(ii) If one considers the electrical charge Q as the relevant physical parameter, one can approximate eq (3.13) by another soliton equation, called the *Boussinesq equation*. That is, in the continuum approximation the charge–voltage relation (3.12) can be rewritten as

$$bV^2 - V + (Q/C_0) = 0.$$

Then, we have

$$V = (1/2b) [1 \pm (1 - (4bQ/C_0)^{1/2}].$$

For $4bQ << C_0$, one can expand the square root to get

$$V \approx (Q/C_0) + b (Q^2/C_0^2).$$

Note that we keep the minus sign in order to obtain the correct relation $Q = C_0 V$ in the linear approximation (b=0) of (3.12). Setting $bv_0^2/C_0 = \alpha$ and $v_0^2/12 = \beta$ in the above charge–voltage relations and ignoring the high-order nonlinear term, one can approximate (3.13) by the so-called Boussinesq equation,

$$\frac{\partial^2 Q}{\partial t^2} - v_0^2 \frac{\partial^2 Q}{\partial x^2} = \alpha \frac{\partial^2 Q^2}{\partial x^2} + \beta \frac{\partial^4 Q}{\partial x^4} .$$

However, unlike the voltage V, which can be easily measured in an experiment, it is difficult to measure the electrical charge Q.

3.4 Laboratory experiments on pulse solitons

To allow the reader to become familiar with the essential properties of solitons, we now describe simple experiments on an electrical network or nonlinear transmission line of the type described in Sect.2.4, where dissipation can be neglected for short-distance propagation. Thus, in the continuum and low-amplitude-approximation (3.9) can be reduced to (3.13), or to (3.21), which describes the unidirectional propagation of dispersive nonlinear waves.

3.4.1 Experimental arrangement

This electrical network is constructed with available commercial components, it contains 144 unit Sects. each with a linear inductor L = 300μH and a nonlinear capacitor C which consists of a reverse-biased diode, as shown in Fig.3.6. Note in

46

passing that the transmission lines described in the literature are constructed with a large number of sections: from 50 to 1000. The components are carefully selected and adjusted for minimal ohmic losses and reflection effects caused by inhomogeneities. The diodes are biased by a d.c. voltage which is kept constant at $V_0 = 1$ Volt for any series of experiments. In practice, the d.c. voltage source biases each diode as shown in the insert of Fig.3.6. One can also simply connect the d.c. source at the input of the line (point P in Fig.3. 6).

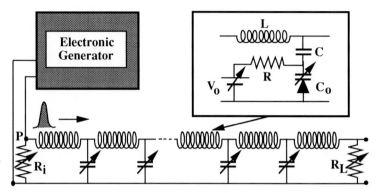

Fig.3.6. Schematic representation of an experimental arrangement. The detailed representation (insert) of a unit section shows that each diode with nonlinear capacitance $C(V)$ is biased through a large resistance R. The linear capacitance $C \gg C(V)$ is used to block the DC bias current corresponding to the bias voltage V_0.

As discussed in previous Sects., if the voltage V is small enough, the capacitance–voltage relation can be approximatively fitted by a relation which is similar to (3.2a):

$$C(V) = C_{o1} (1 + a_1 V + ...).\qquad(3.28)$$

In fact, the coefficients $C_{o1} = C_{o1}(V_o)$ and $a_1 = a_1(V_o)$ depend on the DC voltage V_0 and can be determined from the capacitance–voltage characteristics represented in Fig.3.7 where the straight line corresponds to the above relation. The electrical charge is

$$Q(t) = \int_0^{V_o} C(V)dV + \int_{V_o}^{V_o+V} C(V)dV.\qquad(3.29)$$

Inserting (3.28) into (3.29) and integrating one gets (3.10) if

$$C_o = C_{o1}(1 + a_1 V_o), \qquad b = \frac{a_1}{2(1+a_1 V_o)} .\qquad(3.30)$$

At the d.c. bias $V_0 = 1$ V chosen for experiments, one finds the above quantities, which characterize the nonlinear capacitor and the charge-voltage relations, to be: $C_{o1} = 598.67$ pF, $a_1 = -0.237$ V^{-1}, $C_o = 456.8$ pF, $b = 0.155$ V^{-1}.

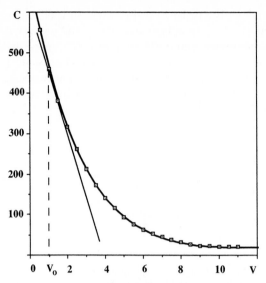

Fig.3.7. Typical capacitance–voltage characteristics for a reverse-biased diode (BB 112 -Philips).

These parameter values hold for signals with amplitudes such as $0 \le V \le 1V$, (see Fig.3.7). In a typical experiment, an initial voltage pulse, with a given profile, is launched at one end of the transmission line, which is terminated by a resistor R_L. This resistor can be adjusted to minimize the reflection effect at the output terminal. The voltage waveforms are examined at various points of the line using an analog or digital oscilloscope. In order to minimize the parasitic reflections introduced by the probes of the oscilloscope during the measurement, these probes must have a high impedance and a low capacitance compared to C_0.

3.4.2 Series of experiments

We first measured the dispersive properties of the line by launching at one end of the line a small amplitude ($0.05 V_{pp}$) sine wave, fed from a variable frequency generator. The experimental dispersion curve is shown in Fig.3.8 and was obtained by measuring the wavelength of the wave with a movable probe.

For small signals (linear approximation) and for very small frequencies, the line is nondispersive and the corresponding long-wavelength phase velocity is given by $c_0 = 1/\sqrt{LC_0}$. For large signals due to the effect of nonlinearity the phase velocity $1/\sqrt{LC(V)}$ depends on the voltage as discussed in Sect.3.1. Consequently, if we launch a large amplitude ($1.80 V_{pp}$) sine wave of frequency F = 15 KHz, we observe a steepening of the leading edge as represented in Fig.3.9.

Having checked separately the dispersive and nonlinear properties of the line, let us consider experiments which illustrate the balance between dispersion and nonlinearity, and show several fundamental properties of the solitons.

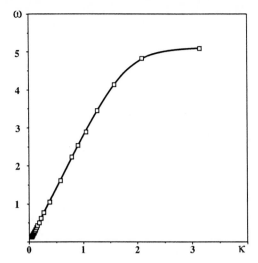

Fig.3.8. Experimentally measured dispersion curve for a d.c. bias : $V_0 = 1V$. Units: ω in 10^{-6} rad s^{-1} and κ in rad $cell^{-1}$.

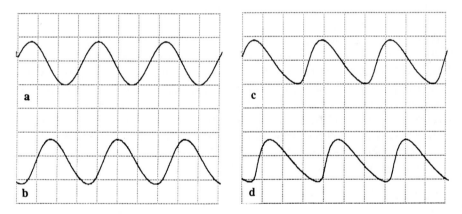

Fig.3.9. Oscillogram showing an initial large-amplitude sine-wave signal at point $n = 1$, and its progresssive steepening at points $n = 48$, $n = 94$, and $n = 132$. Scales : $1Vcm^{-1}$, $20\mu scm^{-1}$.

Decomposition of an initial pulse into solitons and oscillations

If the initial voltage pulse has a rectangular shape we observe that it breaks into a finite train of solitons and a low amplitude (linear) oscillatory tail, as represented in Fig.3.10. One can check that the number of solitons depends on the width and height of the rectangular pulse. These results agree with the empirical results of Scott Russell (see Chap.1) for surface water waves and further theoretical predictions (see Sect.10.5) for the KdV model. When they propagate, the solitons separate from each other, each traveling with its own velocity and a soliton of large amplitude traveling faster than a soliton with low amplitude.

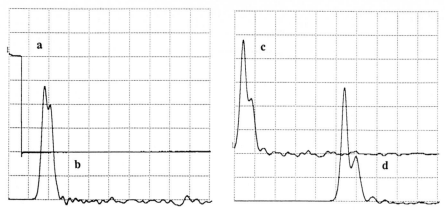

Fig.3.10. Oscillogram showing the decomposition of an input rectangular pulse (**a**) into a finite number of solitons and a low-amplitude oscillatory tail at points n = 23 (**b**), n = 71 (**c**), and n = 142 (**d**). Scales: 0.2Vcm^{-1}, 5μscm^{-1}.

The above experiments can be improved (Kofane et al 1988) by using a function generator (now commercially available), with which the form, amplitude, and width of the input signal can be carefully programmed and adjusted close to the calculated soliton solution, that is, a voltage pulse with a sech-squared initial profile. In this case the number of solitons depends on the amplitude of the initial pulse, and a remarkable theory known as the inverse scattering transform (see Sect.10.5) predicts that one can choose the pulse amplitude in order to get exactly one soliton. In practice one can adjust the initial amplitude in order to get just one soliton and a small oscillatory tail, which can be minimized but not suppressed, as observed in Fig.3.11. This linear tail disperses when progagating along the transmission line and the shape of the KdV soliton, carefully measured with a numerical oscilloscope, agrees very well with the theoretical shape given by (3.22), as compared in Fig.3.12.

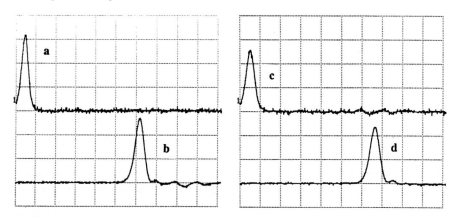

Fig.3.11. Oscillogram showing the evolution of an initial sech-squared pulse (**a**) at point n =1into a KdV soliton observed at points n = 63 (**b**), n = 87 (**c**) and n = 140 (**d**). Scales : 0.1Vcm^{-1}, 5μscm^{-1}.

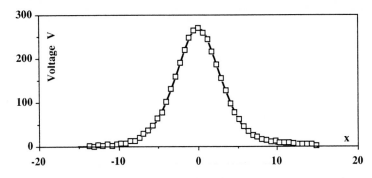

Fig.3.12. Comparison between the theoretical soliton shape (<u>continuous line</u>) and the experimental soliton shape (<u>points</u>) measured as a function of time t at position n = 63, and then expressed as a function of x by using the velocity v = 2.86 cellμs^{-1}.

Table 3.13. Characteristic parameters of KdV solitons measured at cell n = 63.

V_m mV	182	264	368	467	598	691	848	1008
L_s μs	2.7	2.4	2	1.8	1.6	1.45	1.3	1.2
v cellμs^{-1}	2.82	2.86	2.87	2.88	2.90	2.94	2.97	3
L_s cell	7.6	6.85	5.74	5.18	4.64	4.26	3.86	3.6
$V_mL_s^2$	1327	1520	1472	1513	1532	1453	1433	1451

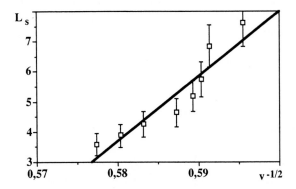

Fig.3.14. Width L_s (cell) of the soliton as a function of the square root of the velocity v (cellμs^{-1})$^{-1/2}$ dots: experiments; continuous line: theory.

By adjusting the initial (sech-squared) voltage pulse one can successively create a soliton with different amplitude V_m, velocity v, and width L_s, as illustrated in Table 3.13. Once the velocity is determined one can express the width L_s in cell units. From these results, as shown in Figs. 3.14 and 3.15, we see that the soliton amplitude increases linearly with its velocity, and that its width is inversely proportional to the square root of its velocity as predicted by theory in Sect.3.4 .

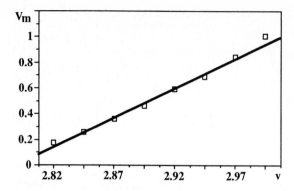

Fig.3.15. Amplitude V_m (Volts) of the soliton as a function of its velocity v (cellμs^{-1}). Dots : experiments; continuous line: theory.

Collision of two solitons traveling in the same direction

As mentioned in Chap.1, Zabusky and Kruskal discovered numerically that two solitons which are solutions of the KdV equation conserve their identities after interacting nonlinearly with each other. This remarkable property can be observed experimentally by using the present transmission line. Two sech-squared pulses of different amplitudes with the larger following the smaller are launched at the input of the line. As shown in Fig.3.16, after some time the larger soliton, which propagates faster, catches the smaller one and the collision occurs. Thereafter, the two solitons separate having kept their identities.

Although we are working with a real transmission line, that is, a system with dissipation and imperfections, all these experiments nicely confirm the remarkable soliton properties predicted by the KdV model.

3.5 Experiments with a pocket version of the electrical network

One can construct a pocket version of the previous electrical network, shown in Fig.3.17. The reader who chooses to make such a transportable system will find the experiment rewarding. The device includes a pulse generator and an electronic commutator. The role of this analog commutator, which can be switched or not, is the following. When it is "on" it allows one to simulate periodic boundary conditions by artificially increasing the length of the line: the pulse at the output terminal is launched again at the input terminal and so forth. As a result, the maximum duration of the propagation of a pulse along the line is no longer limited by the number of sections. When the commutator is "off", one has a line matched (as in Fig.3.6) with a resistance R_L. Details of the electronic circuits are given in Appendix 3D.

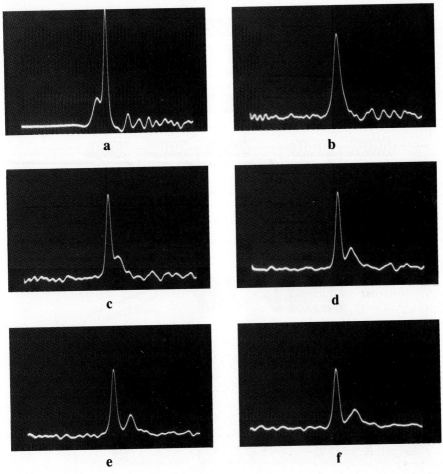

Fig.3.16. Observation of the collision of two solitons moving in the same direction (to the left); their amplitudes are: $V_{m1} = 2V$ and $V_{m2} = 0.4V$. The successive oscillogram traces (a–f) correspond to different positions along the line.

Experiments on pulse solitons are illustrated in Figs 3.18. In Fig.3.18a the evolution of an initial pulse into a soliton and its dispersive tail is represented when the commutator is "off". In Fig. 3.18b the evolution of the same initial pulse is represented when the commutator is "on". In this case, if we neglect the electronic switching effects, which slightly perburb the soliton propagation, we see that the dissipative effects result in a reduction of the soliton amplitude and an increase of the soliton width. They cannot be neglected when the soliton propagates over long distances.

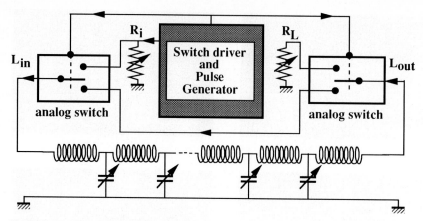

Fig.3.17a. Schematic representation of the pocket version of the line.

Fig.3.17b. Pocket version (dimensions: 14cm x8cm) of the electrical line which includes the generator and the commutator in a separate circuit (dimensions: 4cm x 8cm). The d.c. voltage (9V) is supplied by a battery (Photo R. Belleville).

54

Fig.3.17c. Pocket version of the line connected to an oscilloscope by its probes (Photo R. Belleville).

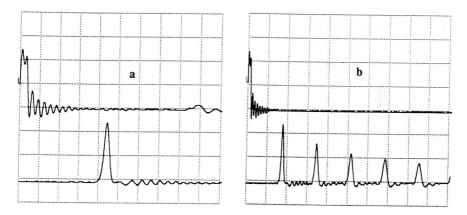

Fig.3.18. (**a**) Evolution of an initial pulse (<u>upper trace</u>) at n = 1 into a soliton and a low amplitude oscillatory tail (<u>lower trace</u>) at point n = 48 when the commutator is "off". Scales: $0.2\,\text{Vcm}^{-1}$, $4\,\mu\text{scm}^{-1}$.
(**b**) The same experiment with the commutator "on". The soliton is depicted at positions n = 1 + 48, 1 + 96, ...

Remark

The effects of dissipation can be taken into account by considering a resistance in parallel to L and a resistance in parallel to C in the circuit of Fig.3.4. Under these conditions, the wave propagation is modeled by a modified KdV equation where the weak dissipative terms can be treated as perturbations (Nagashima and Amigishi 1979).

3.6 Nonlinear transmission lines in the microwave range

The nonlinear transmission line (NLTL for short) is likely to become an important circuit technology for microwaves. In this context, research in ultrafast electronics and optoelectronics explores the limits of very high frequency technologies (Jäger 1985, Jäger et al. 1991). The objectives include the generation and detection of transient electrical signals such as picosecond and subpicosecond pulses and periodic waves with frequencies from 100 Ghz to several terahertz. Nonlinear electrical wave propagation devices play an important role in broadband circuits, instrumentation and the electronic generation of optical signals. In particular, microstrip lines or coplanar waveguides on semiconducting materials can exhibit various interesting wave propagation phenomena (for a review see Rodwell et al. 1994).

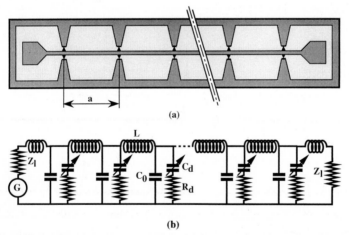

Fig. 3.19. (a) Schematic representation of a planar transmission line periodically loaded by Schottky diodes and (b), its equivalent electrical circuit.

Distributed devices consisting of a planar transmission line periodically loaded by Schottky diodes as depicted in Fig.3.19, have been constructed. Under reverse bias, the Schottky diode is a nonlinear element with a capacitance–voltage relation (compare to (3.12)) of the form $C_d(V) = C_0\sqrt{1 - V/V_0}$ where $C_d(V)$ is the depletion capacitance which depends on the voltage V, and C_0 and V_0 are constants depending on the semiconductor material. The diode has a cutoff frequency $\omega_c = 1/$

56

R_dC_d in the infrared, where R_d is a series parasitic resistance. Due to the spatial distribution of the diodes (Fig.3.19) the transmission line presents a periodic or Bragg cutoff frequency $\omega_b = 2/\sqrt{L(C+C_d)}$, where L and C are the inductance and capacitance per unit cell of the line.

When the frequency ω approaches ω_b the curvature of the dispersion curve increases, that is, the dispersion effect cannot be ignored. However, if ω_b is made very large by reducing both the diode spacings and diode capacitance $C_d(V)$ and, if $R_d \approx 0$ and $\omega_c < \omega_b$, dispersion can be neglected, shock waves (Landauer 1960) will form (see Sect.3.1) and propagate along the line. For example waveforms in the femtosecond regime (480fs, 3.5V transients) have been generated (Van der Weide 1994) by using an integrated NLTL with delta-doped (very localized) Schottky diodes. The application of this particular NLTL is shown in the picture and its accompanying diagram represented in Fig. 3.20. Here, the diode sampling bridge is strobed by one NLTL as it measures the output of another NLTL.

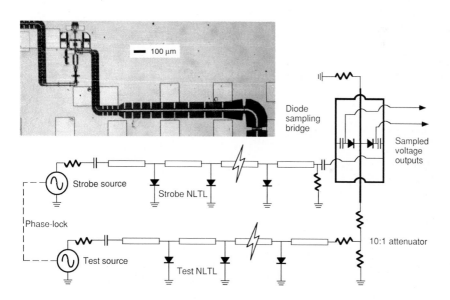

Fig.3.20. Delta-doped Schottky diode nonlinear transmission lines for 480fs, 3.5V transients (shock wave fronts) (Reproduced by kind permission of D. W. Van der Weide)

Now, if $\omega_c >> \omega_b$, wave motion is dominated by the periodic-network dispersion and, soliton propagation can result. In this case the splitting of input pulses (see Sect.3.4) into a finite number of solitons can be used as a method (Carman et al 1992) of second- and third-harmonic generation.

As we have seen above (see Fig.3.20), instead of constructing devices which have a periodic distribution of the diodes one can construct devices in a fully distributed form, e.g., with the integration applied to the semiconductor devices (Schottky elements) themselves. For example, a Schottky contact coplanar transmission line (Dragoman et al. 1993, Rodwell et al. 1994) where the center conductor forms a metal–semiconductor junction can be constructed. In order to implement the required dispersion properties of this nonlinear transmission line, periodic ion implantation or periodic variation of the center conductor can be performed.

Appendix 3A. Calculation of the effect of nonlinearity on wave propagation

Let us first show that (3.3) and (3.4) are correct solutions. For a wave propagating along the positive x direction we define the wave variable s as $s = x - [\ell c(v)]^{-1/2}t$. Note that s is the spatial coordinate moving with velocity $[\ell c(v)]^{-1/2}$, which is itself a function of x and t. Then, from (3.3) we calculate

$$\frac{\partial v}{\partial x} = \frac{\partial v}{\partial s}\frac{\partial s}{\partial x} = \frac{\partial v}{\partial s}\left(1 - \frac{\partial\,[\ell c(v)]^{-1/2}}{\partial x}t\right). \qquad (3A.1)$$

The derivative $\partial\,[\ell c(v)]^{-1/2}/\partial x$ can be calculated from (3.4). One gets

$$\frac{\partial[\ell c(v)]^{-1/2}}{\partial x} \approx \frac{1}{\sqrt{\ell C_o}}\,b\frac{\partial v}{\partial x}. \qquad (3A.2)$$

Solving (3A.1) for $\partial v/\partial x$ yields

$$\frac{\partial v}{\partial x} = \frac{\partial v/\partial s}{1 + b[\ell C_o]^{-1/2}t\,(\partial v/\partial s)}. \qquad (3A.3)$$

Using the same procedure we calculate

$$\frac{\partial v}{\partial t} = -\frac{1}{\sqrt{\ell C_o}}(1 + bv)\frac{\partial v/\partial s}{1 + b[\ell C_o]^{-1/2}t\,(\partial v/\partial s)}. \qquad (3A.4)$$

From these equations, we obtain

$$\frac{\partial v}{\partial t} = -\frac{1}{\sqrt{\ell C_o}}(1 + bv)\frac{\partial v}{\partial x}. \qquad (3A.5)$$

Inserting (3A.5) into (3.1) and using (3A.3), we get

$$\frac{\partial v}{\partial t} = \frac{1}{\sqrt{\ell C_o}}(1 + bv)\ell\frac{\partial i}{\partial t},$$

$$\frac{\partial i}{\partial x} = \frac{c(v)}{\sqrt{\ell C_0}} (1+ bv) \frac{\partial v}{\partial x} .$$

(3A.6)

Then, using (3.2) and (3.4) we see that it is a simple matter of fact to verify that the current is given by (3.5).

Equation (3A.3) expresses the slope of the voltage profile at a point (x, t) in terms of the slope at s. Here, x is the position at time t of the point which was initially at s. Setting $\beta = b (\ell C_0)^{-1/2}$, we then note that $\partial v/\partial x$ is infinite at time $t = -1/\beta(\partial v/\partial s)$ if the quantity $\partial v/\partial s$ is negative. Therefore, if the initial waveform has a negative slope such that $(\partial v/\partial s) = - |\partial v/\partial s|$ at some point s, then for $t > \tau = (1/\beta|\partial v/\partial s|)_{min}$ the solution is no longer single-valued in the neighborhood of the point $x_0 = s_0 + \tau f(s_0)$. Here, s_0 is the point where τ reaches its minimum value. To find out the changes of an initial pulse profile we examine (3A.3) at time

$$t = \tau + \delta = (1/\beta|\partial v/\partial s|)_{min} + \delta,$$

where $\delta <<1$. Then, from (3A.3) we can write

$$\frac{\partial v}{\partial x} = \frac{\partial v/\partial s}{1 + \beta t (\partial v/\partial s)}, \quad \text{at } s = s_0, t = \tau + \delta.$$

(3A.7)

Thus we have

$$\frac{\partial v}{\partial x} = \frac{-|\partial v/\partial s|_{min}}{1 - \beta(\tau+\delta)(|\partial v/\partial s|_{min})} = \frac{1}{\delta\beta} .$$

(3A.8)

This result implies that

$$\frac{\partial v(\tau \pm 0)}{\partial x} \rightarrow \pm \infty .$$

(3A.9)

After some time $t > \tau$, the slope of the wavefront tends to infinity and a physically meaningful solution is the one which contains a propagating jump discontinuity. For the initial pulse we assume the following parabolic shape:

$$v(x, 0) = \begin{cases} V_m (1 - (x/x_0)^2) , & \text{for } |x| \leq 1 \\ 0 , & \text{for } |x| > 1 . \end{cases}$$

(3A.10)

In view of (3A.10) the solution (3.3) becomes

$$V_m [1 - (s/x_0)^2] = V_m\{ 1 - [\frac{x}{x_0} - \frac{(1 + bv)t}{x_0\sqrt{\ell C_0}}]^2 \}, \quad \text{for } |x| \le 1 \qquad (3A.11)$$

$$v(x,t) =$$

$$0 \qquad\qquad\qquad\qquad , \quad \text{for } |x| > 1 .$$

Introducing the dimensionless quantities

$$v' = v/V_m \qquad x' = x/x_0, \qquad t' = t/(x_0\sqrt{\ell C_0}), \qquad b' = bV_m. \qquad (3A.12)$$

and solving explicitly for v, we find that

$$v'(x', t') = \frac{1}{2b't'^2} [(2x't' - \frac{1}{b'}) \pm \sqrt{\frac{1}{b'^2} - \frac{4x't'}{b'} + 4t'^2(1 + \frac{1}{b'})}] - \frac{1}{b'} . \qquad (3A.13)$$

Appendix 3B. Derivation of the solitary-wave solution

To find a solution of (3.13) with a permanent profile , we assume that

$$V(x, t) = V(x - vt) = V(s), \qquad\qquad (3B.1)$$

where s = x - vt represents the position in a coordinate system moving at a constant velocity v for which the wave appears stationary. Accordingly,

$$\partial/\partial x = d/ds, \qquad \partial/\partial t = -v\, d/ds, \qquad\qquad (3B.2)$$

and eq (3.13) becomes the ordinary differential equation

$$(v^2 - v_0^2) V_{ss} = bv^2 (V^2)_{ss} + \delta^2\frac{v_0^2}{12} V_{ssss}. \qquad\qquad (3B.3)$$

Here, we have used the convention of writing a derivative $\partial V/\partial s$ as V_s, $\partial^2 V/\partial s^2$ as V_{ss}, and so forth. Two integrations with respect to s yield

$$(v^2 - v_0^2) V = bv^2 V^2 + \delta^2\frac{v_0^2}{12} V_{ss}, \qquad\qquad (3B.4)$$

where the integration constants have been set to zero because we are looking for a localized solution, and its derivatives tend to zero when $s \to \pm \infty$. After multipliying both sides by 2dV/ds, one can effect another integration to obtain

$$\frac{2bv^2}{3} V^3 - (v^2 - v_0^2) V^2 + \delta^2\frac{v_0^2}{12} (\frac{dV}{ds})^2 = 0, \qquad\qquad (3B.5)$$

60

where the integration constant has been set to zero for the same reasons as above. By setting

$$U = \frac{6(v^2 - v_0^2)}{\delta^2 v_0^2}, \qquad A = \frac{4bv^2}{\delta^2 v_0^2}, \qquad (3B.6)$$

we transform (3B.5) into

$$AV^3 - UV^2 + \frac{1}{2}(\frac{dV}{ds})^2 = 0. \qquad (3B.7)$$

Then, the general traveling wave solution can be written in the form

$$\frac{1}{\sqrt{2U}} \int_{V_0}^{V} \frac{dV}{\sqrt{V^2 - (AV^3/U)}} = \int_{s_0}^{s} ds, \qquad (3B.8)$$

and the integration can be performed in terms of elementary transcendental functions. As a matter of fact, we set

$$W^2 = AV/U \qquad (3B.9)$$

and substitute in (3B.8). We find that

$$\int_{s_0}^{s} ds = \frac{2}{\sqrt{2U}} \int_{W_0}^{W} \frac{dW}{W\sqrt{1 - W^2}} = \frac{2}{\sqrt{2U}} [\text{arc sech}W - \text{arcsech}W_0] \qquad (3B.10)$$

which gives

$$W = \text{sech} [\sqrt{\frac{U}{2}} (s - s_0)]. \qquad (3B.11)$$

if, taking $W_0 = 1$, we disregard arcsechW_0. We then assume that $s_0 = x_0 - vt_0$, which corresponds to the wave at position x_0 and time t_0, is equal to zero. Finally, using (3B.9) and (3B.6) we obtain the solitary wave solution

$$V = \frac{3(v^2 - v_0^2)}{2bv^2} \text{sech}^2 [\frac{\sqrt{3(v^2 - v_0^2)}}{v_0} \frac{(x - vt)}{\delta}]. \qquad (3B.12)$$

Appendix 3C. Derivation of the KdV equation and its soliton solution

Expressing dV in terms of the variables (n,t) and the new variables (s, T) defined by (3.18), we have

$$dV = \frac{\partial V}{\partial n} dn + \frac{\partial V}{\partial t} dt = \frac{\partial V}{\partial s} \varepsilon^{1/2}(dn - c_0 dt) + \frac{\partial V}{\partial T} \varepsilon^{3/2} dt, \tag{3C.1a}$$

and identifying the coefficients of dt and dn we get

$$\frac{\partial}{\partial n} = \varepsilon^{1/2} \frac{\partial}{\partial s}, \qquad \frac{\partial}{\partial t} = -\varepsilon^{1/2} c_0 \frac{\partial}{\partial s} + \varepsilon^{3/2} \frac{\partial}{\partial T}, \tag{3C.1b}$$

Substituting these differential operators and the voltage expansion (3.20) in (3.13) yields

$$(\varepsilon c_0^2 \frac{\partial^2}{\partial s^2} - 2\varepsilon^2 c_0 \frac{\partial^2}{\partial s \partial T} + \varepsilon^3 \frac{\partial^2}{\partial T^2}) \varepsilon V_1 - c_0^2 \varepsilon^2 \frac{\partial^2 V_1}{\partial s^2} = \frac{c_0^2}{12} \varepsilon^3 \frac{\partial^4 V_1}{\partial s^4}$$

$$+ b(\varepsilon c_0^2 \frac{\partial^2}{\partial s^2} - 2\varepsilon^2 c_0 \frac{\partial^2}{\partial s \partial T} + \varepsilon^3 \frac{\partial^2}{\partial T^2}) \varepsilon^2 V_1^2.$$

Keeping terms to lowest order $O(\varepsilon^3)$, we find that

$$c_0 \frac{\partial^2 V}{\partial s \partial T} + b \frac{c_0^2}{2} \frac{\partial^2 V^2}{\partial s^2} + \frac{c_0^2}{24} \frac{\partial^4 V}{\partial s^4} = 0, \tag{3C.2}$$

where we have set $V = V_1$ to simplify matter. Integrating with respect to s and rearranging, we find the KdV equation:

$$\frac{\partial V}{\partial T} + b c_0 V \frac{\partial V}{\partial s} + \frac{c_0}{24} \frac{\partial^3 V}{\partial s^3} = 0. \tag{3C.3}$$

To look for a soliton solution of (3C.3) we first rewrite it in a scaled form

$$V_\tau + a V V_s + V_{sss} = 0, \tag{3C.4}$$

by setting

$$\tau = (c_0/24) T, \qquad a = 24b. \tag{3C.5}$$

Now, we proceed as in Appendix 3A and look for a solution with a permanent profile,

$$V(s, \tau) = V(\xi) = V(s - u\tau),$$ (3C.6)

where $\xi = s - u\tau$, represents the position in a coordinate system moving at a velocity u for which the wave appears stationary. Accordingly,

$$\partial/\partial s = d/d\xi, \qquad \partial/\partial\tau = -u \, (d/d\xi),$$ (3C.7)

and (3C.4) becomes the ordinary differential equation

$$(aV - u)\frac{dV}{d\xi} + \frac{d^3V}{d\xi^3} = 0.$$ (3C.8)

A first integration of eq (3C.8) gives

$$\frac{aV^2}{2} - uV + \frac{d^2V}{d\xi^2} = K_1,$$ (3C.9)

where K_1 is a constant. After multiplying both sides by $dV/d\xi$, one can perform a second integration which yields

$$\frac{aV^3}{6} - \frac{uV^2}{2} + \frac{1}{2}(\frac{dV}{d\xi})^2 = K_1V + K_2.$$ (3C.10)

If, however, one looks for a localized solution, its first and second derivatives must vanish as $\xi \to \pm\infty$. With the above relations this condition implies that $K_1 = 0$ and $K_2 = 0$. Setting

$$A = a/6, \qquad U = u/2,$$ (3C.11)

we obtain an equation similar to (3B.7). Consequently, we can proceed exactly as in Appendix 3B and obtain a solution in the form

$$V = \frac{3u}{a} \, \text{sech}^2 \, [\frac{\sqrt{u}}{2}(s - u\tau)].$$ (3C.12)

In terms of the original laboratory coordinates (x,t) the solution (3C.12) becomes

$$V = \frac{\varepsilon u}{8b} \, \text{sech}^2 \, \{ \frac{\sqrt{\varepsilon u}}{2} \, [n - c_0(1 + \frac{\varepsilon u}{24})t] \} = \frac{\varepsilon u}{8b} \, \text{sech}^2 \, \{ \frac{\sqrt{\varepsilon u}}{2} \, [\frac{x}{\delta} - \frac{v_0}{\delta}(1 + \frac{\varepsilon u}{24})t] \}.$$

(3C.13)

Appendix 3D. Details of the electronics: switch driver and pulse generator

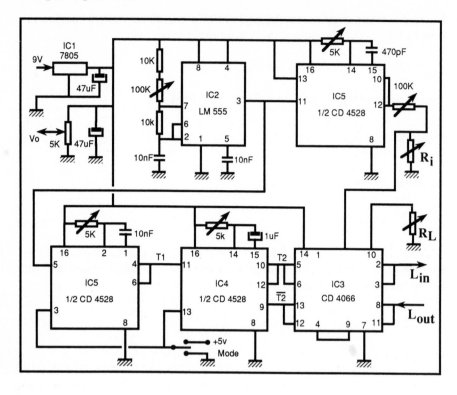

4 More on Transmission-Line Solitons

The study of solitons on discrete lattices dates back to the early days of soliton theory (Frenkel and Kontorova 1939, Fermi et al. 1955) and is of great physical importance. Generally, the discrete nonlinear equations which model these lattices cannot be solved analytically. Consequently, one looks for possible pulse-soliton solutions in the continuum or long wavelength approximation, that is, solitons with a width much larger than the electrical length of a unit section of the electrical network, as described in Chap.3. When this approach is not workable, one has to use numerical approaches (Zabusky 1973, Eilbeck 1991) or simulations. Nevertheless, there exist some lattice models for which the governing equations can be solved exactly. This is the case for the lattice with exponential interactions, introduced in 1967 by Toda (see also Chap.9).

In this Chapter, we first consider *lattice solitons* that can propagate on the remarkable electrical network (Hirota and Susuki 1973) corresponding to the *Toda lattice*. We examine the nonlinear periodic waves that can propagate on transmission lines. Finally, we consider the case of modulated waves or nonlinear wave packets, such as *envelope and hole solitons*, which present soliton properties.

4.1 Lattice solitons in the electrical Toda network

We consider the electrical network represented in Fig.4.1, which is modeled by (3.9) which we rewrite here as

$$\frac{d^2Q_n}{dt^2} = \frac{1}{L}(V_{n+1} + V_{n-1} -2V_n), \qquad n = 1, 2, \dots . \tag{4.1}$$

Then, we assume that over a given voltage range (V_o, V_o+V_n) the differential capacitance $C(V_n)$ can be approximated (Fig.4.2) by

$$C(V_n) = \frac{Q_o}{F_o-V_o+V_n} , \tag{4.2}$$

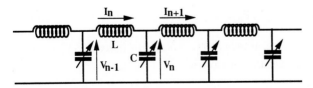

Fig. 4.1. Electrical Toda lattice with linear inductance L and a nonlinear capacitance such a $C(V_n) = Q_o/(F_o-V_o+V_n)$.

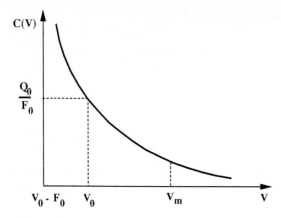

Fig.4.2. Sketch of the nonlinear capacitance $C(V)$ which may be approximated by (4.2) for $V_0 < V < V_m$.

where $Q_0 = Q(V_0)$ and $F_0 = F(V_0)$ depend on the d.c. voltage V_0; for $V_n = V_0$ we have $C(V_0) = Q_0/F_0 = C_0$. The charge $Q_n(t)$ can be divided into two parts:

$$Q_n(t) = \int_0^{V_0} C(V_n)dV_n + \int_{V_0}^{V_0+V_n} C(V_n)dV_n = q_0 + Q_0 \ln[1 + \frac{V_n(t)}{F_0}]. \qquad (4.3)$$

Since q_0 is constant from eqs(4.1) and (4.3) we get the relation

$$LQ_0\frac{d^2}{dt^2} \ln\left(1 + \frac{V_n}{F_0}\right) = (V_{n+1} + V_{n-1} - 2V_n), \qquad n = 1, 2, \dots . \qquad (4.4)$$

This relation yields a Toda lattice equation, as we shall now see. Inverting (4.3) and substituting in (4.1), we get

$$L\frac{d^2Q_n}{dt^2} = F_0 [\exp(\frac{Q_{n+1}-q_0}{Q_0}) + \exp(\frac{Q_{n-1}-q_0}{Q_0}) - 2\exp(\frac{Q_n-q_0}{Q_0})]. \qquad (4.5)$$

Let us make the electromechanical transformation

$$(Q_n - q_0)/Q_0 = -Br_n, \qquad LQ_0/F_0 = LC_0 = m/AB, \qquad (4.6)$$

where r_n is the relative displacement between two points of mass m, and A and B are constants. We then get the equation of motion

$$m\frac{d^2r_n}{dt^2} = A[2\exp(-Br_n) - \exp(-Br_{n-1}) - \exp(-Br_{n+1})]. \qquad (4.7)$$

This equation, which corresponds to a lattice with exponential interactions (see Sect.9.3) was first introduced by M. Toda. It is one of the most interesting models of anharmonic lattices in one dimension (Toda 1967,1978). The Toda-lattice equations (4.4), (4.7) admit exact lattice soliton solutions and rigorous periodic solutions. Here, we shall introduce qualitatively the soliton solution; for details the reader should consult references mentioned in this Chap. and Chap.9.

4.1.1 Lattice solitons

To examine the lattice-soliton solution, we first consider the continuum limit of the Toda lattice. For this purpose we rewrite (4.4) as

$$LQ_o \frac{d^2}{dt^2} \text{Ln} [1 + \frac{V_n}{F_o}] = [\exp(\frac{\partial}{\partial n}) + \exp(\frac{-\partial}{\partial n}) - 2] V_n, \quad n = 1, 2, ... , \qquad (4.8)$$

where

$$V_{n\pm1} = \exp(\pm\frac{\partial}{\partial n}) V_n = 1 \pm \delta\frac{\partial V_n}{\partial x} + \frac{\delta^2}{2}\frac{\partial^2 V_n}{\partial x^2} \pm \frac{\delta^3}{3!}\frac{\partial^3 V_n}{\partial x^3} + \frac{\delta^4}{4!}\frac{\partial^4 V_n}{\partial x^4} +... . \qquad (4.9)$$

Here, as in previous Chaps. in the continuum limit we assume that $V_n(t) = V(x,t)$, with $x=n\delta$, and neglect the higher derivatives with respect to n in expansion (4.9). Assuming that $V_n \ll F_o$, we expand the logarithmic term on the left-hand side of (4.4) and neglect higher powers of V:

$$\text{ln}[1 + V/F_o] \approx V/F_o - V^2/2F_o^2.$$

With these approximations, using: $LQ_o/F_o = LC_o$ and setting

$$\delta/\sqrt{LC_o} = v_o, \quad 1/F_o = 2b, \qquad (4.10)$$

we can reduce (4.4) to

$$\frac{\partial^2}{\partial t^2} (V - bV^2) = \delta^2 \frac{v_o^2}{12}\frac{\partial^4 V}{\partial x^4} + v_o^2\frac{\partial^2 V}{\partial x^2} ,$$

which is similar to (3.13) . Then, proceeding as in Chap.3, we can reduce this equation to the KdV equation (3.21). In laboratory coordinates it has the soliton solution (3.22), which we rewrite here for convenience as:

$$V(t) = \frac{\varepsilon u}{8b} \text{sech}^2 \{ \frac{\sqrt{\varepsilon u}}{2} [n - (1+\frac{\varepsilon u}{24})c_o t] \},$$

where $c_o = v_o/\delta$, as in (3.19). This result, which corresponds to the continuum (long- wavelength) approximation for the lattice, suggests a discrete solution with a sech-squared profile. Thus, we look for a voltage-pulse soliton of the form

$$V_n(t) = F_0\Omega^2 \operatorname{sech}^2 (Pn - \Omega c_0 t), \tag{4.11}$$

where Ω and P are two parameters. Substituting in (4.4) we can check that (4.11) is a solution if

$$\Omega = \sinh P, \qquad c_0 = \sqrt{F_0/LQ_0} = 1/\sqrt{LC_0}. \tag{4.12}$$

This *lattice soliton* has a pulse profile; it has an amplitude $V_m = F_0\Omega^2$ and travels with velocity

$$u_s = \Omega c_0/P = \sinh P/(P\sqrt{LC_0}). \tag{4.13}$$

Its width at half amplitude is proportional to 1/P, which can be calculated by proceeding as for the KDV soliton (see Chap.3). In the small-amplitude limit, that is, for P<<1, one can make the following approximations:

$$\sinh P \approx P + P^3/6, \qquad \sinh^2 P \approx P^2.$$

Substituting in (4.11) and setting $P = \sqrt{\varepsilon u}/2$, we recover the KdV soliton solution, which corresponds to the small amplitude (velocity) limit of the lattice soliton.

4.2 Experiments on lattice solitons

The Toda lattice can be realized experimentally if the nonlinear capacitance satisfies (4.2) which we can rewrite in the form

$$1/C(V) = (F_0 - V_0)/Q_0 + (V/Q_0).$$

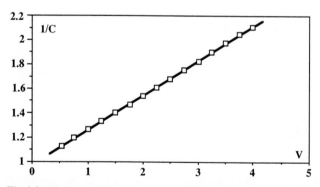

Fig.4.3. Linear variations of $1/C = 1/(C(V) + 330)$ expressed in $(pF)^{-1}$ as a function of the voltage V.

Thus, the electrical circuit represented in Fig.4.1 is equivalent to the Toda lattice if the inverse of the capacitance depends linearly on the voltage. In practice, one has to find an adequate diode which presents such a characteristic; here, we have connected a linear capacitance C_L in parallel with the reverse-biased diode used in Sect.3.5. In this case the variations of $1/(C(V)+ C_L)$ versus the voltage are well fitted by a straight line in the voltage range 0.5V to 4V when $C_L = 330pF$, as depicted in Fig.4.3. Except for this modification, the electrical network and the experimental arrangement are similar to the ones described in Sect.3.5. The remarkable property of solitons can be observed by using the above network.

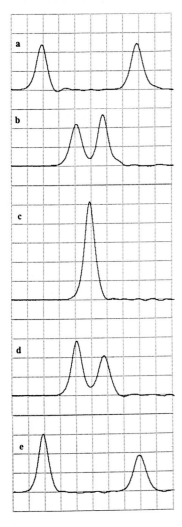

Fig.4.4. Head-on collision of two lattice solitons traveling on an electrical Toda lattice. The successive oscillogram traces correspond to different positions on the network: n=10, n=20, n=24, n=28, and n=38.Scales: 1V/div, 2μs/div.

4.2.1 Collision of two lattice solitons moving in opposite directions

Two narrow pulses of different amplitudes are launched at both ends of the network. These two pulses become two lattice solitons as they travel in opposite directions along the network. The oscillograms represented in Fig.4.4 show the solitons for different positions along the line and their head-on collision at position n = 24. During the overlap the amplitude increases; after the collision both solitons recover their identities. Note that the dissipation effects, which are always present in such a real network, result in a slight decrease of their amplitudes.

4.2.2 The Fermi-Pasta-Ulam recurrence phenomenon

The famous recurrence phenomenon on a one-dimensional lattice discovered in 1955 by Fermi, Pasta, and Ulam (see Chap.1) with computer simulations can be easily observed on the Toda electrical network.

An initial sinusoidal signal of amplitude $V_m = 4V_{pp}$ and frequency F=228.75 kHz is introduced into the input of the line. By observing the waveform at various points along the line, we can examine the influence of nonlinearity and dispersion on the initial signal. These results are shown in Fig.4.5. Fig.4.5a shows the input

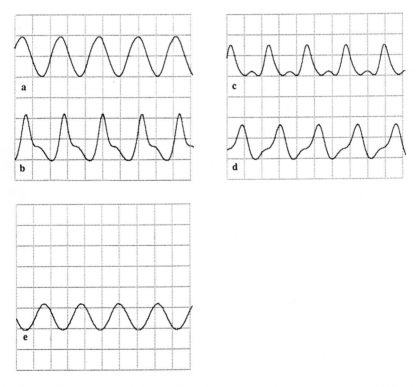

Fig.4.5. Observation of the Fermi–Pasta–Ulam recurrence phenomenon on the Toda electrical network. The oscillograms show the signal at positions: n=1, n=12, n=24, n=36, and n=48. Scales: 2V/div, 5μs/div.

70

sine wave. In Fig.4.5 b and c the signal is progressively decomposed into higher harmonics and two solitons, respectively. In Fig.4.5d the decomposition is reversing itself. Finally, as shown in Fig.4.5e, the original sinusoidal signal is reconstructed. One can observe that the recurrence period N_r (here N_r=47) depends on the amplitude and frequency of the input sine wave. The higher the frequency and amplitude, the shorter is the recurrence period.

4.3 Periodic wavetrains in transmission lines

In this Sect. we examine the nonlinear periodic waves that can propagate on nonlinear transmission lines. For example such waves can propagate in the above Toda lattice. Here, for simplicity we consider only the KdV model to which the above Toda lattice equations and (3.9) (which describes the simple electrical network in Sect.3.2) can be reduced. Under these conditions, we look for periodic solutions of (3.21) with a permanent profile rather than solitary waves. In terms of the laboratory coordinates (x= $n\delta$, t) the cnoidal-wave solution, first found by Korteweg and de Vries (see Chap.1) for surface water waves, is given by

$$V(x,t) = V_2 + (V_3 - V_2)\, cn^2\{\sqrt{2b(V_3-V_1)}\,[\frac{x - v_s t}{\delta}], m\}. \qquad (4.14)$$

Here, V_1, V_2 and V_3 are quantities that depend on the line considered and can be calculated (see Appendix 4A), cn denotes the Jacobi elliptic function of parameter m (see Appendix 4B), and v_s is a velocity as defined in (3.23). In Fig.4.6 the cnoidal wave is sketched as a function of x for t =0; it represents a nonlinear voltage wavetrain. The amplitude of the cnoidal wave is given by (V_3 -V_2)/2, the value V_2 corresponds to a trough and the value V_3 corresponds to a peak. The wavelength λ, or spatial period of the wave, depends on the parameter m and is given by the relation

$$\lambda = \frac{2K(m)}{\sqrt{2b(V_3 - V_1)}}\,\delta, \qquad (4.15)$$

where K(m) is the complete integral of the first kind (see Appendices 4A and 4B).

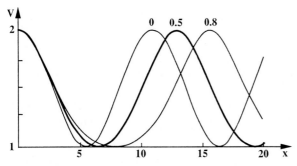

Fig 4.6. Voltage wavetrain represented by a cnoidal wave which oscillates between V_2 =1 and V_3=2, for m =0 (sinusoidal limit), m=0.5, and m=0.8.

4.3.1 The solitary wave limit and sinusoidal limit of the cnoidal wave

It is instructive to examine the two limits of (4.14), which depend on the parameter m. When m = 1, that is, when $V_1 = V_2$, the cnoidal wave degenerates to the soliton solution of the KdV equation,

$$V(x,t) = V_1 + (V_3 - V_1) \text{ sech}^2 [\sqrt{2b(V_3-V_1)} \frac{x - v_s t}{\delta}]. \qquad (4.16)$$

Specifically, we recover (3.22) if we set

$$V_2 = 0, \qquad V_3 = \varepsilon u/8b.$$

In the limit m → 0, we have the linear limit cn (y, m) → cos (y), and the cnoidal-wave solution (4.14) takes the sinusoidal limiting form

$$V(x,t) \approx V_2 + (V_3 - V_2) \cos^2 [\sqrt{2b(V_3-V_1)} \frac{x - v_s t}{\delta}]. \qquad (4.17a)$$

To simplify this we can set $V_2 = 0$ and transform (4.17a) into

$$V(x,t) \approx \frac{V_3}{2} \{ 1 + \cos [2\sqrt{2b(V_3-V_1)} \frac{x - v_s t}{\delta}]\}, \qquad (4.17b)$$

which represents a sinusoidal electrical wave, sketched in Fig.4.6, traveling along the transmission line.

4.4 Modulated waves and the nonlinear dispersion relation

In this Sect. we consider modulated waves or wavepackets which can exist when both dispersion and weak nonlinearity are present. In the analysis of the behavior of such waves the concept of the nonlinear dispersion relation, where the frequency depends both on the wavenumber of the carrier wave and on the voltage (or current amplitude), was first introduced by Stokes (1847) for deep water waves (see Chaps 1 and 5). Such a concept is useful for the physicist; it brings out clearly the origins of the dispersive and nonlinear terms in the slow-evolution equation of the wave envelope compared to that of the rapid carrier wave.

Let us show that one can obtain a nonlinear dispersion relation for a simple transmission line. Specifically, we consider a nonlinear network modeled by (4.1), with a capacitance–voltage relation similar to (3.2a). We obtain

$$Q_n(V_n) = C_0[V_n + (a_1/2)V_n^2 + (a_2/3)V_n^3 + ...].$$

To simplify the calculations, we then assume that $a_1 = 0$ and that the voltage (or current) waveform is sinusoidal with frequency ω and wavenumber κ:

$$V_n = \psi(\varepsilon n, \varepsilon t)\, e^{\,i\,[\omega t - \kappa n]} + \text{cc}.$$

Here, the voltage envelope ψ is assumed to vary slowly in space εn and time εt compared to the variations of the carrier wave; and to a first approximation it can be considered constant in the differentiations with respect to time. As in Sect.2.4 one has $n = x/\delta$ in order to ensure that κ has the correct units. Inserting the above relations into (4.1) and neglecting the third harmonics we get

$$\omega \approx \frac{2}{\sqrt{LC_0}} \sin\frac{\kappa}{2} \left[1 - 3\,\frac{a_2}{2}\, |\psi|^2 \right],$$

or

$$\omega = \omega\,(\kappa, |\psi|^2). \tag{4.18}$$

Nonlinear dispersion relations of the form of (4.18) where the frequency depends on both the wavenumber of the carrier wave and the voltage (or current amplitude) were obtained for different kinds of transmission lines (Mizumura and Noguchi 1975).

Next, proceeding as in Sect.2.5, we can derive (Craik 1985; Infeld and Rowlands 1990) the equation describing the evolution of the slow-envelope wave. We consider expansion (2.41) with an additional nonlinear term:

$$\omega - \omega_0 = \left(\frac{\partial\omega}{\partial\kappa}\right)(\kappa - \kappa_0) + \frac{1}{2}\left(\frac{\partial^2\omega}{\partial\kappa^2}\right)(\kappa - \kappa_0)^2 + \left(\frac{\partial\omega}{\partial|\psi|^2}\right)(|\psi|^2 - |\psi_0|^2), \tag{4.19}$$

where the derivative $(\partial\omega/\partial|\psi|^2)$ is evaluated at ψ_0, which represents a constant amplitude. Thus, when the nonlinear effects are taken into account, the dispersion relation (2.45) of the envelope wave is now modified to

$$\Omega = v'_g\, K + P' K^2 - Q\, |\psi|^2, \tag{4.20}$$

where $\Omega = \omega - \omega_0$ and $K = \kappa - \kappa_0$. Here, the linear group velocity v'_g and the linear dispersion coefficient P' (see Sect.2.5) and the nonlinear coefficient Q evaluated at $\psi = \psi_0 = 0$ are given by

$$v'_g = \frac{\partial\omega}{\partial\kappa}, \text{ at } \kappa = \kappa_0; \quad P' = \frac{1}{2}\frac{\partial^2\omega}{\partial\kappa^2}, \text{ at } \kappa = \kappa_0; \quad Q = -\frac{\partial\omega}{\partial|\psi|^2}, \text{ at } \psi_0 = 0; \tag{4.21}$$

and can be calculated if we know explicitly the dispersion relation. The nonlinear coefficient is assumed to be of the same order: $O(\varepsilon^2)$ (see (2.46)) as the dispersive term P. Thus, proceeding as in Sect.2.5, instead of (2.46) we get

$$i\varepsilon[\frac{\partial\psi}{\partial T} + v_g\frac{\partial\psi}{\partial X}] + \varepsilon^2 P\frac{\partial^2\psi}{\partial X^2} + \varepsilon^2 Q|\psi|^2\psi = 0, \tag{4.22a}$$

where

$$X = \varepsilon x = \varepsilon n\delta, \qquad T = \varepsilon t, \qquad v_g = \delta v'_g, \qquad P = \delta^2 P'.$$

Equation (4.22a) indicates that, to leading order, the wave envelope propagates at the group velocity of the carrier wave. In a frame of reference moving at the group velocity, that is using transformation (2.47a), $\xi = X - v_g T$, $\tau = \varepsilon T$, this equation is cast into the standard form

$$i\frac{\partial\psi}{\partial\tau} + P\frac{\partial^2\psi}{\partial\xi^2} + Q|\psi|^2\psi = 0. \tag{4.22b}$$

Equation (4.22b) (and also (4.22a)) is called the nonlinear Schrödinger (NLS) equation because it is formally similar to the Schrödinger equation of quantum mechanics

$$i\hbar\frac{\partial\psi}{\partial\tau} + \frac{\hbar^2}{2m}\frac{\partial^2\psi}{\partial\xi^2} - U\psi = 0.$$

Here, the potential U is proportional to the absolute square of the wave envelope (or wave function ψ) and $\hbar = h/2\pi$, where h is the Planck constant. It represents the self-trapping of the wave energy or, equivalently the local density $|\psi|^2$ of the voltage (of the quasi-particle) described by the Schrödinger equation. In (4.22) the linear dispersive term and the nonlinear term, which are of the same order, can balance and yield *envelope and hole-soliton* solutions, as we shall see now.

4.5 Envelope and hole solitons

When $Q/P > 0$, that is, when P and Q have the same sign, the NLS equation (4.22) admits solutions (see Appendix 4C) of the form

$$\psi = V_m \, \mathrm{sech} \, [\sqrt{\frac{Q}{2P}} \, V_m \, \varepsilon(x - v_g t)] \exp(i\varepsilon^2\frac{Q}{2} V_m^2 t). \tag{4.23}$$

We note that the envelope propagates at the group velocity v_g, and the phase velocity depends on the amplitude V_m squared. Proceeding as in Sect.3.3 we can

74

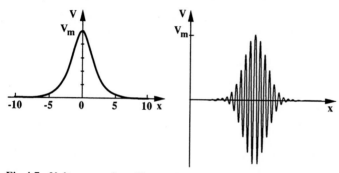

Fig.4.7. Voltage envelope V as a function of x for t =0, P = Q =1, and V_m=1 and the corresponding modulated waveform, a soliton wavepacket.

calculate the width of the envelope at half height $V_m/2$. We find that

$$L = \frac{1}{V_m} \sqrt{\frac{2P}{Q}} \text{ sech}^{-1}(\tfrac{1}{2}).$$

Zakharov and Shabat (1972) have shown that such a sech-shaped pulse, depicted in Fig.4.7, is an *envelope soliton;* moreover, they have shown that any initial localized disturbance eventually evolves into a number of envelope solitons and a dispersive tail (see Sect.10.6). The bulk of the energy is contained in the solitons, which propagate with permanent form once produced.

Since the nonlinear Schrödinger equation describes the envelope of waves with a carrier frequency $\omega=\omega_0$, the wavepackets obtained by combining the envelope waveform (4.23) and the sinusoidal carrier wave,

$$V_n(t) = 2\varepsilon^2 V_m \text{sech} [\sqrt{\frac{Q}{2P}} V_m \varepsilon(n - v'_g t)] \cos [\kappa_0 n - (\omega_0 + \varepsilon^2 \frac{|Q|}{2} V_m^2)t], \quad (4.24)$$

will have soliton properties.

When Q/P < 0, in terms of the laboratory coordinates (x,t), the solution (see Appendix 4C) has the form

$$\psi = a (x,t) e^{i\phi(x,t)}, \qquad\qquad\qquad (4.25)$$

where the amplitude function a(x,t) is given by

$$a = V_m\{ 1 - m^2 \text{ sech}^2 [\sqrt{\frac{-Q}{2P}} mV_m \varepsilon(x - v_{gt})] \}^{1/2}, \qquad (4.26)$$

where m is a parameter that controls the depth of the modulation of the amplitude $(0 \le m \le 1)$. For the phase function $\phi(x,t)$ one finds (see Appendix 4C) the following expression

75

$$\phi = \sqrt{\frac{-Q}{2P}} \; [\; V_m(1-m^2)^{1/2}\varepsilon(x-v_gt)$$

$$+ \tan^{-1}\{ \frac{m}{\sqrt{1-m^2}} \tanh [\sqrt{\frac{-Q}{2P}} \; mV_m\varepsilon(x-v_gt)]\; \}] \; - \varepsilon^2 \frac{Q}{2}(3-m^2)\; V_m^2 t. \quad (4.27)$$

The amplitude or envelope function propagates at group velocity v_g. Such a soliton solution, which was originally found by Zakharov and Shabat (1973), is called a *hole soliton* (a *dark soliton* in optics, see Sect.8.3) because it corresponds to a hole in the amplitude or envelope function, as represented in Fig.4.8. For $m = 1$, the modulation depth is zero and the amplitude function of the soliton solution becomes

$$a = V_m \tanh \{ \; [\sqrt{\frac{-Q}{2P}} \; V_m \; \varepsilon(x - v_gt)\;) \} \qquad (4.28)$$

and the amplitude of the hole is maximum, as represented in Fig.4.8 (*dotted line*). Combining the amplitude function (4.26) and the sinusoidal carrier, and ignoring the phase term (4.27), we obtain a wave of the form

$$V(n,t) = 2V_m\{1 -m^2 \; \text{sech}^2 \; [\sqrt{\frac{-Q}{2P'}} \; mV_m \; (n - v'_gt)]\}^{1/2} \cos (\kappa_0 n - \omega_0 t). \quad (4.29)$$

If the electrical network contains linear capacitances and nonlinear inductances rather than nonlinear capacitances and linear inductances, the physics of the problem can also be modeled by an NLS equation that describes the evolution of the current $I(n,t)$ rather than the voltage $V(n,t)$.

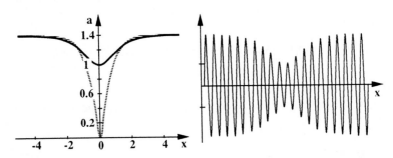

Fig.4.8. Amplitude function of the hole soliton versus x, for m = 0.7 (<u>continuous line</u>) and m = 1 (<u>dotted line</u>) and the corresponding modulated waveform

4.5.1 Experiments on envelope and hole solitons

Experiments on envelope or hole solitons are not so easy to perform as those on pulse solitons. Nevertheless, several authors have reported interesting results concerning nonlinear modulated waves. It was shown experimentally by Yagi and Noguchi (1976, 1977) that envelope solitons could be generated on a transmission line with nonlinear inductances. Despite the dissipation due to hysteresis losses, the form and the width of these solitons were qualitatively similar to the characteristics predicted from solution (4.23), as sketched in Fig.4.7.

Experimental observations of a hole soliton in a nonlinear electrical network, with nonlinear capacitances, were made by Muroya et al (1982). They showed that the hole soliton generated from an initial wave propagates stably in the circuit. The width and modulation depth of this soliton agreed with the theoretical predictions based on the NLS model (4.22).

4.6 Modulational instability

In Sect.4.5 we showed that envelope solitons can exist in an electrical transmission line, and can be created by an input with a pulse shape. Now, if a sine wave applied at one end of a transmission line, is slowly modulated, qualitatively, we might expect some nonlinear distortion of the envelope to occur, since larger-amplitude waves travel faster than smaller ones. As discussed in Chap. 1, this remarkable phenomenon, which has been called *modulational instability*, was discovered by Lighthill (1965). It is often referred as *Benjamin−Feir instability*, for it was Benjamin and Feir (1967) who first studied hydrodynamic waves by using a small-time instability analysis.

In this section we show that the nonlinear Schrödinger equation admits a uniform wave-train solution that is unstable to a small perturbation. The starting point is the nonlinear Schrödinger equation in the form (4.22), where, to simplify the writing, we set t for τ and x for ξ. We look for a solution with constant amplitude a_0 of the form

$$\psi = a_0 \exp[\, i\theta_0(t)], \tag{4.30}$$

where the phase $\theta(t)$ depends only on time. Replacing (4.30) in (4.22) yields

$$\partial\theta_0/\partial t = Qa_0{}^2.$$

After integrating this equation we get

$$\psi = a_0 \exp(iQa_0{}^2 t), \tag{4.31}$$

which represents a continuous wave solution with a phase or frequency which depends on the square of the amplitude.

We now ask if this solution is stable against small perturbations. For this purpose we assume that ψ has the form

$$\psi = a\,(x,t)\, \exp[i\theta(x,t)]. \tag{4.32}$$

Inserting this into (4.22) we get

$$Pa_{xx} - a\theta_t - aP(\theta_x)^2 + Qa^3 = 0, \tag{4.33a}$$

and

$$Pa\theta_{xx} + 2Pa_x\theta_x + a_t = 0. \tag{4.33b}$$

We then consider small perturbations εa_1 and $\varepsilon\theta_1$ of a_o and θ_o, such that

$$a = a_o(t) + \varepsilon a_1(x,t), \qquad \theta = \theta_o(t) + \varepsilon\theta_1(x,t). \tag{4.34}$$

Inserting (4.34) into (4.33), to zero order we obtain

$$\theta_o = (Qa_o^2)t, \qquad a_o = C^{te}$$

that is, we recover (4.31). Then, to order ε, we find a set of linear equations

$$a_o\theta_{1t} - 2Qa_o^2 a_1 - Pa_{1xx} = 0, \qquad a_{1t} + Pa_o\theta_{1xx} = 0 \tag{4.35}$$

We now assume that the perturbations have the form of sinusoidal modulations with wavenumber K and frequency Ω:

$$a_1 = A \exp[i(Kx - \Omega t)] + A^* \exp[-i(Kx - \Omega t)], \tag{4.36a}$$

$$\theta_1 = B \exp[i(Kx - \Omega t)] + B^* \exp[-i(Kx - \Omega t)]. \tag{4.36b}$$

Substituting relations (4.36) in (4.35) provides a set of two homogeneous equations

$$i\Omega A + Pa_o K^2 B = 0, \qquad (PK^2 - 2Qa_o^2)A - i\Omega a_o B = 0.$$

These equations have a nontrivial solution when the determinant is zero, that is, only when K and Ω satisfy the dispersion relation

$$\Omega^2 = (K^2 - 2\frac{Q}{P}a_o^2)P^2K^2. \tag{4.37}$$

We see that the stability of the continuous wave depends critically on the sign of Q/P in the transmission line. If, for example $Q=-|Q|$ and $P>0$, one has $\Omega^2>0$. Consequently, the frequency σ is real for all K and the continuous wave is stable against small perturbations. By contrast, when $P<0$, Ω becomes imaginary if the perturbation wavenumber lies in the range

78

$$0 < |K| < |K_c| = a_0\sqrt{2Q/P} \ . \tag{4.38}$$

In this case, the perturbation (see (4.36)) grows exponentially in time, and the continuous wave solution (4.30) is unstable. The maximum instability occurs at

$$K_m = a_0\sqrt{Q/P} = K_c/\sqrt{2}. \tag{4.39}$$

It is calculated from the condition $d\Omega/dK=0$. The corresponding frequency or maximum growth rate is then calulated to be

$$\Omega_m = a_0^2 Q. \tag{4.40}$$

It is independent of P but increases linearly with the amplitude squared, that is, with the power of the signal. A plot of the normalized growth rate (Ω/Ω_m) versus normalized wavenumber (K/K_m) is given in Fig.4.9.

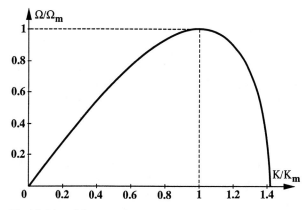

Fig.4.9. Plot of the normalized growth rate (Ω/Ω_m) versus normalized wavenumber (K/K_m).

Many systems with nonlinearity exhibit such an instability, which leads to modulation of the steady state as a result of the interplay between nonlinearity and dispersion. In Fig.4.10 we have sketched the evolution of a sinusoidal wave initially modulated by a weak perturbation. As time goes on, the modulation increases and the continuous wave breaks into a periodic pulse train. In the Fourier spectrum the modulational instability manifests itself as a pair of sideband components around the carrier wave component. There is a relationship between these envelope pulses and the envelope-soliton solution (4.23): the quantity $1/K_m$ corresponds approximatively to the width of a pulse envelope soliton (4.23) with amplitude $V_m = a_0$.

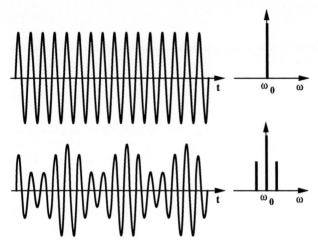

Fig.4.10. Sketch of the time evolution of an initial plane wave which is weakly modulated at point x=0 and the same wave at point x, where the instability has developed.

Let us now introduce characteristic parameters which may be evaluated in terms of expectations for the practical observations of envelope solitons and modulational instability in transmission lines or other physical systems (see also Sects 8.2 and 5.7). Noting that in (4.31) the coefficient of t has the dimension of the inverse of time, we can define a nonlinear time

$$T_{nl} = 1/Qa_o^2. \tag{4.41}$$

Moreover, in the linear approximation (4.22b) reduces to (2.48) and the dispersive broadening of a pulse is governed (see (2.53) with $\varepsilon=1$) by the dispersion time

$$T_d = L_o^2/2P, \tag{4.42}$$

where L_o is the initial half-width of the envelope pulse. Thus, we can introduce a nonlinear length and a dispersive length,

$$L_{nl} = v_g T_{nl}, \qquad L_d = v_g T_d, \tag{4.43}$$

which provide the two characteristic length scales over which respectively the nonlinear effects and the dispersive effects become important. Here, we have assumed that to order ε (see (4C.16)) the linear group velocity v_g is approximatively the velocity of propagation in the nonlinear line. Using (4.40) and (4.41) we can obtain the modulation frequency F_m at which the growth rate is maximum:

$$2\pi F_m = 1/T_{nl}. \tag{4.44}$$

80

Mizumura and Noguchi (1975) have observed experimentally that an input modulated wave on a transmission line exhibited self-modulational instability: the modulational ratio grew exponentially in the initial cells. Moreover, as the modulated wave was increased in amplitude, an envelope soliton train propagated from the rear of the wave flux.

Remarks

(i) Modulational instability can develop spontaneously due to electrical noise which is present in the electrical circuit; it is called *spontaneous* modulational instability. Otherwise, when it is induced by a weak external amplitude or phase perturbation it is called *induced* modulational instability

(ii) Qualitatively, the essential idea of Benjamin–Feir instability or resonance mechanism is as follows. Consider a periodic wavetrain (Fig.4.10), moving in the x direction, whose fundamental simple-harmonic component has amplitude a_0, wavenumber κ_0 and frequency ω_0. It is represented by $a_0\cos(\kappa_0 x - \omega_0 t)$. Now, suppose that a weak disturbance of the wavetrain is introduced. It consists of two progressive wave modes: a lower "sideband" with wavenumber $\kappa_1=(1-\varepsilon)\kappa_0$ and frequency $\omega_1=(1-\delta)\omega_0$ and a upper sideband with wavenumber $k_2=(1+\varepsilon)k_0$ and frequency $\omega_2=(1+\delta)\omega_0$. They are represented by $a_1\cos(\kappa_1 x - \omega_1 t)$ and $a_2\cos(\kappa_2 x - \omega_2 t)$. Here ε and δ are small parameters: $\varepsilon<<1$, $\delta<<1$ and the amplitudes a_1 and a_2 of these modes are assumed to be much smaller than a_0. In a linear situtation, as a result of dispersion, the modes within the packet would simply spread out, moving with their individual phase velocities. However, when their amplitudes become sufficiently large these modes undergo nonlinear interactions as a result of cubic nonlinearity. Under this condition harmonics of the primary wavetrain must also be present with phase $2(\kappa_0 x - \omega_0 t)$, $3(\kappa_0 x - \omega_0 t)$,... Thus the second harmonic $a_0^2\cos 2(\kappa_0 x - \omega_0 t)$ will interact with the lower (upper) sideband to give

$$a_0^2\, a_1\cos 2(\kappa_0 x - \omega_0 t)\,\cos(\kappa_1 x - \omega_1 t) = a_0^2 a_1\cos(\kappa_2 x - \omega_2 t) + ...$$

or

$$a_0^2\, a_2\cos 2(\kappa_0 x - \omega_0 t)\,\cos(\kappa_2 x - \omega_2 t) = a_0^2\, a_2\cos(\kappa_1 x - \omega_1 t) + ...$$

Each side-band mode combines with the second harmonic, bringing about a mutual reinforcement. This synchronous or resonant forcing effect felt by a sideband is proportional to the amplitude of the other, and so the two amplitudes can grow exponentially. The primary wavetrain is unstable to the above form of disturbance, it is the so-called modulational instability considered in Sect.4.6. As is clearly seen, the resonant mechanism requires

$$\kappa_1 + \kappa_2 = 2\kappa_0, \qquad \omega_1 + \omega_2 = 2\omega_0,$$

a condition which must be consistent with the dispersion relation. Thus the modulational instability is a special case $(\kappa_1 + \kappa_2 = \kappa_0 + \kappa_0,\ \omega_1 + \omega_2 = \omega_0 + \omega_0)$, of the resonant four-wave interaction (Phillips 1981, Craik 1985) process.

4.7 Laboratory experiments on modulational instability

In this section we present an experimental analysis (Marquié et al 1994) of modulational instability in a nonlinear electrical network with N cells, represented in Fig.4.11 (compare to Fig.2.6.). Each cell contains a linear inductance L_1 and a linear inductance L_2 in parallel with a nonlinear capacitance C. As in Sect. 3.4 this capcacitance consists of a reversed biased diode with the typical characteristics given in Fig. 3.7. It is biased by a constant voltage V_0 and depends on the voltage V_n at cell n so that the charge–voltage relation has the form (see Sects.3.4 and 4.4):

$$Q_n(V_n) = C_0[V_n - \alpha V_n^2 + \beta V_n^3 + ...],\qquad(4.45)$$

where the nonlinear coefficients $\alpha = -a_1/2$ and $\beta = a_2/3$ are positive.The polynomial approximation (4.45) of the C-V curve (see Fig. 3.7) is justified if $V_n < 2V_{pp}$.

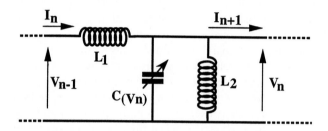

Fig.4.11. Unit cell of an electrical network with an inductance L_2 in parallel with the nonlinear capacitance C (V_n).

4.7.1 Model equations

The capacitance–voltage relation (4.45) is similar to (3.2a) (see also Sect.4.4). If, to a first-order approximation, we ignore the dissipative effects, we can model the network by a set of first-order equations (see Remark (iii) in Sect.2.4.3) which corresponds to the discrete version of (2.25). Combining these equations yields

$$\frac{d^2Q_n}{dt^2} = \frac{1}{L_1}(V_{n+1} + V_{n-1} - 2V_n) - \frac{V_n}{L_2}, \qquad n = 1, 2, ...,N. \qquad(4.46)$$

From (4.45, 4.46) we obtain the set of N nonlinear discrete equations

$$\frac{d^2V_n}{dt^2} + \omega_0^2 V_n + u_0^2(2V_n - V_{n+1} - V_{n-1}) = \alpha\frac{d^2V_n^2}{dt^2} - \beta\frac{d^2V_n^3}{dt^2}, \qquad(4.47)$$

with $u_0^2 = 1/L_1C_0$, $\omega_0^2 = 1/L_2C_0$ and n = 1, 2, ...,N. In the linear approximation ($V_n \ll 1$) the r.h.s of (4.47) can be ignored. Thus, considering a solution with constant amplitude V_1 of the form $V_n = V_1 Re[e^{i(\omega t - \kappa n)}]$, from (4.47) we obtain the linear dispersion relation (compare to 2.27) of a typical pass-band filter

$$\omega^2 = \omega^2{}_l = \omega_o{}^2 + 4u_o{}^2 \sin^2 \frac{\kappa}{2}, \qquad\qquad\qquad (4.48)$$

where ω_l stands for the angular frequency in the linear regime. The linear dispersion curve corresponding to (4.48) is represented in Fig.4.12, where $f_o = \omega_o/2\pi$ is the lower cutoff or gap frequency introduced by the parallel inductance L_2, and

$$f_{max} = \frac{\omega_{max}}{2\pi} = \frac{1}{2\pi}\sqrt{\omega_o{}^2 + 4u_o{}^2}, \qquad\qquad (4.49)$$

is the cutoff frequency introduced by the lattice effects. The corresponding group velocity and dispersion coefficient are

$$v'_g = \frac{\partial\omega}{\partial\kappa} = \frac{u_o{}^2\sin\kappa}{\omega}, \qquad\qquad P' = \frac{1}{2}\frac{\partial v_g}{\partial\kappa} = \frac{u_o{}^2\cos\kappa - v'_g}{2\omega}. \qquad (4.50)$$

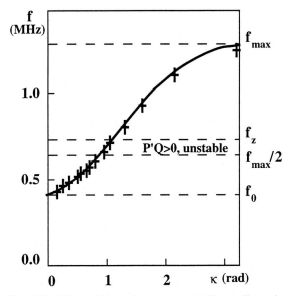

Fig.4.12. Theoretical and experimental linear dispersion curve. The instability region $f_{max}/2 < f_c < f_z$ where P'Q>0 in the nonlinear regime, is indicated.

The dispersion curve of Fig.4.11 is S shaped and presents a zero dispersion point P=0 for $f=f_z$ (see (4.50)), because two kinds of dispersion compete. The first dispersion effect, which dominates (P'>0) for $f<f_z$, arises from the term ω_o (existence of inductance L_2, see Sect.2.4.1) and the second one, which dominates (P'<0) for $f>f_z$, originates from lattice effects.

Next, we focus on nonlinear waves with a slowly varying envelope in time and space with regard to a given carrier wave (see Sect.4.4) with angular frequency $\omega=\omega_c=2\pi f_c$ and wave number $\kappa=\kappa_c$. By inspection of the r.h.s of (4.47) we see that, due to the presence of quadratic and cubic nonlinearity, the d.c., first harmonic, second harmonic and third harmonic will be created if an initial sinusoidal signal is launched into the electrical lattice. Nevertheless, in order to simplify the problem we restrict our investigation to frequencies such as $f_c>f_{max}/2$. In this case all the harmonics have their frequencies lying above the cutoff frequency f_{max} and, to a first approximation, such small-amplitude signals can be ignored. Moreover, the d.c. term vanishes as a result of the existence of a low frequency gap. Under these conditions we can assume that the voltage has the form

$$V_n = V_1(\varepsilon n, \varepsilon t)\, e^{i(\omega t - \kappa n)} + cc. \tag{4.51}$$

Inserting (4.51) in (4.47) we get a nonlinear dispersion relation

$$\omega^2 = \omega^2_1(1 - 3\beta|V_1|^2). \tag{4.52}$$

Then, proceeding as in Sect.4.4, we find a nonlinear Schrödinger equation

$$i\frac{\partial V_1}{\partial \tau} + P'\frac{\partial^2 V_1}{\partial s^2} + Q|V_1|^2 V_1 = 0, \tag{4.53}$$

where $s = \varepsilon(n - v'_g t)$, $\tau = \varepsilon^2 t$ and $Q = \frac{3}{2}\beta\omega_1$. As discussed in Sect.4.6 a uniform wave-train solution of (4.53) will be unstable to a small perturbation if $P'Q>0$. In the present case this condition is fulfilled if $f_{max}/2<f=f_c<f_z$, as shown in Fig.4.12, because both P' and Q are positive. The modulational instability conditions are given by (4.38) and (4.39).

4.7.2 Experiments

Experiments were carried out on a nonlinear electrical network with N=45 identical cells (see Fig.4.11). The nonlinear capacitances are similar to those used previously (see Sect.3.4.1); with a constant bias voltage $V_0=2V$ one has $C_0=320\pm10pF$, $\alpha=0.21V^{-1}$ and $\beta=0.0197V^{-2}$. The linear inductances are $L_1= 220\pm5\mu H$ and $L_2= 470\pm10\mu H$. The general experimental arrangement is similar to the one represented in Fig.3.6. The network is matched by a variable resistor and the input signal is created by a function generator. The waveforms are observed and stored by using a numerical oscilloscope with fast-Fourier-transform processing.

In the very-small-amplitude approximation the linear dispersion curve is measured as follows. At the input of the line one applies a sinusoidal voltage of low amplitude V=0.05V. The results are plotted in Fig. 4.12. The measured gap frequency is $f_0= 435\pm10kHz$ and the cutoff frequency is $f_{max}= 1280\pm30kHz$.

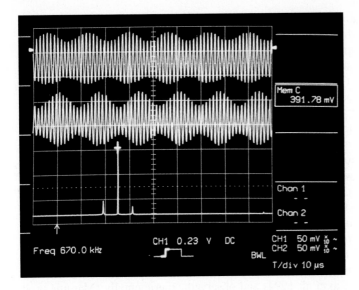

Fig.4.13. Induced modulational instability. (a) Initial signal at the input of the network with a 19% modulation rate. The carrier frequency is f_p= 670kHz while the modulation frequency is F=60kHz. Scales: 0.5V/div, 10µs/div. (b) Self modulated wave at cell n=12 with a 34% modulation rate. Scales: 0.5V/div, 10µs/div. (c) Fourier spectrum of the signal represented at cell n=12. Note the sidebands due to self-modulation, which lie at ± 60kHz of the carrier (central peak) frequency f_c=670kHz. Vertical scale: 0.165V/div. (Photo B. Michaux).

In order to observe the modulational instability that can be induced (compare to Sect.5.8) by a coherent and weak external modulation, a weakly amplitude modulated (19%, as shown in Fig. 4.13) sinusoidal wave with frequency f_c=670kHz is launched. For an appropriate value of the modulation frequency, lying in the range predicted by theory ($f_{max}/2 < f_c < f_z$), instability develops, leading to an increase in the modulation. Namely, at cell number n=12, and for amplitude a_0=0.19V, the modulation develops and becomes maximum (34%) when the envelope frequency measured from the Fourier spectrum (see Fig.4.13) is F = 60±5kHz. This experimental value agrees quantitatively with the theoretical value F_m= v'$_g$ $K_m/2\pi$ = 61kHz, in spite of the approximations (no second harmonic and no dissipation) that have been made.

Remarks

(**i**) As we have seen in Sect. 4.6 the modulational instability can also be induced by a weak phase modulation rather than an amplitude modulation.
(**ii**) Spontaneous modulational instability can also develop along the transmission line (Marquié et al. 1994). However, in order to observe it one has to increase the amplitude of the input wave because the noise is weak. In that case the nonlinear

Schrödinger equation, which is valid for relatively small amplitudes, does not correctly describe the physics of the system

(iii) When modulation instability occurs spontaneously it is initiated by the noise present in the physical system. The amplitude and phase fluctuations associated with this noise act as a seed for modulation instabilities and are likely to be amplified in the process (Cavalcanti et al. 1995), resulting in significant changes in the coherence and spectral properties of the wave. Such theoretical predictions, obtained in the context of optical fibers, have not yet been verified experimentally in any physical systems.

(iv)The investigation of Sect.4.7 is based on the continuum–limit approximation for the envelope wave, that is, the physics is governed by a continuous NLS equation. However, this approach cannot be used to describe the dynamics of nonlinear waves in the region of strong lattice effects. In this case the dynamics of modulated waves in an electrical lattice can be modeled (Marquié et al. 1995) by *a generalized discrete NLS equation,* which was introduced by Salerno (1992). In this case, contrary to the NLS model, modulational instability can occur for any value of the wave number K of the perturbation, as predicted by theory (Kivshar and Salerno 1994, Cai et al. 1994) and observed experimentally (Marquié et al. 1995).

4.8 Modulational instability of two coupled waves

When two electrical waves copropagate along an electrical transmission line, they can interact nonlinearly with each other. In fact, in addition to the phenomenon of self-interaction of a single wave, studied in the previous section, the same nonlinearity provides a coupling between the waves through a phenomenon referred to as a cross interaction. In this context let us consider two distinct propagating waves with different carrier frequencies and wavenumbers

$$V_{1,n}= V_1(\varepsilon n,\varepsilon t)\, e^{i\theta_1} + cc, \qquad V_{2,n}= V_2(\varepsilon n,\varepsilon t)\, e^{i\theta_2}+ cc, \qquad (4.54)$$

where the phases are given by $\theta_1=\omega_1 t-\kappa_1 n$ and $\theta_2=\omega_2 t-\kappa_2 n$. Then, by inserting (4.54) in (4.47), keeping terms in $e^{i\theta_1}$ or in $e^{i\theta_2}$ and, neglecting higher–order terms, we obtain a nonlinear dispersion relation of the form

$$\omega_j^2= \omega^2_{1,j}(1 - 3\beta|V_j|^2+ 6\beta|V_{3-j}|^2), \qquad (4.55)$$

with j=1or 2 and

$$\omega_j^2=\omega^2_{1,j}= \omega_0^2+ 4u_0^2\sin^2\frac{\kappa_j}{2} \;.$$

A straightforward generalization of the procedure in Sect. 4.4 leads to the following set of coupled NLS equations

$$i[\frac{\partial V_1}{\partial T} + v'_{g,1}\frac{\partial V_1}{\partial X}] + P'_1\frac{\partial^2 V_1}{\partial X^2} + Q(|V_1|^2 + 2|V_2|^2)V_1 = 0, \qquad (4.56a)$$

$$i[\frac{\partial V_2}{\partial T} + v'_{g,2}\frac{\partial V_2}{\partial X}] + P'_2\frac{\partial^2 V_2}{\partial X^2} + Q(|V_2|^2 + 2|V_1|^2)V_2 = 0, \qquad (4.56b)$$

where, to simplify the analysis, the parameters ε and δ (see 4.22a) have been dropped. Here, $v'_{g,1}$, P'_1 and $v'_{g,2}$, P'_2 are the group velocities and dispersion coefficients, evaluated at the carrier frequencies, of each wave. Note that in the derivation of (4.56) terms of the form $3V_1^2V_2^* e^{i(2\theta_1-\theta_2)}$ and $3V_2^2V_1^* e^{i(2\theta_2-\theta_1)}$ have been neglected because a large-phase *mismatch* is assumed. In (4.56) the cross-interaction terms have a factor of two, indicating that cross-interaction, often called *cross-phase modulation* (XPM), is twice as effective as self-interaction, called *self-phase modulation* (SPM) (see Sect.8.2).

Equations (4.56) admit solutions (see (4.30)) with constant amplitudes a_{o1} and a_{o2}

$$V_1 = a_{o1} \exp[i(Q|a_{o1}|^2 + 2Q|a_{o2}|^2)t], \qquad (4.57a)$$

$$V_2 = a_{o2} \exp[i(Q|a_{o2}|^2 + 2Q|a_{o1}|^2)t]. \qquad (4.57b)$$

A perturbation analysis of (4.57) similar to the one presented in Sect.4.6 allows one to obtain the modulational instability conditions, which are extremely rich and have been studied independently by many authors (see Mc Instrie and Luther 1990, and references therein). The modulational growth rate associated with a single modulationally unstable wave is increased by the presence of a second, modulationally unstable, wave. A wave that is modulationally stable in isolation can be destabilized by the presence of a second modulationally unstable wave. Two waves that are both modulationally stable by themselves are often modulationally unstable in the other's presence. Although the modulation growth rates depend on the relative amplitudes of the waves, existence of modulational instability does not. Thus, the evolution of two waves can differ qualitatively, as well as quantitatively, from the evolution of a single wave.

As a particular case, modulational instability of two counterpropagating waves has been observed (Bilbault et al. 1995) in an electrical transmission line. In this case of a forward and a backward wave with the same amplitudes $a_{o1} = a_{o2} = a$ and same carrier frequency $\omega = \omega_1 = \omega_2$, but opposite wavenumbers $k_1 = k$ and $k_2 = -k$, one has $v'_{g,1} = v'_g$, $v'_{g,2} = -v'_g$ and $P'_1 = P'_2 = P'$. The experimental results confirm the theoretical predictions mentioned above. The maximum growth rate is larger for two counterpropagating waves than for either wave alone. Moreover, a single wave that is modulationally stable becomes unstable in the other's presence.

4.9 Microwave solitons in magnetic transmission lines

In the last decade considerable progress has been made in theoretical studies of nonlinear magnetization waves or spin waves in media. The *magnetic soliton concept* (Akhiezer.and Borovik 1967) has stimulated a number of new experiments on the formation, propagation and collisions of microwave envelope solitons in magnetic films or transmission lines, as we shall discuss now.

4.9.1 Nonlinear spin waves

Traditionally, in the theory of solid state magnetism, nonlinearities were considered small and were described in terms of linear theory. In the linear approximation eigen-excitations in the magnetic medium were considered as an ideal gas of non-interacting spin waves or magnons. A propagating linear packet of spin waves is spread by dispersion because individual spin waves in the packet are totally independant of each other.

As discussed throughout this book for other physical systems, it is obvious that in the case of strong excitation in a magnetic medium the procedure of linearization of the equation of motion for the magnetization, which leads to the notion of spin waves as magnetic excitations, is not correct anymore. Nonlinear equation of motion for the magnetization should be solved from the very beginning and a new system of localized nonlinear excitations or *magnetic solitons* should be considered. The magnetic soliton is a cluster or a bound state of large number of spin waves where the attraction between spin waves compensate the spreading effect of dispersion.

The existence of spin wave envelope solitons was predicted theoretically (Lukomskii 1978, Zvezdin and Popkov 1983). In 1983, magnetic films were chosen, by Kalinikos and coworkers, as magnetic transmission lines, for investigation of nonlinear properties of spin waves and for observation of spin wave envelope formation in the microwave range. The relative simplicity of the techniques for excitation and detection of spin waves, the richness of their dispersion characteristics, the accessibility of waves from the surface of the film, and the weak attenuation per unit length of the line, all make films or magnetic transmission lines a convenient medium for investigating nonlinear spin wave processes.

4.9.2 NLS model equation for spin waves

It can be shown (Slavin et al. 1994a) that the nonlinear dispersion relation for spin waves in the slow envelope approximation has the general form (see (4.18)):

$$\omega = \omega(\kappa, |\psi|^2). \tag{4.58}$$

Because the envelope ψ varies slowly with time $T = \varepsilon t$ and space $X = \varepsilon x$, the spin wave packet is spectrally narrow in frequency and wave vector. Proceeding as in

Sect 4.4 one can expand (4.58) in terms of κ near the point $\kappa=\kappa_0$. As a result, one gets a nonlinear Schrödinger equation (see Zhang et al. 1998) similar to (4.22a) for the envelope $\psi(X, T)$

$$i[\frac{\partial \psi}{\partial T} + v_g\frac{\partial \psi}{\partial X}] + P\frac{\partial^2\psi}{\partial X^2} + Q|\psi|^2\psi = 0, \qquad (4.59)$$

where

$$v_g= \frac{\partial \omega}{\partial \kappa}, \text{ at } \kappa = \kappa_0, \; P= \frac{1}{2}\frac{\partial^2\omega}{\partial \kappa^2}, \text{ at } \kappa = \kappa_0, \text{ and } Q = -\frac{\partial \omega}{\partial|\psi|^2}, \text{ at } \psi_0 = 0;$$

are respectively the linear group velocity, the dispersion and the nonlinear coefficient of the magnetic waves. These coefficients depend on the constant external constant magnetic field H_0 that is applied, tangentially or perpendicularly to the film. As in Sect 4.4, eq (4.59) can be reduced to a standard NLS equation. Here, dissipation has been neglected but it can be taken into account (Slavin et al. 1994a).

If the condition $P/Q>0$ is fulfilled, one can predicts, like in Sect 4.5, envelope solitons with a form similar to (4.24), as solutions of (4.59) and also modulational instability (see Sect 4.6). Thus, equation (4.59) is a model for describing the nonlinear dynamics of envelope spin waves. In recent years microwave envelope solitons have been observed in ytrium garnet (YIG) films as we shall discuss in the next section.

4.9.3 Observation of magnetic envelope solitons

Ferromagnetic films were chosen for the investigation of nonlinear spin waves processes. The relative simplicity of the techniques for the excitation and detection of solitons, the richness of their dispersion characteristics, the accessibility of waves from the surface of the film, and the weak attenuation per unit wavelength, all make these ferromagnetic films convenient transmission lines, in the microwave range, for investigating nonlinear processes. Single-crystal ytrium garnet (YIG) films stand out as materials with the necessary homogeneity properties and low magnetic losses.

Figure 4.14 shows a scheme of the typical laboratory instrumentation which has been used for the envelope soliton experiments (Kalinikos et al. 1988, 1990). The YIG film placed between the poles of an electromagnet is the key element of the experiments (see Figs 4.14 and 4.15).

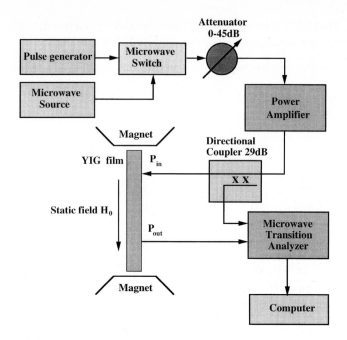

Fig. 4.14. Schematic diagram of the typical laboratory instrumentation used for the microwave magnetic envelope soliton measurements (Reproduced by kind permission of B. A. Kalinikos, C. E. Patton and A.N. Slavin)

A microwave signal (P_{in}) reaching the input of a delay transmission line (YIG film strip), via an antenna, as sketched in Fig. 4.14, is used to excite nonlinear magnetic waves that can propagate along the transmission line. The output antenna is movable, so that the propagation distance can be easily varied. The microwave signal (P_{out}) at the output antenna can be stored and analyzed (for details see Slavin et al. 1994a, 1994b, Kovshikov et al. 1996 and Patton et al. 1998).

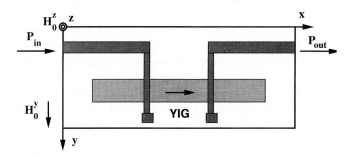

Fig. 4.15. Schematic representation of a YIG film or magnetic transmission line (parallel to the x direction) which is placed between two antennas (parallel to the y direction). The output antenna is movable, so that it allows to vary the propagation distance.

90

An external constant magnetic film H_o, tangential ($H_{o,y}$) or perpendicular ($H_{o,z}$) to the film, allows to bias the film.

Under pulsed excitation, the results show that, as predicted by (4.59), envelope magnetic solitons can form and propagate. The soliton parameters can be determined. In the case of continuous excitation, the modulation instability of an initial plane wave can be observed.

Remarks

(i) In order to simplify the presentation, dissipation effects have been ignored. However, in such real magnetic films dissipation inevitably occurs and must be fully taken into account (Slavin et al. 1994a).

(ii) Microwave magnetic envelope soliton pulse trains with no decay in amplitude have been generated in magnetic films (Kalinikos et al. 1997). The solitons were formed from magnetostatic spin waves with negative dispersion. They were propagating in a long and narrow 5.2 μm thick YIG transmission line biased by an external magnetic field H_O =1096 Oe. In order to compensate the effects of losses, an interrupted feedback arrangement was used to generate a train of soliton pulses from an initial 25ns wide input pulse at carrier frequency of 5 GHz. The train extended for over 40 μs, 2 orders of magnitude longer than the single soliton lifetime. Measurements of the soliton width and phase profiles confirmed the soliton nature of the pulses over the entire train.

(iii) Very recently, continuous self-generation of microwave magnetic soliton trains was reported (Kalinikos et al. 1998). Solitons with a width of 20 ns and a carrier frequency of about 5 GHz were generated in a 5.2 μm thick YIG transmission line. The measured pulse amplitude-width characteristics and phase profiles demonstrated the soliton character of these self-generated pulse signal, produced whithout any input pulse whatsoever.

4.10 Solitons and signal processing

Traditional signal processing relies on linear systems and linear techniques. Nevertheless, nonlinear systems look like very attractive to produce more efficient algorithms for a variety of signal processing problems. For these reasons the exploration of the properties of solitons as signals, e.g. for optical fiber communications systems considered in Chap. 8, is important and useful. In this regard, nonlinear transmission lines or electrical networks may be efficient to both generate and process signals that will be transmitted over linear or nonlinear (see Sect. 8.5.2) communication channels.

A pulse soliton modulation technique, based on the Fermi-Pasta-Ulam recurrence phenomenon described in Sect 4.2.2, was proposed by Hirota and coworkers (1973). A sinusoidal signal is introduced (see Fig.4.16) into a nonlinear network (A) (Toda lattice) and after propagation over some distance it decomposes into a train of two pulse-solitons (only a two-soliton period is represented in Fig.4.16).

At this point the amplitudes of the solitons signal are individually modulated by the information signal. Then, an appropriate length of nonlinear network (B) is used to allow the soliton to recombine into a sinusoidal signal which can propagate on a linear communication channnel (LCC). Finally, a nonlinear network (C) with appropriate length allows to recover the two modulated solitons from which the information signal can be extracted.

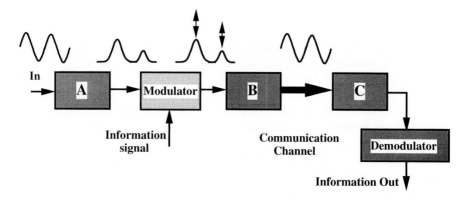

Fig.4.10. Amplitude modulation and demodulation of soliton trains

Another technique has been proposed (Singer 1996) by considering a soliton train. Such a nonlinear periodic wave, described mathematically by a Jacobi elliptic function (see Sect. 4.3) can play the role of a carrier wave. Thus, the amplitude or width of the soliton carrier wave can be modulated and demodulated after propagation on a communication channel. In this regard, the techniqes developed for sinusoidal carrier demodulation have to be adapted, at least from a signal generation wiewpoint, and a variety of sinusoidal carrier modulation and demodulation techniques can be directly tranposed to soliton carriers.

Signal processing using envelope-solitons is also interesting. Such solitons, which are solutions of NLS equation or its generalizations, can carry information in their envelope, amplitude, width and position; in their group and phase velocities and in their carrier phase. This information can be exchanged in collisions with other envelope solitons, as opposed to colliding waves in a linear medium, which do no interact because of linear superposition, and therefore cannot have information interaction. Altough, information transfer (Jakubowski et al. 1997) and secure communication (Lonngren et al. 1991) using envelope-soliton collisions have been considered in the context of nonlinear optics, it would be interesting to explore such processes by using electrical networks.

Appendix 4A. Periodic wavetrain solutions

To look for periodic solutions of (3.21) we first proceed as in Appendix 3C: we rewrite it in scaled form (3C. 4) and reduce it to (3C.10). Then we take nonzero values for the constants K_1 and K_2 in (3C.10). For simplicity, we set

$$6K_2/a = E, \qquad 6K_1/a = B, \qquad 3u/a = C. \tag{4A.1}$$

Then (3C.10) is rewritten as

$$\frac{3}{a}(\frac{dV}{d\xi})^2 = E + BV + CV^2 - V^3 \tag{4A.2}$$

or

$$\pm \sqrt{\frac{3}{a}} \int_{V_3}^{V} \frac{dV}{\sqrt{E + BV + CV^2 - V^3}} = \int_{\xi_3}^{\xi} d\xi. \tag{4A.3}$$

The solution for V oscillates between V_2 and V_3 where $V_1 < V_2 < V_3$ are the real roots (Fig.4A.1) or three distinct zeros, if they exist, of the cubic equation

$$E + BV + CV^2 - V^3 = (V_1-V)(V_2 - V)(V_3 - V) = P(V) = 0. \tag{4A.4}$$

By using the transformation $V = -f+C/3$, we can transform (4A.4) into

$$f^3 + (B - \frac{C^2}{3})f + \frac{C}{3}(\frac{2C^2}{9} - B) + E = 0,$$

and calculate the solutions of this cubic equation. Using the substitution

$$V = V_3 + (V_2 - V_3) \sin^2\theta = V_2 + (V_3 - V_2) \cos^2\theta, \tag{4A.5}$$

we can rewrite (4A.3) as

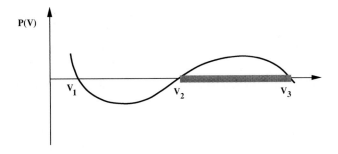

Fig. 4A.1. Sketch of the three possible roots of P(V)

93

$$\sqrt{\frac{3}{a}} \int_0^\theta \frac{2\sin\theta\cos\theta d\theta (V_2-V_3)}{\sqrt{[(V_3-V_1)-(V_3-V_2)\sin^2\theta][(V_2-V_3)\sin\theta\cos\theta]^2}} = \pm (\xi - \xi_3), \quad (4A.6)$$

which gives a standard elliptic integral (see Appendix 4B) :

$$\sqrt{\frac{12}{a(V_3-V_1)}} \int_0^\theta \frac{d\theta}{\sqrt{(1-m\sin^2\theta)}} = \pm (\xi - \xi_3), \quad (4A.7)$$

where $m = k^2 = (V_3-V_2)/(V_3-V_1)$ is the square of the modulus k. Thus, we have $0<m<1$ and

$$cn \ [\ \sqrt{\frac{a(V_3-V_1)}{12}} \ (\xi-\xi_3), \ m] = \cos\theta, \quad (4A.8)$$

where the sign $\pm$ has been suppressed since cn is an even function (see Appendix 4B). The solution for $V(\xi)$ can then be expressed as the so-called *cnoidal wave*

$$V(\xi) = V_2 + (V_3 - V_2) \ cn^2 \ [\ \sqrt{\frac{a(V_3-V_1)}{12}} \xi, \ m], \quad (4A.9)$$

where ξ_3 is assumed to be zero. The function $cn^2(\alpha\xi, m)$ has a period $2 K(m)$ i.e

$$cn^2(\alpha\xi,m) = cn^2[\alpha\xi + 2K(m), m \ ], \quad (4A.10)$$

where $K(m)$ is the complete elliptic integral of the first kind,

$$K(m) = \int_0^{\pi/2} \frac{d\theta}{\sqrt{(1-m\sin^2\theta)}}. \quad (4A.11)$$

Hence, the wavelength λ, which is the spatial period of the function $V(\xi)$, is given by the relation

$$\alpha \ (\xi +\lambda) - \alpha\xi = 2K \ (m), \quad (4A.12)$$

94

which yields

$$\lambda = \frac{2K(m)}{\alpha} = 2K(m) \sqrt{\frac{12}{a(V_3 - V_1)}}. \tag{4A.13}$$

Appendix 4B. The Jacobi elliptic functions

In this Appendix we give some information about Jacobi elliptic functions, which should facilitate the reading of this book. For more information the reader should consult the book of Lang (1973).

Looking for solutions of a nonlinear equation, one can often transform it into an equation of the form

$$(\frac{d\theta}{dt})^2 = (1 - k^2 \sin^2\theta), \tag{4B.1}$$

where $0 \leq k \leq 1$. The solutions of (4B.1) can be written as

$$u = \int_0^\varphi \frac{d\theta}{\sqrt{1 - k^2 \sin^2\theta}}. \tag{4B.2}$$

We note that by setting $q = \sin\theta$, (4B.2) can be transformed into

$$u = \int_0^{\sin\varphi} \frac{dq}{\sqrt{(1 - q^2)(1 - k^2 q^2)}}.$$

The inverse function $\varphi(u)$ is called the amplitude:

$$\varphi = amu. \tag{4B.3}$$

Thus, the three Jacobi elliptic functions are defined:

$$snu = \sin(amu) = \sin\varphi, \tag{4B.4a}$$
$$cnu = \cos(amu) = \cos\varphi, \tag{4B.4b}$$
$$dnu = (1 - k^2 sn^2 u)^{1/2}. \tag{4B.4c}$$

These functions also depend on the number k, called *the modulus*; they are written as

$$am(u, k), \quad sn(u,k), \quad cn(u,k), \quad dn(u,k). \tag{4B.4d}$$

Note that one often defines the number m = k^2, called *the parameter*, and *the complementary parameter* m_1 = 1 - m.

Relations (4B4) imply that

$$sn^2u + cn^2u = 1, \quad dn^2u + k^2sn^2u = 1.$$

In (4B.2) the integrand is periodic with period π. As a consequence, when φ increases from φ to $\varphi + \pi$, u increases from u to u + 2K, where

$$K = \int_0^{\pi/2} \frac{d\theta}{\sqrt{1 - k^2\sin^2\theta}} , \qquad (4B.5)$$

is the complete integral of the first kind. Under these conditions one has

$$am (u + 2K) = amu + \pi, \quad am (- u + 2K) = - amu + \pi, \qquad (4B.6)$$

because amu is an odd function of u. From (4B.4) and (4B.6) we obtain

$$sn (2K + u) = - snu, \quad cn (2K + u) = - cnu, \quad dn (2K + u) = dnu, \quad (4B.7)$$

$$sn (2K - u) = snu, \quad cn (2K - u) = - cnu, \quad dn (2K - u) = dnu. \quad (4B.8)$$

These results show that the functions sn and cn have the period 4K and the function dn has the smaller period 2K.

4B.1 Asymptotic limits

We now give the asymptotic values for special values of the modulus k .

For k =0, from (4B.2) and (4B.3) we get

$$u = \varphi = am (u,0). \qquad (4B.9)$$

Inserting (4B.9) into (4B.4) shows that dn (u,0)=1 and the elliptic functions defined by (4B.4a,b) reduce to sinusoidal functions:

$$sn (u, 0) = sinu, \quad cn (u, 0) = cosu. \qquad (4B.10)$$

For k<< 1, (4B.2) can be expanded in powers of k:

$$u \approx \int_0^\varphi [1+ \frac{k^2}{2}\sin^2\theta + O(k^4)]d\theta. \qquad (4B.11)$$

Integrating (4B.11) yields

$$u \approx \varphi\, (1 + \frac{k^2}{4}) - \frac{k^2}{8}\sin2\varphi + O(k^4). \qquad (4B.12)$$

From (4B.12) we obtain

$$\varphi \approx u\, (1 - \frac{k^2}{4}) + \frac{k^2}{8}\sin[2\,(1 - \frac{k^2}{4})u\,]. \qquad (4B.13)$$

Using (4B.13) the elliptic functions defined by (4B.4) are expressed as follows:

$$snu \approx \sin(1 - \frac{k^2}{4})u + \frac{k^2}{4}\sin(1 - \frac{k^2}{4})u\,\cos^2(1 - \frac{k^2}{4})u, \qquad (4B.14a)$$

$$cnu \approx \cos(1 - \frac{k^2}{4})u - \frac{k^2}{4}\sin^2(1 - \frac{k^2}{4})u\,\cos(1 - \frac{k^2}{4})u, \qquad (4B.14b)$$

$$dnu \approx 1 - \frac{k^2}{2}\sin^2(1 - \frac{k^2}{4})u. \qquad (4B.14c)$$

Relations (4B.14) are transformed into

$$snu \approx \sin u - (k^2/4)\,(u - \sin u\,\cos u\,)\cos u, \qquad (4B.15a)$$

$$cnu \approx \cos u + (k^2/4)\,(u - \sin u\,\cos u\,)\sin u, \qquad (4B.15b)$$

$$dnu \approx 1 - (k^2/2)\,\sin^2 u. \qquad (4B.15c)$$

For $k = 1$, the elliptic integral K, which enters in the periods of the elliptic functions, becomes infinite. In this case (4B.2) can be integrated to give

$$u = \ln[\tan(\frac{\varphi}{2} + \frac{\pi}{4})] \text{ or } \varphi = 2\tan^{-1}[\tanh(\frac{u}{2})\,], \qquad (4B.16)$$

which yields

$$sn(u,1) = \tanh u, \qquad cn(u,1) = \operatorname{sech} u, \qquad dn(u,1) = \operatorname{sech} u. \qquad (4B.17)$$

4B.2 Derivatives and integrals

Considering the definition (4B.3) of amu and using (4B.2), we can write

$$\frac{d\varphi}{du} = \sqrt{1 - k^2 \sin\varphi}$$
(4B.18a)

or

$$\frac{d\ amu}{du} = dnu.$$
(4B.18b)

Taking into account (4B.4) we get

$$\frac{d\ snu}{du} = \cos\varphi\frac{d\varphi}{du} = cnu\ dnu,$$
(4B.19a)

$$\frac{d\ cnu}{du} = -\sin\varphi\frac{d\varphi}{du} = -snu\ dnu,$$
(4B.19b)

$$\frac{d\ dnu}{du} = -\frac{k^2\sin\varphi\cos\varphi}{\sqrt{1 - k^2\sin\varphi}}\frac{d\varphi}{du} = -k^2 snu\ cnu.$$
(4B.19c)

We can rewrite (4B.19a),

$$dnu = \frac{d\ (snu)/du}{\sqrt{1 - sn^2 u}},$$

and obtain

$$\int dnu\ du = \sin^{-1}(snu).$$
(4B.20)

Two other relations are deduced from (4B.19b,c):

$$\int cnu du = \frac{1}{k}\cos^{-1}(dnu), \qquad \int snu du = \frac{1}{k}\ln(dnu - kcnu).$$
(4B.21)

Appendix 4C. Envelope and hole soliton solutions

Setting $\tau P = \tau'$ and $Q/P = q$ we rewrite (4.22b) as

$$i\frac{\partial\psi}{\partial\tau'} + \frac{\partial^2\psi}{\partial\xi^2} + q|\psi|^2\psi = 0.$$
(4C.1)

Let us first consider the pulse soliton solution of (4C.1) when P and Q have the same sign, i.e when q > 0. Note that when both P and Q are negative it only changes the sign, of the time τ. We write ψ in the form

$$\psi = a(\xi,\tau') \, ei[\theta(\xi,\tau') + n\tau'], \qquad\qquad (4C.2)$$

where a and θ are real functions, and n is a real constant. Then, we plug (4C.2) into (4C.1) and separate the real and imaginary parts to obtain

$$a_{\xi\xi} - a(\theta_{\tau'} + n) - a(\theta_\xi)^2 + qa^3 = 0 \qquad\qquad (4C.3)$$

and

$$a\theta_{\xi\xi} + 2a_\xi\theta_\xi + a_{\tau'} = 0. \qquad\qquad (4C.4)$$

To simplify the writing we used the notation $a_\xi = \partial a/\partial\xi$, $a_{\xi\xi} = \partial^2 a/\partial\xi^2$, and so on, for the other derivatives. We now look for traveling solutions of (4C.4) of the form

$$a\,(\xi,\, \tau) = a\,(\xi - c\tau'), \qquad \theta(\xi,\, \tau) = \theta\,(\xi - c\tau'), \qquad\qquad (4C.5)$$

where the real constant c represents a velocity. Setting $s = \xi - c\tau'$, from (4C.5) we get $\partial/\partial\xi = \partial/\partial s$ and $\partial/\partial\tau' = -c\partial/\partial s$, and (4C.3) and (4C.4) become

$$a_{ss} + a\,[c\theta_s - n - (\theta_s)^2] + qa^3 = 0 \qquad\qquad (4C.6a)$$

and

$$a\theta_{ss} + 2a_s\theta_s - ca_s = 0 \qquad\qquad (4C.6b)$$

Upon integration of (4C.6b) we find that

$$a^2\,(2\theta_s - c) = A, \qquad\qquad (4.C.7)$$

where A is an integration constant. Solving equation (4C.7) for θ_s and substituting into (4C.6a) yields

$$a_{ss} - \frac{a}{4}\,(c + \frac{A}{a^2})^2 + a\frac{c}{2}\,(c + \frac{A}{a^2}) - na + qa^3 = 0$$

which is rearranged to give

$$a_{ss} = (n - \frac{c^2}{4})a + \frac{A^2}{4a^3} - qa^3. \qquad\qquad (4C.8)$$

99

Multiplying both sides of (4C.8) by $(2a_s)$ and integrating, we find that

$$(a_s)^2 = (n - \frac{c^2}{4}) a^2 - \frac{A^2}{4a^2} - q\frac{a^4}{2} + K, \qquad (4C.9)$$

where K is an integration constant. Then, multiplying (4C.9) by $4a^2$ and setting S = a^2, we get

$$(S_s)^2 = 4(n - \frac{c^2}{4}) S^2 - A^2 - 2qS^3 + 4KS. \qquad (4C.10)$$

Dividing by 2q and putting

$$E = -A^2/2q, \quad B = 4K/2q, \quad V_m^2 = 2(n - c^2/4)/q, \qquad (4C.11)$$

we obtain an equation similar to (3B.5) and (4A.2):

$$(1/2q)S_s^2 = E + BS + V_m^2 S^2 - S^3 = P(S), \qquad (4C.12)$$

where the polynomial P(S) may have three real roots (see Appendix 4A): S_1, S_2 and S_3, such that

$$P(S) = (S_1 - S)(S_2 - S)(S_3 - S).$$

One has a localized pulse solution if the function a, that is, S and its derivatives, tend to zero for $|\xi| \to \infty$. This implies that E =0 and B =0. Under these conditions integrating (4C.12) as in Appendix 3B yields

$$S = V_m^2 \operatorname{sech}^2 (\frac{\sqrt{2q}}{2} V_m s). \qquad (4C.13)$$

Then, using $S = a^2$ and (4C.7), and solving for ψ when A =0, we find that

$$\psi = V_m \operatorname{sech} [\frac{\sqrt{2q}}{2} V_m(\xi - c\tau')] \exp[i\frac{c}{2}(\xi - c\tau') + in\tau'], \qquad (4C.14)$$

where V_m, given by (4C.11), is the amplitude of the pulse envelope. In terms of the original space and time variables defined by $\xi = \varepsilon(x - v_g t)$ and $\tau' = \tau P = \varepsilon^2 Pt$, we obtain

$$\psi = V_m \operatorname{sech} [\sqrt{\frac{Q}{2P}} V_m \varepsilon(x - u_e t)] \exp\left\{ i\varepsilon\frac{c}{2}(x - u_c t) \right\}, \qquad (4C.15)$$

where

$$u_e = v_g + \varepsilon cP, \quad u_c = (v_g + \varepsilon cP) - \varepsilon\frac{2}{c}(\frac{Q}{2}V_m^2 + \frac{c^2}{4} P), \qquad (4.C.16)$$

represent the envelope and phase velocity, respectively. In the expression for u_e the second term is of order ε compared to v_g. We can choose $c = 0$, as assumed in Sect. 4.5. In this case the envelope function depends only on ξ and the phase function depends only on τ'; in other words the wave is stationary in the frame of reference moving at group velocity.

If the nonlinear coefficient Q and the dispersion P are of different signs, their ratio q previously defined is negative. In this case (4C.10) can be written as

$$(S_s)^2 = 4(n - \frac{c^2}{4})\, S^2 - A^2 + 2|q|S^3 + 4KS, \qquad (4C.17)$$

and we look for a solution such that $S \to$ constant but its derivatives tend to zero for $|\xi| \to \infty$. Unlike the case of the pulse soliton the constants A^2 and K are not zero and (4C.17) can be cast into the form

$$(S_s)^2 = 2|q|(S - S_1)^2(S - S_3), \qquad (4.C.18)$$

where

$$2\,(n - c^2/4)/|q| = -S_3 - 2S_1, \qquad 2K/|q| = S_1^2 + 2S_1 S_3, \qquad A^2 = 2|q|\, S_1 S_3^2.$$

Here, S_1 and S_3 correspond to the double root and the single root of (4C.18), respectively. By putting $S = S_1 - f^2$ and $S_1 - S_3 = f_1$, (4C.18) is transformed into

$$\sqrt{2|q|} \int ds = \pm\, 2 \int \frac{df}{f\sqrt{f_1 - f^2}}. \qquad (4C.19)$$

This equation can be integrated by putting $f = \sqrt{f_1}\, Z$, which gives

$$\sqrt{2|q|} \int ds = \pm\, \frac{2}{\sqrt{f_1}} \int \frac{dZ}{Z\sqrt{1 - Z^2}} = \pm\, \frac{2}{\sqrt{f_1}} \int d\,(\text{arc sech}Z).$$

Then, taking account of the above transformations, one obtains the so-called hole or dark soliton

$$S = S_1\{\, 1 - m^2 \,\text{sech}^2\, [\sqrt{\tfrac{|q|}{2}}\, mV_m\, s]\, \}, \qquad (4C.20)$$

where we have set

$$S_1 = V_m^2, \qquad m^2 = (S_1 - S_3)/S_1. \qquad (4C21)$$

From (4C.20) we then express the amplitude function in terms of the original space and time variables defined by $s = (\xi - c\tau') = \varepsilon(x - v_g t) - \varepsilon^2 c P t$:

$$a = V_m\{\, 1 - m^2\, \text{sech}^2\, [\sqrt{\tfrac{-Q}{2P}}\, mV_m\, \varepsilon(x - v_g t - \varepsilon c P t)]\, \}^{1/2}, \qquad (4C.22)$$

From (4C.7) we have

$$\theta = \frac{A}{2} \int \frac{ds}{S} + \frac{c}{2} s.$$ (4C.23)

Using (4C.20) one integrates (4C.22) and obtains the total phase $\phi = \theta + n\tau'$:

$$\phi = \sqrt{\frac{-Q}{2P}} \left[V_m (1 - m^2)^{1/2} s + \tan^{-1} \left\{ \frac{m}{\sqrt{1 - m^2}} \tanh \left[\sqrt{\frac{-Q}{2P}} m V_m s \right] \right\} \right]$$

$$+ \frac{c}{2} \varepsilon(x - v_g t) - \left[\frac{c^2}{4} + \frac{Q}{2P} V_m^2 (3 - m^2) \right] \varepsilon^2 P t.$$ (4C.24)

For $c = 0$ (4C.22) and (4C.24) simplify as assumed in the text.

5 Hydrodynamic Solitons

The behavior of water waves can be observed in nature by each of us. For example, the wave that most people first encounter may be one which approaches a beach. Water waves can be appreciated in a descriptive way without any technical knowledge (The Open University course team 1989). This is a fascinating subject (Lamb 1932), because the phenomena are familiar and the mathematical problems are various (Witham 1974). The literature is vast and the more modern literature concentrates on nonlinear problems (Lighthill 1978; Crapper 1984).

In this Chapter we first review the basic properties of shallow-water surface waves and deep-water waves with very small amplitude, that is, in the linear approximation. When the amplitude of the waves increases, the nonlinearity contained in the equations of fluid mechanics cannot be neglected, and we examine successively the hydrodynamic pulse soliton and envelope soliton, which appear as shallow and deep waves, respectively.

5.1 Equations for surface water waves

Waves can appear on the initially unperturbed plane surface of a liquid under the action of some external perturbation. Waves on the surface of water are of two main kinds: *gravity waves* and *capillary waves*.

The former have wavelengths from about half a meter to several hundred meters and they result from the action of the gravitational field **g** which tends to keep the water surface at its lowest level, as depicted in Fig.5.1.These waves occur mainly on the surface of the liquid, but their properties depend on the depth.

Capillary waves are ripples of fairly short wavelength, no more than a few centimeters. They result from the surface tension T which acts as a restoring force and tends to keep the surface flat (see Fig.5.1). For example, these waves with

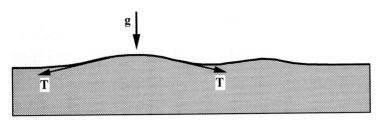

Fig.5.1. The gravitation g and surface tension T act as restoring forces which tend to keep the water surface flat.

short wavelengths are created when raindrops fall on a pond. By contrast, when one generates waves by dropping a large stone into a pond, the effects of gravity predominate, on account of the longer wavelengths involved. From this Sect.to Sect.5.3 we will deal exclusively with surface gravity waves. The surface tension effects will be examined in Sect.5.4.

5.1.1 Reduced fluid equations

We consider the motion of waves on the surface of water, with air above it, in the constant gravitational field of modulus g, which is first taken as the dominant restoring force: the effects of surface tension are first neglected and will be discussed in Sect.5.4. For water waves this is a good assumption if the wavelength is substantially longer than $\lambda_m = 1.74$ cm. The liquid is assumed to be bounded from below by a hard horizontal bed; the depth of the water is h (Fig.5.2). We choose a coordinate system such that the z axis is vertically upwards and the xy plane coincides with the unperturbed fluid surface. Under these conditions, if $\mathbf{k}$ is a unit vector in the z direction the gravitational field is $\mathbf{g} = - g\mathbf{k}$. At the upper free surface we set

$$z = \eta \ (x,y,t), \tag{5.1}$$

where η is the z coordinate of a point on the surface, it corresponds to the vertical displacement of the surface. At equilibrium the surface is unperturbed and one has $\eta = 0$.

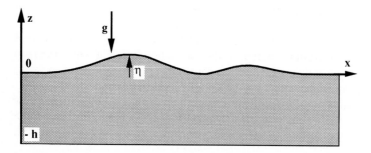

Fig.5.2. Sketch of the water surface for gravity waves.

The following reasonable assumptions are made. Water is incompressible, i.e., its density ρ is supposed to be constant throughout the volume of the fluid and throughout its motion. Water is inviscid: since viscosity is effective only for small-scale motions, it is negligible for most of the phenomena considered in the following. Under these conditions, the equations taken for the study of gravity waves on water are the Euler equations (see Appendices 5A and 5B) for an irrotational flow of an incompressible fluid with a free surface:

$$\Delta\phi = \frac{\partial^2\phi}{\partial x^2} + \frac{\partial^2\phi}{\partial y^2} + \frac{\partial^2\phi}{\partial z^2} = 0, \qquad \text{for } -h < z < \eta \ (x, y, t);$$ (5.2a)

$$\phi_z = \eta_x\phi_x + \eta_y\phi_y + \eta_t, \qquad \text{at } z = \eta \ (x, y, t);$$ (5.2b)

$$\phi_t + \frac{1}{2}(\phi_x^2 + \phi_y^2 + \phi_z^2) + g\eta = 0, \quad \text{at } z = \eta \ (x, y, t);$$ (5.2c)

$$\phi_z = 0, \qquad \text{at } z = -h.$$ (5.2d)

Here, ϕ is the velocity potential (see Appendix 5B). These equations are dispersive owing to g and nonlinear owing to the terms $\eta_x\phi_x + \eta_y\phi_y$ and $(\phi_x^2 + \phi_y^2 + \phi_z^2)$. Generally one looks for solutions by first solving the linear Laplace equation (5.2a); then one applies the highly nonlinear boundary conditions (5.2b,c).

5.2 Small-amplitude surface gravity waves

If we consider small disturbances on water otherwise at rest, the free surface elevation above mean η and the velocity potential ϕ are both small on the scale of the wavelength and the wave period. Thus, the free surface boundary conditions can be linearized. Neglecting the nonlinear terms, (5.2b) and (5.2c) are approximated by

$$\eta_t - \phi_z = 0, \qquad \text{at } z = \eta,$$ (5.3a)

and

$$\phi_t + g\eta = 0, \qquad \text{at } z = \eta.$$ (5.3b)

Since the quantity η is small, condition (5.3a) may be simplified by expanding ϕ_z in a Taylor series around $z = 0$:

$$(\phi_z)_\eta = (\phi_z)_0 + \eta\,(\phi_{zz})_0 + \dots.$$

Thus, to the first order of approximation, ϕ_z in (5.3a) can be evaluated at $z = 0$ instead of $z = \eta$. Then elimination of η after differentiating (5.3b) with respect to time yields

$$\phi_{tt} + g\phi_z = 0, \qquad \text{at } z = 0.$$ (5.4)

This process reduces the nonlinear free boundary problem (5.2) to a linear fixed-boundary problem for the velocity potential:

$$\Delta\phi = 0, \qquad \text{for } -h < z < 0; \tag{5.5a}$$

$$\phi_{tt} + g\phi_z = 0, \qquad \text{at } z = 0; \tag{5.5b}$$

$$\phi_z = 0, \qquad \text{at } z = -h. \tag{5.5c}$$

From now on we restrict ourselves to *the one-dimensional problem , that is, we consider waves propagating along the x axis which are uniform in the y direction.* In this case, all quantities are independent of y. Experimentally, waves on the surface of water look approximatively sinusoidal some time after their generation by a disturbance. This suggests looking for a separable solution of (5.5a), with frequency ω and wavenumber k, of the form

$$\phi = q(x, z) \sin (kx - \omega t). \tag{5.6}$$

Substituting (5.6) into Laplace's equation (5.5a) we have

$$(\frac{\partial^2 q}{\partial x^2} + \frac{\partial^2 q}{\partial z^2} - k^2 q) \sin (kx - \omega t) + k \frac{\partial q}{\partial x} \cos (kx - \omega t) = 0.$$

This equation is satisfied if the coefficients of the cosine and sine terms are both zero. This yields $\partial q/\partial x = 0$, which means that q must be chosen independent of x (there is no dissipation). We then have

$$\frac{\partial^2 q}{\partial z^2} - k^2 q = 0. \tag{5.7}$$

The general solution of (5.7) is

$$q(z) = A e^{kz} + B e^{-kz},$$

with two arbitrary constants A and B . The velocity potential can be expressed as

$$\phi = (A e^{kz} + B e^{-kz}) \sin (kx - \omega t).$$

Boundary condition (5.5c) yields

$$B = A e^{-2kh}.$$

Inserting this result into the expression for ϕ gives

$$\phi = 2Ae^{-kh} \cosh [k(z + h)] \sin (kx - \omega t). \tag{5.8}$$

From (5.3a) one can calculate the free surface displacement:

$$\eta = 2 \frac{kA}{\omega} e^{-kh} \sinh kh \cos (kx - \omega t). \tag{5.9}$$

There is no integration constant because the mean value of the surface is zero. Note that one can redefine a new amplitude

$$a_m = 2e^{-kh} \frac{kA}{\omega} \sinh kh \qquad (5.10)$$

such that

$$\eta = a_m \cos (kx - \omega t). \qquad (5.11)$$

The velocity potential then becomes

$$\phi = \omega a_m \frac{\cosh [k(z + h)]}{k \sinh kh} \sin (kx - \omega t). \qquad (5.12)$$

Taking into account (5.5b), which must also be satisfied by (5.8) (or (5.12)), we get the dispersion relation for small amplitude gravity waves at the surface of water:

$$\omega^2 = gk \tanh (kh), \qquad (5.13)$$

which is valid when the surface tension effects are neglected. From (5.13) we calculate the phase velocity $v = \omega/k$:

$$v = \sqrt{\frac{g}{k} \tanh(kh)} = \sqrt{\frac{g\lambda}{2\pi} \tanh \left(\frac{2\pi h}{\lambda}\right)}, \qquad (5.14)$$

and the group velocity v_g:

$$v_g = \frac{\sqrt{g}}{2\sqrt{k\tanh(kh)}} [\tanh(kh) + kh \, \mathrm{sech}^2(kh)] = \frac{v}{2} [1 + \frac{2kh}{2\sinh(2kh)}]. \qquad (5.15)$$

Equations (5.13–15) tell us that ω, v, and v_g depend on kh or equivalently on the dimensionless ratio h/λ. Then it is natural to consider the two following interesting asymptotic cases .

5.3 Linear shallow- and deep-water waves

If $h/\lambda \ll 1$, the wavelength λ is large compared to the depth h and one has *shallow-water waves*. In the opposite case if $h/\lambda \gg 1$, one has *deep-water waves*.

5.3.1 Shallow-water waves

In the first case, kh $\ll 1$, the term tanh (kh) can be expanded in powers of kh:

$$\tanh (kh) \approx kh - \frac{(kh)^3}{3} \,,$$

and the dispersion relation (5.13) is approximated by

$$\omega = c_0 k \, [\, 1 - (kh)^2/6 + ...], \tag{5.16}$$

where $c_0 = \sqrt{gh}$. This relation corresponds to *shallow-water waves* with weak dispersion. Setting $h/\lambda = \delta$, where δ is a small parameter, we see that in (5.16) the leading-order dispersive term is of order δ^2. When this term is neglected, that is, for very long shallow-water waves, (5.13) and (5.14) may be approximated by

$$\omega \approx c_0 \, k, \qquad v \approx \sqrt{gh} = c_0. \tag{5.17}$$

This gives the important result that for a harmonic wave with a very long wavelength compared to the depth, the dispersion can be neglected and the phase velocity c_0 is independent of the wavelength but varies as the square root of the depth. The group velocity is

$$v_g \approx c_0.$$

The approximation (5.17) gives better than 3% accuracy if $h < 0.07\lambda$. Surface waves are therefore considered to be shallow water waves if the depth h is less than 7% of the wavelength. Expression (5.11) for the free surface displacement remains unchanged; it describes a simple harmonic progressive wave of small amplitude as represented in Fig.5.3. On the other hand, one has the approximations

$$\cosh [\, k(z+h)] \approx 1, \qquad \sinh kh \approx kh.$$

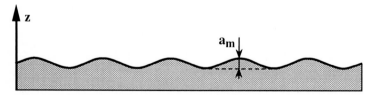

Fig. 5.3. Sketch of sinusoidal shallow-water waves

Then the expression (5.12) for the velocity potential can be approximated by a simple harmonic progressive wave of constant amplitude. From (5.12), the horizontal and vertical components of the velocity:

$$u = dx/dt = \phi_x, \qquad w = dz/dt = \phi_z, \tag{5.18}$$

are calculated, and we find that

$$u = \frac{a_m \omega}{kh} \cos (kx - \omega t) = c_0 \frac{\eta}{h}, \qquad w = a_m \omega \left(1 + \frac{z}{h}\right) \sin (kx - \omega t). \tag{5.19}$$

This relation shows that the horizontal component is larger than the vertical component.

5.3.2 Deep-water waves

Let us now consider the case of *deep-water waves*, where the wavelength λ is small compared to the depth h: $h/\lambda \gg 1$. The dispersion relation (5.13) and the phase velocity relation (5.14) are approximated by

$$\omega \approx \sqrt{gk}, \qquad v \approx \sqrt{g\lambda/2\pi}. \tag{5.20}$$

The dispersion is nonzero and the wave speed is proportional to the square root of the wavelength for a progressive wave with small wavelength compared to the depth. With $g = 9.81$ ms^{-2}, one gets $v \approx 1.25 \sqrt{\lambda}$. For example, for waves in the sea 50m $< \lambda <$ 150m which gives: 8.8 ms$^{-1} < v <$ 15.3ms^{-1} for the velocity range and 5.7 s $< T <$ 9.8s for the period $T = \lambda/v$.

For deep-water waves the form of η given by (5.11) remains unaffected. On the other hand, the velocity potential expression is modified. Recalling that the fluid occupies the region z<0, for kh $\gg 1$ we have $(\cosh k(z+h))/\sinh kh \approx e^{kz}$. With this approximation (5.12) becomes

$$\phi = \frac{a_m \omega}{k} e^{kz} \sin (kx - \omega t). \tag{5.21}$$

This means that ϕ decreases with z as we go into the interior of the fluid. The horizontal and vertical components of the velocity in the moving water are:

$$u = \phi_x = a_m \omega e^{kz} \cos (kx - \omega t), \qquad w = \phi_z = a_m \omega e^{kz} \sin (kx - \omega t). \tag{5.22}$$

From (5.22) we see that the amplitude of the motion $A_s = a_m \omega e^{kz}$ of the fluid particles diminishes exponentially with depth. At depth $- z = \lambda/2$, it is $(1/e^\pi)A_s \approx (1/22)A_s$ where A_s is the amplitude at the surface (z = 0). At depth $-z = \lambda$, it is

$(1/e2\pi)A_s \approx (1/500)A_s$. These simple calculations tell us that *the surface waves would not feel the bottom until the depth becomes less than about half a wavelength.*

From (5.22) one can show (Lamb 1932; Kundu 1990) that the trajectories or paths of the individual water particles, regarded as points, in the wave are described by circles whose radii diminish exponentially with depth.

5.4 Surface-tension effects: capillary waves

Up to now we have assumed that the pressure on the free surface of the water was identical to the atmospheric pressure p_a. In fact, a density discontinuity exists between the two fluids, air and water, in contact , and a pressure difference results. In this case the interface is found to behave as if it were under tension, in other words it behaves like a stretched membrane. Like gravitation g, surface tension T acts as a restoring force which also contributes to the dispersion of the waves. Analytically, when the surface tension effects are taken into account (see Appendix 5B), the nonlinear dynamic boundary condition (5.2c) is replaced by

$$\frac{\partial \phi}{\partial t} + \frac{1}{2}[(\frac{\partial \phi}{\partial x})^2 + (\frac{\partial \phi}{\partial y})^2 + (\frac{\partial \phi}{\partial z})^2] + g\eta - \frac{T}{\rho}(\frac{\partial^2 \eta}{\partial x^2} + \frac{\partial^2 \eta}{\partial y^2}) = 0, \qquad (5.23)$$

at $z = \eta(x,y,t)$.

For small amplitude waves (5.23) is approximated by a linear boundary condition

$$\frac{\partial \phi}{\partial t} + g\eta - \frac{T}{\rho}(\frac{\partial^2 \eta}{\partial x^2} + \frac{\partial^2 \eta}{\partial y^2}) = 0, \qquad \text{at } z = 0, \qquad (5.24)$$

which replaces (5.3b). Differentiating this equation with respect to time t and using (5.3a) yields

$$\frac{\partial^2 \phi}{\partial t^2} + g\frac{\partial \phi}{\partial z} - \frac{T}{\rho}\frac{\partial}{\partial z}(\frac{\partial^2 \phi}{\partial x^2} + \frac{\partial^2 \phi}{\partial y^2}) = 0, \qquad \text{at } z = 0 \qquad (5.25)$$

Now, as previously, we restrict ourselves to waves propagating along the x axis that are uniform in the y direction. Under this condition, solution of the small-amplitude wave problem including surface-tension effects is identical to the one for gravity waves presented previously, except that (5.5b) is replaced by (5.25). Thus, we look for sinusoidal solutions of the form (5.12) and obtain the dispersion relation

$$\omega = \sqrt{gk (1+ \frac{Tk^2}{\rho g}) \tanh kh} \ . \qquad (5.26)$$

The phase velocity relation (5.14) becomes

$$v = \sqrt{\frac{g}{k}\left(1 + \frac{Tk^2}{\rho g}\right)\tanh kh} = \sqrt{\frac{g\lambda}{2\pi}\left(1 + \frac{4\pi^2 T}{\lambda^2\rho g}\right)\tanh\frac{2\pi h}{\lambda}}. \qquad (5.27)$$

The importance of surface tension is measured by the parameter

$$\mathcal{T} = Tk^2/\rho g = 4\pi^2 T/\lambda^2\rho g. \qquad (5.28)$$

When $\mathcal{T}$ is small (long wavelengths) the effect of surface tension may be neglected, (5.26) reduces to (5.13), and we have *pure gravity waves*. For example if we consider a air–water interface at 20°C the surface tension measured in newton per meter is $T = 0.074 Nm^{-1}$, with $\rho = 10^3 kgm^{-3}$ and $g = 9.81 ms^{-2}$. Under these conditions one obtains $4\pi^2 T/\rho g = 3.02 \times 10^{-4} m^2$. Thus, for waves with wavelength $\lambda = 10 cm$ one gets $\mathcal{T} = 0.03$, which corresponds to a modification of only 1.5 % of the phase velocity of *pure gravity waves* given by (5.14).

In the opposite case of short wavelengths, i.e., when $\mathcal{T}$ becomes large, the phase velocity v increases above its value for pure gravity waves at all wavelengths. The effects of gravity may be neglected and the waves are called *capillary waves* or *ripples*. Intermediates cases correspond to *capillary gravity waves*.

A plot of (5.27) giving phase velocity versus wavelength is shown in Fig.5.4. We note that surface tension dominates for $\lambda < \lambda_m$, where λ_m corresponds to a minimum phase velocity v_m. For these small wavelengths we can assume the deep water approximation $\tanh(2\pi h/\lambda) \cong 1$, which is valid if $h > \lambda/2$. Relation (5.27) is then approximated by

$$v \cong \sqrt{\frac{g\lambda}{2\pi}(1 + \mathcal{T})}. \qquad (5.29)$$

Setting $dv/d\lambda = 0$, we find that

$$\lambda_m = 2\pi\sqrt{T/\rho g}, \qquad v_m = (4gT/\rho)^{1/4}. \qquad (5.30)$$

For the air–water interface we have: $\lambda_m = 1.74$ cm and $v_m = 23.3$ cm s^{-1}. As seen above, only waves with wavelength $\lambda < 10 cm$ are affected by surface tension. For wavelengths $\lambda < 5 cm$ the behavior of small waves is dominated by surface tension. In this case, assuming again $\tanh(2\pi h/\lambda) \cong 1$, the dispersion relation, the phase velocity and group velocity of *pure capillary waves* are obtained:

$$\omega = \sqrt{Tk^3/\rho}, \qquad v = \sqrt{2\pi T/\rho\lambda}, \qquad v_g = 3v/2. \qquad (5.31)$$

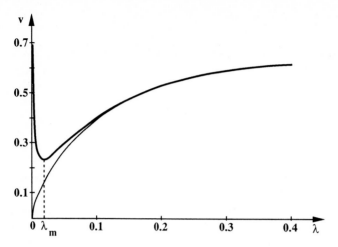

Fig 5.4. Phase velocity v versus wavelength λ when both gravitation and surface tension effects are taken into account. The <u>thin curve</u> represents the phase velocity for $\mathcal{T} = 0$.

5.5 Solitons in shallow water

Up to now we have considered linear surface-water waves, that is, waves with small amplitude. When the amplitude of the waves increases, the nonlinearity cannot be neglected, as expected from the boundary conditions (5.2b) and (5.2c). Thus, in the following we show that the propagation of weakly nonlinear and dispersive shallow water waves can be modeled by the KdV equation. To proceed with this problem we adopt a phenomenological approach.

In Sect.5.3, we saw that when the wavelength decreases, the waves become weakly dispersive. In this case, according to (5.16) the phase velocity is

$$v \approx c_0 (1 - k^2 h^2 / 6).$$

It deviates from the nondispersive phase velocity c_0 by a term which is proportional to the square of the wavenumber k; that is, the leading-order dispersive term is of order $(h/\lambda)^2 = \delta^2$.

Next, when the wavelength is very large compared to the depth, we have seen that the so-called shallow-water waves do not disperse. Thus, if the free surface displacement η, though small compared with the wavelength, is not regarded as infinitely small, a closer approximation to the wave velocity may be used if in (5.17) we replace h by h + η. This suggests a velocity, relative to the fluid, of the form

$$v = \sqrt{g(h + \eta)} \approx c_0 (1 + \frac{1}{2}\frac{\eta}{h}).$$

112

Moreover, since from (5.19) the horizontal velocity of fluid particles is $u = c_0 \eta/h$, the velocity of propagation of the wave is approximatively

$$c \approx c_0 \left(1 + \frac{1}{2} \frac{\eta}{h}\right) + c_0 \frac{\eta}{h} = c_0 \left(1 + \frac{3}{2} \frac{\eta}{h}\right). \qquad (5.32)$$

Consequently, the phase velocity deviates from the dispersionless phase velocity c_0 by a term which is proportional to the wave height η. This result, which predicts that nonlinear waves of elevation will steepen and eventually break, is due to Airy (1845). This steepening effect of a wave resulting from the amplitude dependence of the wave velocity (see Chaps.1 and 3) can be observed at the surface of the ocean. It is enhanced when the wave approaches a beach: the amplitude becomes so large (at constant energy) that the wave breaks down , as depicted in Fig.5.5. The final state of the wave when it hits the beach is catastrophic; this is the case for a *Tsunami* (The Open University course team 1989). The word Tsunami originates from the Japanese indicating ocean waves of great wavelength, of the order of hundreds of kilometers, and of small height, usually 1meter, generated by a major initial disturbance such as an earthquake.

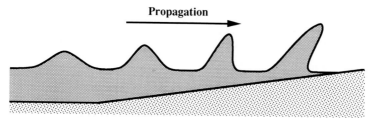

Fig.5.5. The steepening and breaking of a wave is enhanced when it approaches a beach.

Next, *when both weak nonlinearity and weak dispersion are present, they can balance.* We then combine the above equations into

$$v = \frac{\omega}{k} \approx c_0 \left(1 - \frac{k^2 h^2}{6} + \frac{3}{2} \frac{\eta}{h}\right). \qquad (5.33)$$

The dimensionless parameter

$$(3\eta/2h)/(h^2 k^2/6) \sim a_m/h^3 k^2, \qquad (5.34)$$

where a_m is the amplitude of η, is called the *Ursell number*. It gives a measure of nonlinearity compared to dispersion and was introduced by Ursell (1953). Then, for (5.11) we note that the nth time and space derivatives of the free surface elevation η (wave function) correspond to the nth powers of ω and k, respectively. Under these conditions, in the dispersion relation (5.33) the terms ω and k are replaced by the operators $i\partial/\partial t$ and $-i\partial/\partial x$, and we let the resulting expression operate on the function $\eta(x, t)$. We obtain

$$i\frac{\partial}{\partial t}\eta = -ic_0 (\frac{\partial}{\partial x} + \frac{1}{6}h^2\frac{\partial^3}{\partial x^3} + \frac{3}{2}\frac{\eta}{h}\frac{\partial}{\partial x})\eta \tag{5.35a}$$

or

$$\frac{\partial\eta}{\partial t} + c_0\frac{\partial\eta}{\partial x} + \frac{3}{2}\frac{c_0}{h}\eta\frac{\partial\eta}{\partial x} + \frac{1}{6}c_0 h^2\frac{\partial^3\eta}{\partial x^3} = 0. \tag{5.35b}$$

This KdV equation describes the propagation of lossless shallow-water waves. It can be cast in its standard form (see Sect.3.3) by eliminating the term $c_0\eta_x$. Using the transformation $s = x - c_0 t$, that is, by considering a frame moving at velocity c_0, we obtain

$$\frac{\partial\eta}{\partial t} + \frac{3}{2}\frac{c_0}{h}\eta\frac{\partial\eta}{\partial s} + \frac{1}{6}c_0 h^2\frac{\partial^3\eta}{\partial s^3} = 0. \tag{5.36}$$

In the equation originally found by Korteweg and de Vries the surface-tension effects were taken into account. By expanding (5.27) instead of (5.14), we find that

$$v \approx c_0 [1 + k^2 (\frac{T}{2\rho g} - \frac{h^2}{6})]. \tag{5.37}$$

Under these conditions the coefficient h^2 of η_{sss} in (5.36) is replaced by $(h^2 - 3T/\rho g)$. Now, proceeding as in Appendix 3C, we get the soliton solution

$$\eta = a_m \operatorname{sech}^2 [\sqrt{\frac{3a_m}{8h^3}} (x - Vt)], \tag{5.38}$$

which is shown in Fig.5.6. Here a_m is the amplitude, and the profile or width depends on the quantity $\sqrt{8h^3/3a_m}$. As for transmission-line solitons, the propagation velocity is amplitude dependent; it is given by

$$V = c_0 (1 + \frac{a_m}{4h}). \tag{5.39}$$

As we have seen in Sect.4.3 a set of periodic solutions, expressed in terms of elliptic functions, may also be found. These so-called cnoidal waves have sharpened crests and flattened troughs, as sketched in Fig.5.6b.

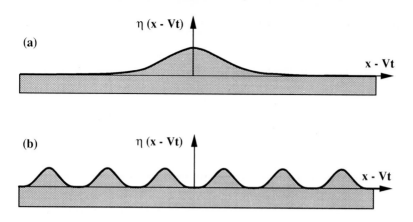

Fig.5.6a,b. Dispersive nonlinear waves in shallow water: (a) soliton; (b) cnoidal waves.

Remark

To proceed more systematically with the nonlinear problem a perturbative approach can be adopted (Witham 1974; Leblond and Mysak 1978). We first identify two small dimensionless parameters. One, δ, was introduced earlier (see (5.16)); it measures dispersion and is the ratio of the undisturbed depth to a typical wavelength. For disturbances in which the dominant wavelengths λ are of magnitude ℓ, we can write $\delta = h/\ell$. The other parameter $\varepsilon = a_m/h$ is the ratio of the maximum wave amplitude a_m to the undisturbed depth h; it measures nonlinearity. Thus, a simultaneous perturbation expansion in the two dimensionless parameters δ and ε is performed. The KdV equation is derived (see Appendix5C) under the hypothesis that $\varepsilon = \delta^2$, which represents the appropriate balance between nonlinearity and dispersion to get a wave of permanent form.

5.6 Experiments on solitons in shallow water

The most striking aspects of soliton behavior (see Chap.1) were seen by Scott Russell in a water tank which he described as being "a foot wide, eight or nine inches deep and twenty of thirty feet long". Following Russell's observations, a series of experiments were performed by Hammack and Segur (1974) in a big wave tank. Since then, simple water-tank experiments, which allow one to illustrate the important features of solitons in shallow water, have been described by Bettini et al. (1983) and Olsen et al. (1984). Following these authors we have constructed a water tank which is suitable for simple experiments on linear dispersive waves and solitons, which we now describe.

5.6.1 Experimental arrangement

This tank, shown in Fig.5.7, is 5m long, 0.4 m wide and 0.6m high. It consists of two elements, connected via flanges with O-rings. The vertical side walls and the bottom of the tank are constructed from transparent perspex screwed and glued together. The whole tank is on wooden bars, which provide the necessary horizontal alignement. Two types of wavemakers A and B are employed. Wavemaker A consists of a rectangular wooden block floating in water at one end of the tank; it is used to generate low-amplitude or linear waves. Wavemaker B consists of a perspex plate that is guided by two vertical grooves cut in the walls, at a given distance from the nearest end of tank. To generate waves, a certain amount of water is poured in the small space separated by the movable perspex plate and the end of the tank. In order to give a contrast with the backside of the tank, the water is colored with a small amount of fluoreiscein dissolved in ethanol.

At the end of the tank opposite to where waves are created, one has pure reflections. Otherwise, for some experiments one can place a smoothly degrading pebble beach which absorbs the incoming waves and allows one to limit parasitic reflections.

In the experiments we describe, two probes at different positions along the tank are used to measure the position of a water surface wave. They are of the resistance type, consisting of two parallel conductive wires with radius a=3mm separated by a distance d=10mm. The resistance between the two electrodes dipped into the water of conductivity σ at a depth z is given by

$$R(z) = \frac{1}{\pi \sigma z} \log \frac{d}{a} \; .$$

Fig.5.7. Sketch of the water tank with wavemaker B and two probes.

One part of this resistance is connected to an AC generator; the other part is connected to an amplifier used as an inverter in order to get an output voltage proportional to the depth z. The details of the electronics are given in Appendix5E. A numerical oscilloscope and a printer enable one to vizualise, store, and plot the profiles of the waves.

5.6.2 Experiments

First, wavemaker A is employed. It is slightly pushed down by hand to create a small-amplitude wavepacket. As the wave propagates, the changes in its profile are

116

recorded as a function of time. In the experiment we present, the level in the tank is
h = 4.5cm and the distance between the two probes is L_1 = 110cm. As shown in
Fig.5.8, the shape of the linear wavepacket spreads out; it is dispersive, the longer
wavelengths leading the shorter.

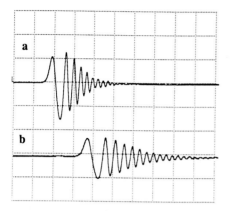

Fig.5.8a,b. Oscillograms showing the spreading out of a linear wavepacket when it propagates
along the tank; the velocity of the wave front is 64cm/s. Waves are generated with wavemaker A.
Scales: 1V/div, 1s/div.

Second, for experiments on solitons we use wavemaker B and the basic level
in the tank remains the same as before. By quickly removing the perspex plate, the
step in the small space can create one or several waves propagating along the tank.
In this case, the nearest closed end of the tank acts as a reflection plane for the
wave, which immediately after the removal of the plate is approximatively
rectangular (see Hammack and Segur 1974).

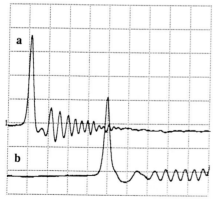

Fig.5.9a. Soliton profile recorded at x = 66cm and x = 296cm. Scales: 1V/div,0.9s/div.

For a first experiment the initial level in the small space is H = 6.5cm, the first probe is at a distance L_2 = 66cm from the movable plate, and the distance between the probes is L_1 = 230cm. Fig.5.9a shows records of this soliton. At the first probe its linear oscillatory tail is present; at the second it has already begun to disperse. The soliton velocity is v=66.6 cm/s; its experimental profile is well fitted by the theoretical sech-squared profile, as shown in Fig.5.9b.

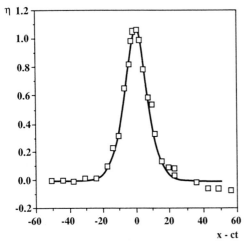

Fig.5.9b. Comparison between the experimental profile: (dots) and the theoretical profile (continuous line) given by (5.38).

Fig.5.10a. Picture showing the initial condition with a rectangular shape.

118

In a second experiment where H > 6.5cm, we can observe (just as for electrical transmission lines) that an initial rectangular wave breaks into a finite number of solitons and a low-amplitude (linear) oscillatory tail. The number of solitons depends on the width and height of the rectangular pulse (see Sect.10.5). Fig.5.11 shows the results for H = 8cm; the second probe is at a distance L_1= 2.87m.

Fig.5.10b. Picture showing the generated soliton moving past a probe (Photo R. Belleville).

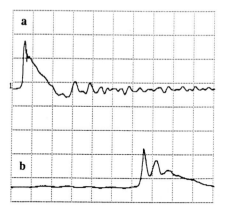

Fig.5.11. Evolution of an initial rectangular disturbance. Wave profile recorded at positions x = 66cm and x = 353cm. Scales: 5V/div, 1s/div.

The water tank we have described can be easily constructed and can be used for a number of other experiments on linear and nonlinear waves. The interested reader will find more details in the references mentioned at the beginning of this section (see also Kuwabara et al 1986).

5.7 Stokes waves and soliton wavepackets in deep water

We now turn our attention to weakly nonlinear deep-water waves whose characteristic length k^{-1} is comparable to or greater than the depth h. In 1847, Stokes showed, through the use of a small amplitude expansion of a sinusoidal wave, that periodic waves of finite amplitude are possible in deep water.

5.7.1 Stokes waves

Following Lamb (1932) (see also Kinsman 1965) we can calculate (Appendix 5D) the surface-wave profile and phase speed by using the stream function (defined in Appendix 5B) and a trick due to Rayleigh (1876) whereby the progressive waves are rendered steady. Thus, to third order the surface elevation of waves in deep water is given by

$$\eta = \frac{1}{2} k a_m^2 + a_m \cos k(x - ct) + \frac{1}{2} k a_m^2 \cos 2k(x - ct)$$
$$+ \frac{3}{8} k^2 a_m^3 \cos 3k(x - ct) + \dots , \tag{5.40}$$

where a_m is the amplitude, and the propagation velocity is given by (5D.13),

$$c = \sqrt{\frac{g}{k}(1 + k^2 a_m^2)}. \tag{5.41}$$

From (5.40), where the additive constant can be eliminated by an appropriate vertical translation of the coordinates, we see that the wave profile of the so-called Stokes wave contains components of different wavelengths propagating at the same velocity c. It is no longer exactly sinusoidal, as shown in Fig.5.12. Moreover, when the waves are not infinitesimal, the phase velocity increases with the amplitude a_m; this is Stokes original result (see Chap. 1) for deep-water waves.

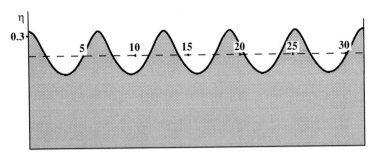

Fig.5.12. Stokes wave profile, as given by (5.40) for k=1 and a_m =0.3, which is a finite-amplitude wave in deep water.

5.7.2 Soliton wavepackets

In fact, as discussed in Chap.1, the Stokes waves are unstable to infinitesimal modulation perturbations. This instability was first discovered by Lighthill (1965) and then theoretically and experimentally confirmed by Benjamin and Feir (1967). These findings generated great interest in the problem of unsteady time evolution of deep-water wavetrains and wavepackets.

Thus, we further confine our attention to the weakly nonlinear deep-water gravity waves and look for the model equation that governs the evolution of a wavetrain envelope. The dispersion relation, deduced from (5.41), is

$$\omega^2 = gk \ (1 + k^2 a_m^2). \tag{5.42}$$

It has the general form $\omega = f(k, |A|^2)$ that we have already found for transmission lines (Chap. 4) and that we will find for mechanical transmission lines and optical fibers (Sects. 6.9 and 8.2). Consequently, proceeding exactly as in Chap.4 , we consider a wavetrain where the envelope A is assumed to vary slowly in time and space compared to the variations of the carrier wave:

$$\eta = \text{Re} \ [A \ (X, T) \ e^{i \ (\omega_0 t - k_0 x)}]. \tag{5.43}$$

Here, $X = \varepsilon x$ and $T = \varepsilon t$, where ε is a small parameter, $\varepsilon \ll 1$, represent the slow space and time variables . As the envelope of the wavetrain is slowly varying it contains a large number of crests of the carrier wave, and the amplitude distribution is concentrated in wavenumbers close to the dominant value $k = k_0$. Under these conditions we can expand the frequency ω as a function of k in Taylor series about the wave number k_0 of the sinusoidal carrier wave:

$$\omega - \omega_0 = (\frac{\partial \omega}{\partial k})(k - k_0) + \frac{1}{2} (\frac{\partial^2 \omega}{\partial k^2}) \ (k - k_0)^2 + (\frac{\partial \omega}{\partial |A|^2}) \ (|A|^2 - |A_0|^2). \tag{5.44}$$

Here, the first two derivatives, which represent the group velocity and the dispersion in the linear approximation, are evaluated at $k = k_0$. The third derivative, which represents the nonlinearity, is evaluated at $|A| = A_0 = 0$. Proceeding as in Sect.4.4, we can transform (5.44) into

$$i \ (\frac{\partial A}{\partial T} + \frac{\omega_0}{2k_0} \frac{\partial A}{\partial X}) - \frac{\omega_0}{8k_0^2} \frac{\partial^2 A}{\partial X^2} - \frac{1}{2} \ \omega_0 k_0^2 |A|^2 = 0. \tag{5.45}$$

The nonlinear Schrödinger equation (5.45) indicates that, to leading order, the wave envelope propagates at the group velocity

$$v_g = \omega_0 / 2k_0 \tag{5.46}$$

of the carrier wave. The evolution is governed by the balance of an envelope dispersion term and a nonlinear term

$$P = - \omega_0/8k_0^2, \quad Q = - \omega_0 k_0^2/2. \tag{5.47}$$

Historically, this equation was first discussed by Benney and Newell (1967), but it was first derived for deep water waves by Zakharov in 1968. Moreover, Zakharov and Shabat (1971) were the first to show that (5.45) admits exact envelope-soliton solutions and that an initial wavepacket eventually evolves into a number of envelope solitons and a dispersive tail (Sect.10.6). Here, the envelope-soliton solutions (see 4C.15) take the form

$$A = a_m \, sech \, [\sqrt{2} \, a_m \, k_0^2 \, \varepsilon(x - \frac{\omega_0}{2k_0} t \,)] \, exp \, (i \, \frac{\omega_0 k_0^2}{4} \, a_m^2 \, \varepsilon t \,), \tag{5.48}$$

where a_m is the amplitude. The corresponding wave packet is

$$\eta \, (x, t) = a_m \, sech \, [\sqrt{2} \, a_m \, k_0^2 \, \varepsilon(x - \frac{\omega_0}{2k_0} t \,)] \, cos \, [\, (\omega_0 + \varepsilon \, \frac{\omega_0 k_0^2}{4} \, a_m^2)t - k_0 x)]. \tag{5.49}$$

Moreover, (5.45) admits a uniform wavetrain (see Sect.4.6) solution with amplitude a_0 and wavenumber k_0,

$$A = a_0 \, exp \, (\, i \, \frac{\omega_0 k_0^2}{2} \, a_0^2 \, \varepsilon t \,), \tag{5.50}$$

which is modulationally unstable if the product of dispersion by nonlinearity is positive, a situation which is obtained in the present example. As time passes the wavetrain evolves into a series of pulses or envelope solitons. The quantity $a_0 k_0 = 2\pi(a_0/\lambda_0)$, where λ_0 is the wavelength, defines the slope of the wave or *wave steepness*.

From relations (4.41–43) the characteristic nonlinear length and dispersion length, which provide the length scales over which the nonlinear and dispersion effects become important, respectively, are given by

$$L_{nl} = v_g/|Q|a_m^2 = 1/k_0^3 a_m^2, \quad L_d = v_g L_0^2/2|P| = 2k_0 L_0^2.$$

Here L_0 is the half width of the pulse (see Sect.2.6). Thus for practical applications one can evaluate these parameters (compare to (8.20) and (8.24) for optical fibers).

5.7.3 Experiments on solitons in deep water

The predictions for the properties of soliton in deep water were tested by carefully controlled experiments (Yuen and Lake 1975, 1978) and the agreement between experiment and theory was very good. Specifically, the experiments were performed in a water tank 13.33m long, 1m wide, and 1m deep. As in the

122

experiments described in Sect.5.6, waves were generated at one end by a wavemaker (hinged paddle) and absorbed at the other end by a shallow beach. The typical frequencies of the waves generated were in the range 2–4Hz, and their profiles were recorded by probes and displayed on an oscillograph.

Experiments confirmed the theoretical prediction that an initial disturbance with carrier frequency F_0 =2Hz disintegrates into a finite number of solitons and a tail. The number of solitons depends on the initial profile. The head-on collision of two wavepackets with the same carrier frequency F_0 = 3Hz was observed. The results verify the stability properties predicted by the theory for envelope solitons. A series of experiments on modulational instability showed that slightly modulated initial wavetrain evolves into a strongly modulated wavetrain.

5.8 Experiments on modulational instability in deep water

Wave modulations are not just theoretical curiosities but have important consequences in oceanography and meteorology, as well as naval architecture and ocean engineering (Longuet-Higgins 1980). Several experiments (Feir 1967; Lake et al. 1977; Melville 1982; Su 1982) were performed in a water tank. Following these authors we present laboratory experiments on *induced* modulational instability (compare to Sect.4.7.2) of surface gravity waves in deep water.

Fig.5.13. Picture showing the long water tank, which consists of 10 elements of 2m connected via flanges with O rings. (Photo B. Michaux).

The tank is similar to the one described in Sect.5.6.1; it is a longer channel of dimensions 0.4m x 0.6m x 20m which consists of ten elements. It is filled to a typical depth of about 50cm (see Fig.5.13). At one end of the tank a wavemaker has been installed, which consists of a vertical, computer-controlled paddle that occupies the entire tank cross section. The movement of the wavemaker is fully programmable through a control system. A desktop computer provides intelligent control of the paddle motion. The probes are similar to those described in Sect. 5.6.1 and Appendix 5E. A numerical oscilloscope with fast-Fourier-transform

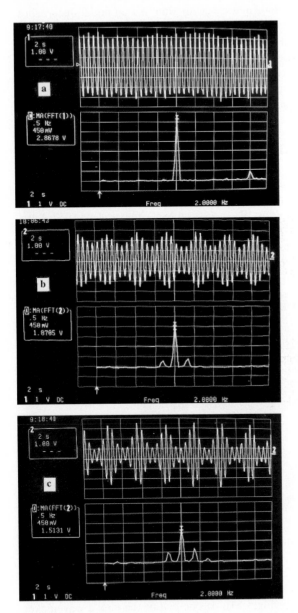

Fig.5.14. Successive oscillograms traces (a–c) showing the waveform modulation and its corresponding Fourier spectrum, with sidebands at ± (0.33±0.02) Hz from a central frequency at 2Hz, measured at the three probes situated at 0.8 m, 10m and 13m of the wave maker. (Photo B. Michaux)

processing and a printer enables one to observe, store, and plot the wave forms and their Fourier spectrum.

Consistent with the assumptions implicit in the derivation of the NLS equation (5.45), we have conducted several experiments for which the input wave amplitude a_o is small but finite with respect to the depth $h=h_o$, and the input frequency $f_o=\sqrt{g/2\pi\lambda_o}$ is chosen in order to have $\tanh(k_oh)\approx 1$ (see Sect.5.3.2). In the experiment we present, an initial periodic wave train with amplitude $a_o=1.1$cm and frequency $f_o=2$Hz is generated at the input end of the channel of depth $h_o=46$cm. Under these conditons the wave steepness, defined in Sect.5.7.2, is $k_oa_o=0.18$. The initial (carrier) wave is weakly amplitude modulated at the frequency $F_m=0.33$Hz; the modulation rate is 5%. This value of the modulation frequency is chosen because, guided by theory (see Sects.4.6 and 5.7.2) that predicts a maximum growth rate of the instability at $F_m=2\pi f_oa_o/\lambda_o=0.36$Hz, we first performed several experiments with different modulation frequencies ranging from 0.4 Hz to 0.3Hz. We have found that for $F_m=0.33\pm0.02$Hz the growth rate is maximum.

The evolution of the wave train and its progressive self-modulation, in time and space, as it propagates along the channel can be simply observed. Figure 5.14 shows records of the waveform and its Fourier spectrum at the three probes that are at distances $L_1=0.8$m, $L_2=10$m and $L_3=13$m from the wavemaker, respectively. At the first probe (Fig.5.14a), if one ignores the apparent fluctuations of the signal amplitude, which arise from digitization (numerical oscilloscope), one cannot detect the weak modulation of the periodic wave form. The Fourier spectrum shows the initial wave frequency and its weak second harmonic. At the second probe (Fig.5.14b) modulation instability can be observed on the wave-form oscillogram and its spectrum, characterized by sidebands at frequencies ±0.33Hz. At the third probe (Figs.5.14c) the amplitude of the sidebands has increased at the expense of the central peak (carrier wave) amplitude, the self-modulation (rate$\approx70\%$) can be clearly observed. The initial wave train has evolved into a series of pulses or envelope solitons. We have used a relatively short tank and, in order to accelerate the nonlinear development, the wave slope (steepness) is typically rather high. Nevertheless, the above results show that the agreement between theory and experiments is quite reasonable.

Remarks

(i) In the laboratory experiments presented here, the physics is modeled by the NLS equation whose derivation is based on three assumptions:
– weak nonlinearity and thus small wave steepness
– slow variations of wave amplitude and phase (narrow spectrum)
– the existence of a constant carrier-wave frequency
However, when the steepness of the initial wave becomes sufficiently large and its spectrum becomes broader, analyses of nonlinear dynamics must be carried to a higher order than in the case of the weakly nonlinear description in Sect.5.7. The nonlinear Schrödinger equation is no longer a good model. Other modulation

125

equations were found, as for example by Dysthe (1979) (for a review see Peregrine 1983); Crawford et al (1980) proposed that an integral equation, also first derived by Zakharov (1968), be used instead. Nevertheless, some problems related to the long time evolution of the modulated wave trains remain open (Pierce and Knobloch 1995). Aside from these numerous theoretical investigations several experimental results on instabilities of water waves were reported (for a review see Hammack and Henderson 1993).

(ii) In the experiments described in Sect.5.8 the data were analyzed by using a fast-Fourier-transform (FFT) procedure. In fact, the Fourier method is perhaps the most common procedure for analyzing experimental time series in a wide variety of physical situations. Linear Fourier analysis provides the experimenter with a powerful tool for probing wave propagation in the frequency and wavenumber domains. Even when the motion is known to be nonlinear, the linear Fourier transform is often applied, simply because it has been proven to be one of the most valuable of all data analysis tools. We apply this method in spite of the fact that it is based upon linear, dispersive wave motion. An alternative data analysis approach, based upon the inverse scattering transform (see Chap.10) used to solve nonlinear wave equations, has been proposed (see Osborne and Petti 1994, and references therein). The procedure of the inverse scattering transform can be thought of as a generalization (Sect.10.1) to certain nonlinear problems of the Fourier transform technique for solving linear problems. Nonlinear Fourier analysis, like its linear counterpart, has several features that make it useful for the analysis of experimental data. These ideas have been exploited (Osborne et al. 1991) to find solitons in complex wave trains in ocean surface wave data and to study (Osborne and Petti 1994) shallow-water laboratory wave trains.

5.9 Some applications of the KdV model

5.9.1 Blood-pressure pulse propagation

Theoretical studies involving modeling of blood pressure propagation have their motivation in the desire to understand one the most fascinating secrets of life. From experiments it is known that blood-pressure pulses propagate along the major arteries with characteristic shape changes (Mc Donald 1974, Pedley 1980). Indeed, the blood motion is accompanied by an amplitude increase with rapid development of a steep front for the pressure pulse known in the literature as peaking and steepening. These two phenomena are combined with an increase of the pulse-wave velocity and are in accordance with the development of a wave called a dichrotic wave.

The majority of theoretical investigations on blood motion deals with linearized models, even though there has been increasing evidence of the presence of strong nonlinear phenomena. Nevertheless, nonlinear models were proposed (Hashizume 1985, 1988; Yomosa 1987) in order to study the motion of weakly nonlinear pressure waves in a thin nonlinear elastic tube filled with an incompressible fluid. It was shown that the dynamics of such pulses may be governed by the KdV equation. Later, a quasi-one-dimensional model (Paquerot and Remoissenet 1994) which describes blood pressure propagation in large arteries, was studied. In the limit of an ideal fluid and for slowly varying arterial parameters a Boussineq-like equation (see remark in Sect.3.3) was obtained which, under certain conditions, may be approximated by a KdV equation . The two-soliton solution (see Sect. 10.6) of the KdV equation is interesting because the "in-vivo profile" of the arterial pulse is composed of a main pulse and an associated "dichrotic wave", that is, the pulse wave in vessels may be regarded as a two-pulse wave. In fact, the results show that the blood-pressure pulse can be seen as a wave whose shape evolves between a solitary wave and a shock wave.

Such oversimplified models represent the first approaches to the nonlinear dynamics of blood pressure waves which represent a very complex but exciting problem. Further explorations, taking into account for example the viscosity and branching, are needed.

5.9.2 Nonlinear modes of liquid drops

Recently, it has been shown (Ludu and Draayer 1998) that the nonlinear equations which describe the dynamics of the surface of a liquid drop may be approximated by a KdV equation, giving traveling solutions that are cnoidal waves (see Sect. 4.3). Specifically, these traveling deformations can range from small oscillations (normal modes), to cnoidal oscillations, and to solitary waves.

Appendix 5A. Basic equations of fluid mechanics

We derive the basic equations of fluid mechanics (Landau and Lifschitz 1954; Batchelor 1967) by considering *the ideal fluid model*. We consider an inviscid fluid, i.e., a fluid whose viscosity can be neglected. Moreover, the motion of the fluid is supposed to be adiabatic: there is no heat exchange between different parts of it. We begin with the equation which expresses the conservation of mass.

5A.1 Conservation of mass

In the fluid we consider (Fig.5A.1) an arbitrary volume V_0 bounded by a closed surface S_0. If we define by ρ the fluid density, the mass M of fluid contained in V_0 is given by

$$M = \int_{V_0} \rho \, dV . \qquad (5A.1)$$

The total mass of fluid flowing per unit time dt into the whole volume V_0, which corresponds to the flux through S_0 of the mass current or mass flux density $\rho \mathbf{v}$, is given by

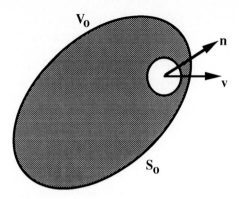

Fig. 5A.1. An arbitrary volume of fluid

$$\frac{\partial M}{\partial t} = - \int_{S_0} \rho \mathbf{v} \mathbf{n} \, dS. \tag{5A.2}$$

Here $\mathbf{v} = \mathbf{v}(x, y, z, t)$ is the velocity of a particle of fluid and $\mathbf{n}$ is a unit vector which defines the normal to the surface element dS. Their scalar product is positive if the fluid is flowing out of the volume and negative if the flow is into the volume. We now write that the incoming flux through S_0 is equal to the increase per unit time of fluid; i.e., from (5A.1) and (5A.2) we get

$$\frac{\partial M}{\partial t} = \frac{\partial}{\partial t} \int_{V_0} \rho \, dV = - \int_{S_0} \rho \mathbf{v} \mathbf{n} \, dS. \tag{5A.3}$$

In (5A.3) the surface integral can be transformed to a volume integral by using the Green's theorem

$$\int_{V_0} [\frac{\partial \rho}{\partial t} + \mathrm{div}(\rho \mathbf{v})] dV = 0. \tag{5A.4}$$

Since the volume V_0 is arbitrary, this equation must hold for any volume, and the integrand must vanish, yielding an equation that expresses the continuity of mass, or *continuity equation*:

$$\frac{\partial \rho}{\partial t} + \mathrm{div}(\rho \mathbf{v}) = 0. \tag{5A.5}$$

128

5A.2 Conservation of momentum

We consider again the arbitrary volume V_0 (Fig.5A.1) in the fluid. We assume that the only force acting on the fluid is due to the pressure $p = p(x, y, z, t)$. Under this condition, the total force $\mathbf{F}$ acting on V_0 is equal to the integral of the pressure over the surface S_0. After transforming this integral into a volume integral, we have

$$\mathbf{F} = - \int_{S_0} p\mathbf{n}dS = - \int_{V_0} (\mathbf{grad}p)dV. \tag{5A.6}$$

Equation (5A.6) tells us that the fluid exerts a force $d\mathbf{F} = - (\mathbf{grad}\ p)\ dV$ on any volume element dV of the fluid. We now use Newton's second law of Mechanics, which states that the time rate of change of the linear momentum of a body is equal to the resultant external force acting on the body, to derive the equation of motion of a volume element dV of mass dM in the fluid. We write

$$\rho \frac{d\mathbf{v}}{dt} = - \mathbf{grad}\ p, \tag{5A.7}$$

where $\rho = dM/dV$ is the density. The velocity $\mathbf{v}$ is a function of the space coordinates and time t. The change of velocity $d\mathbf{v}$ of a particle of fluid can be written as

$$d\mathbf{v} = \frac{\partial \mathbf{v}}{\partial t} dt + \frac{\partial \mathbf{v}}{\partial x} dx + \frac{\partial \mathbf{v}}{\partial y} dy + \frac{\partial \mathbf{v}}{\partial z} dz, \tag{5A.8}$$

where the derivative $\partial \mathbf{v}/\partial t$ is taken at constant x, y, and z, i.e., at a constant point in space. Dividing each term of (5A.8) by dt and then noting that the last three terms can be written as $(\mathbf{v}.\ \mathbf{grad})\ \mathbf{v}$, we obtain

$$\frac{d\mathbf{v}}{dt} = \frac{\partial \mathbf{v}}{\partial t} + (\mathbf{v}.\ \mathbf{grad})\ \mathbf{v}. \tag{5A.9}$$

Substituting (5A.8) ... (5A.7) yields

$$\frac{\partial \mathbf{v}}{\partial t} + (\mathbf{v}.\ \mathbf{grad})\ \mathbf{v} = - \frac{1}{\rho}\ \mathbf{grad}\ p. \tag{5A.10a}$$

Equation (5A.10a) is *Euler's equation of motion* of fluid mechanics, first derived by L. Euler in 1757. It can be generalized to account for an external force $\rho\mathbf{f}$ other than that due to the pressure p. This force must be added to the right-hand side of (5A.7), so that (5A.9) takes the form

$$\frac{\partial \mathbf{v}}{\partial t} + (\mathbf{v}.\ \mathbf{grad})\ \mathbf{v} = - \frac{1}{\rho}\ \mathbf{grad}\ p + \mathbf{f}. \tag{5A.10b}$$

If, for example, the external force is the gravitational field $\mathbf{g}$, one has $\mathbf{f} = \mathbf{g}.$ From standard vector analysis one has

$$\frac{1}{2} \mathbf{grad} \; v^2 = \mathbf{v} \times \mathbf{curlv} + (\mathbf{v}. \; \mathbf{grad}) \; \mathbf{v}.$$

Using this relation one can rewrite (5A.10b) in the form

$$\frac{\partial \mathbf{v}}{\partial t} + \frac{1}{2} \mathbf{grad} \; v^2 + \omega \times \mathbf{v} = - \frac{1}{\rho} \mathbf{grad} \; p + \mathbf{f}, \qquad (5A.10c)$$

where the vector $\omega = \mathbf{curlv}$ is defined as the *vorticity*, which is a concept of central importance in fluid dynamics .

5A.3 Conservation of entropy

The equation for entropy per unit mass s follows from our assumption that we neglect effects which lead to production of entropy due to viscous heating and thermal diffusion. In this adiabatic approximation the entropy of any particle of fluid remains constant as that particle moves about in space. This is expressed by

$$ds/dt=0. \qquad (5A.11)$$

Proceeding as in (5A.8) for the total time derivative, condition (5A.11) can be transformed into

$$\frac{\partial s}{\partial t} + \mathbf{v}. \; \mathbf{grad} \; s = 0. \qquad (5A.12)$$

By using (5A.5) one can write (5A.12) as a continuity equation for entropy:

$$\frac{\partial(\rho s)}{\partial t} + \mathrm{div}(\rho s \mathbf{v}) = 0. \qquad (5A.13)$$

Note that in many cases the entropy can be assumed to be constant in space and time, which considerably simplifies the problem. The entropy equation is then reduced to $s = C^{te}$.

An essential aspect of the above basic equations is that they are *nonlinear*. In the text we apply them to the specific problem of surface-water waves.

Appendix 5B. Basic definitions and approximations

5B.1 Streamline

At time t in a flow, every fluid particle has a given velocity $\mathbf{v}$ with a definite direction. The instantaneous lines everywhere parallel to velocity are called the *streamlines* of the flow. For a *steady flow*, that is, a flow in which the velocity is constant in time at any point occupied by the fluid, the streamline pattern remains

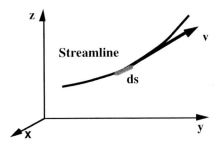

Fig.5B.1. An arbitrary stream line.

constant. By contrast, for an *unsteady flow* the streamline pattern changes with time.

Next, let dx, dy and dz be the components of an element of arc length ds along a streamline (Fig.5B.1). We denote by u = dx/dt, v=dy/dt, and w = dz/dt the components of the local velocity vector. Along a streamline one then has

$$\frac{dx}{u} = \frac{dy}{v} = \frac{dz}{w} .$$ (5B.1)

Thus the knowledge of the velocity components allows one to integrate (5B.1) and to find the streamline equation.

5B.2 Irrotational and incompressible flow

In many cases the density of fluid particles does not change appreciably along the fluid motion and may be considered to be constant throughout the volume of the fluid and throughout its motion. In these cases, Euler's equation (5A.10) remains unchanged. However, the continuity equation (5A.5) simply reduces to

$$\text{div } \mathbf{v} = \frac{\partial u}{\partial x} + \frac{\partial v}{\partial y} + \frac{\partial w}{\partial z} = 0.$$ (5B.2)

Moreover, in the main problems of water waves the flow can be assumed to be *irrotational*, that is, the individual fluid particles do not rotate. Consequently, when the vorticity $\omega = 0$, (5A.10c) reduces to

$$\frac{\partial \mathbf{v}}{\partial t} + \frac{1}{2} \text{grad } v^2 = -\frac{1}{\rho} \text{grad } p + \mathbf{f}.$$ (5B.3)

We then note that, like any vector field having zero curl, the velocity of an irrotational flow can be expressed as the gradient of some scalar. This scalar is called the *velocity potential*, it is usually denoted by ϕ:

$$\mathbf{v} = \mathbf{grad}\ \phi = \mathbf{i}\ \frac{\partial \phi}{\partial x} + \mathbf{j}\ \frac{\partial \phi}{\partial y} + \mathbf{k}\ \frac{\partial \phi}{\partial z} \tag{5B.4}$$

and the fluid flow is called a *potential flow*. In (5B.4) $\mathbf{i}$, $\mathbf{j}$, and $\mathbf{k}$ denote the unit vectors along the directions x, y, and z. With the substitution of (5B.4), the continuity equation (5B.2) for an incompressible fluid becomes Laplace's equation:

$$\Delta\phi = \frac{\partial^2\phi}{\partial x^2} + \frac{\partial^2\phi}{\partial y^2} + \frac{\partial^2\phi}{\partial z^2} = 0, \tag{5B.5}$$

which must be satisfied everywhere in the fluid. One can substitute (5B.4) in (5B.3) to get

$$\mathbf{grad}\ [\ \frac{\partial \phi}{\partial t} + \frac{1}{2}\ (\mathbf{grad}\ \phi)^2 + \frac{p}{\rho}\ ] = \mathbf{f}. \tag{5B.6}$$

5B.3 Two-dimensional flow: the stream function

We now restrict ourselves to an irrotational and incompressible fluid flow in two dimensions. In this case the flow is characterized by certain analytical peculiarities and the solutions of several interesting problems can be obtained. In a two dimensional flow the velocity v depends on only two coordinates, say x and z as in the following for surface-water waves. In this case the continuity equation (5B.2) reduces to

$$\frac{\partial u}{\partial x} + \frac{\partial w}{\partial z} = 0. \tag{5B.7}$$

It is then convenient to define a *stream function* ψ (x, y, t) such that

$$u = \frac{\partial \psi}{\partial z}\ , \qquad w = -\ \frac{\partial \psi}{\partial x}\ . \tag{5B.8a}$$

Moreover, from the definition (5B.4) of the velocity potential we can also write

$$\frac{\partial \phi}{\partial x} = \frac{\partial \psi}{\partial z}\ , \qquad \frac{\partial \phi}{\partial z} = -\ \frac{\partial \psi}{\partial x}\ . \tag{5B.8b}$$

Differentiating the first equation with respect to y and the second with respect to x, and subtracting, we obtain Laplace's equation for the stream function:

$$\Delta \psi = 0. \tag{5B.9}$$

For a steady flow the stream function is independent of time and we can write

132

$$d\psi = \frac{\partial \psi}{\partial x} dx + \frac{\partial \psi}{\partial z} dz = -w\, dx + u\, dz. \qquad (5B.10)$$

If we make

$$d\psi = 0 \qquad (5B.11)$$

along a streamline, we see that the instantaneous streamlines are given by the curves $\psi = C^{te}$. On the other hand the curves $\phi = C^{te}$ are the curves of equal velocity potential. Since from (5B.8) we have $\psi_x \phi_x + \psi_z \phi_z = 0$, the two systems of curves are orthogonal to each other (compare to electric field and potential).

For a steady flow the free surface of a liquid does not change in time and therefore has to be a streamline with flow along it. Without restricting the generality one can assume that $\psi = 0$. Thus, if one can use the stream function as an independent variable, the surface of the liquid will have a known position.

In the notation of complex functions one can introduce a complex potential V which is a function of a complex variable Z such that

$$V = \phi + i\psi, \qquad Z = x + iz. \qquad (5B.12)$$

Mathematically, (5B.12) represent the Cauchy–Riemann conditions. Those tell us that the complex potential V is an analytic function of Z. In the language of analytic function theory, the real and imaginary parts of V(Z) are conjugate harmonic functions, since they enjoy an orthogonality property and both satisfy the two-dimensional Laplace equation. Under these assumptions, for gravitational forces

$(\mathbf{f} = \mathbf{g})$ one can rewrite (5B.6) as

$$\mathbf{grad}\ [\ \frac{\partial \phi}{\partial t} + \frac{1}{2}\ (\mathbf{grad}\ \phi)^2 + \frac{p}{\rho} + gz\ ] = 0. \qquad (5B.13)$$

Equation (5B.13) may be integrated to yield

$$\frac{\partial \phi}{\partial t} + \frac{1}{2}\ (\mathbf{grad}\ \phi)^2 + \frac{p-p_0}{\rho} + gz = C(t), \qquad (5B.14)$$

where p_0 is a constant and C(t) is an arbitrary function of time which can be put to zero without loss of generality. Since the velocity is the space derivative of the potential ϕ, one can add to ϕ any arbitrary function of time. Clearly, replacing ϕ by a new potential $\phi + \int [C(t) + p_0/\rho]dt$ gives zero on the right-hand side of (5B.14). Equation (5B.14) then becomes

$$\frac{\partial \phi}{\partial t} + \frac{1}{2}(\mathbf{grad}\ \phi)^2 + \frac{p}{\rho} + gz = 0. \tag{5B.15}$$

5B.4 Boundary conditions

Laplace's equation (5B.5) and equation (5B.15) must be supplemented by the boundary conditions at the free surface as well as at the horizontal bed. At the upper boundary, that is at the free surface, we set

$$z = \eta\ (x,\ y,\ t), \tag{5B.16}$$

where η is the z coordinate of a point of the surface, corresponding to the vertical displacement of the surface. At equilibrium (unperturbed surface) $\eta = 0$. Differentiating both parts of (5B.16) and proceeding as in (5A.8),

$$w = \frac{\partial \eta}{\partial x} u + \frac{\partial \eta}{\partial y} v + \frac{\partial \eta}{\partial t}, \qquad \text{at } z = \eta\ (x,\ y,\ t), \tag{5B.17}$$

where $u = dx/dt$, $v = dy/dt$, and $w = dz/dt$ are the components of the velocity. Then, taking (5B.4) into consideration, we obtain

$$\phi_z = \eta_x \phi_x + \eta_y \phi_y + \eta_t, \qquad \text{at } z = \eta\ (x,\ y,\ t). \tag{5B.18}$$

This first boundary condition on the surface expressed in either of the forms (5B.17) or (5B.18) is often called *the kinematic boundary condition*, in the literature. At the interface between the surface of the water and the external fluid, surface tension is neglected and the pressure in both fluids must be equal; it is the constant atmospheric pressure p_a. Then, at the surface $z = \eta$, (5B.15) yields

$$\frac{\partial \phi}{\partial t} + \frac{1}{2}(\mathbf{grad}\ \phi)^2 + \frac{p_a}{\rho} + g\eta = 0. \tag{5B.19}$$

The constant atmospheric pressure can be eliminated from (5B.19) by the potential transformation $\phi \rightarrow \phi + (1/\rho)p_a t$, as already mentioned above (see (5B.14)) this makes no difference since $\mathbf{v} = \mathbf{grad}\ \phi = \mathbf{grad}\ [(\ \phi + (1/\rho)p_a t]$. Assuming this is done, and by using (5B.4), (5B.19) is reduced to the so-called *dynamic boundary condition*,

$$\phi_t + \frac{1}{2}(\phi_x{}^2 + \phi_y{}^2 + \phi_z{}^2) + g\eta = 0, \qquad \text{at } z = \eta\ (x,\ y,\ t). \tag{5B.20a}$$

Differentiating (5B.20a) with respect to x, one gets an equivalent expression in terms of the velocities:

$$u_t + uu_x + vv_x + ww_x + g\eta_x = 0. \qquad (5B.20b)$$

At a solid fixed boundary, the normal velocity of the water must vanish because the interface is defined by the property that the fluid does not cross it. Here, the velocity must be zero in the z direction, but not, however, in the x and y directions, owing to our neglecting the viscous effects. Consequently,

$$w = \frac{\partial\phi}{\partial z} = 0 \quad \text{at } z = -h. \qquad (5B.21)$$

5B.5 Surface tension

Let us consider (see Fig.5B.2) two elements of arcs PQ and P'Q', on an element ABCD of the free surface, which are contained in the planes normal to this surface. At point O the local radii of curvature of the surfaces are denoted by R and R', respectively.

The free surface element experiences, at both its ends AD and BC (and similarly at both its ends AB and DC), a surface tension T, per unit length, directed tangentially to the surface (see Fig.5B.2). The atmospheric pressure is p_a and the pressure in the water just inside the interface is p. Therefore, the equilibrium of the free surface element requires that the sum of surface tension forces,

$$dF + dF' = Tdy\frac{dx}{R} + Tdx\frac{dy}{R}, \qquad (5B.22)$$

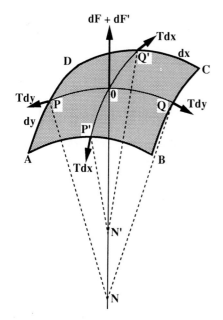

Fig. 5B.2. A free surface element experiencing a tension T per unit length.

balances the resultant $(p_a - p)$ dxdy of the pressure forces. It follows that the pressure difference is related to the curvatures by the Laplace relation:

$$p_a - p = T(\frac{1}{R} + \frac{1}{R'}). \tag{5B.23}$$

At time t, the free surface equation is $z = \eta(x,y,t)$ and the curvatures are

$$\frac{1}{R} = \frac{\partial^2\eta/\partial x^2}{[1+(\partial\eta/\partial x)^2]^{3/2}}, \qquad \frac{1}{R'} = \frac{\partial^2\eta/\partial y^2}{[1+(\partial\eta/\partial y)^2]^{3/2}}.$$

For waves with small slopes it may be approximated by the relations $1/R = \partial^2\eta/\partial x^2$ and $1/R' = \partial^2\eta/\partial y^2$. Therefore, one gets the condition

$$p_a - p = T\left(\frac{\partial^2\eta}{\partial x^2} + \frac{\partial^2\eta}{\partial y^2}\right) \qquad \text{at } z = \eta(x,y,t). \tag{5B.24}$$

Appendix 5C. Derivation of the KdV equation: the perturbative approach

The motion of the free surface is described by obtaining a solution of (5.2a) that satisfies the boundary conditions (5.2b–d). Since the total depth $(z + h)$ is small, we attempt to obtain a solution of the two dimensional Laplace equation

$$\phi_{xx} + \phi_{zz} = 0, \tag{5C.1}$$

by using an expansion of the form

$$\phi = \phi_0 + (z+h)\phi_1 + (z+h)^2\phi_2 + (z+h)^3\phi_3 + (z+h)^4\phi_4 + \dots , \tag{5C.2}$$

where $\phi_0, \phi_1, \dots$ are functions of (x,t). Substituting (5C.2) in (5C.1) and identifying terms according to the same powers of $(z + h)$ yields

$$\phi_{0xx} + 2\phi_2 = 0, \qquad \phi_{1xx} + 6\phi_3 = 0, \qquad \phi_{2xx} + 12\phi_4 = 0. \tag{5C.3}$$

The expansion (5C.2) for the velocity potential ϕ must satisfy the boundary condition (5.2d) at the bottom $(z = -h)$. Thus, we first obtain $\phi_1 = 0$ and using the second relation (5C.3) we find that $\phi_3 = 0$ and so on, all the terms with odd n vanishing. Then, from (5C.2) and (5C.3) we get

$$\phi = \phi_0 - \frac{1}{2}(z+h)^2\phi_{0xx} + \frac{1}{24}(z+h)^4\phi_{0xxxx}. \tag{5C.4}$$

From (5C.4) we calculate the horizontal and vertical components of the velocity:

$$u = \phi_x = \phi_{ox} - \frac{1}{2}(z+h)^2\phi_{oxxx} + O[(z+h)4], \qquad (5C.5a)$$

$$w = \phi_z = -(z+h)^2\phi_{oxxx} + \frac{1}{6}(z+h)^3\phi_{oxxxx} + O[(z+h)4]. \qquad (5C.5b)$$

Then, on substitution of (5C.4) and (5C.5) in the nonlinear boundary conditions (5.2b) and (5.2c) at the free surface ($z = \eta$) we obtain

$$\eta_t + [(h + \eta)\,\phi_{ox}]_x - \frac{1}{6}(h+\eta)^3\phi_{oxxxx} - \frac{1}{2}(h+\eta)^2\eta_x\phi_{oxxx} + ... = 0, \qquad (5C.6a)$$

$$g\eta + \phi_{ot} + \frac{1}{2}\phi_{ox}^2 - \frac{1}{2}(h+\eta)^2[\,\phi_{otxx} + \phi_{ox}\phi_{oxxx} - \phi_{oxx}^2\,] + ... = 0. \qquad (5C.6b)$$

If we linearize (5C.6) i.e., if we neglect product terms, we obtain

$$\eta_t + h\phi_{oxx} - \frac{1}{6}h^3\phi_{oxxxx} = 0, \qquad (5C.7a)$$

$$g\eta + \phi_{ot} - \frac{1}{2}h^2\phi_{otxx} = 0. \qquad (5C.7b)$$

Differentiating (5C.7b) with respect to time and using (5C.7a) yields the linear dispersive equation

$$\phi_{ott} - c_o^2\phi_{oxx} - \frac{1}{2}h^2\phi_{oxxtt} + \frac{g}{6}h^3\phi_{oxxxx} = 0, \qquad (5C.8)$$

where c_o is given by (5.17). By assuming a solution of the form $\phi_o = A\cos(\omega t - kx)$ we get the dispersion relation

$$\omega^2 = c_o^2 k^2(1 + \frac{h^2k^2}{6})(1 + \frac{h^2k^2}{2})^{-1}.$$

Keeping in mind that $kh \ll 1$, we see that this relation can be approximated by the dispersion relation (5.16) obtained for linear and weakly dispersive shallow water waves. The simplest linear approximation is obtained if we neglect the dispersive terms (of order h^2 and h^3) in (5C.7). Under these assumptions, we can combine equations (5C.7) so as to eliminate ϕ_o (or η); we find that η (or ϕ_o) satisfies the simple linear wave equation $\eta_{tt} - c_o^2\eta_{xx} = 0$ (or $\phi_{tt} - c_o^2\phi_{xx} = 0$). These equations admit simple-harmonic-wave solutions .

We now proceed with the nonlinear equations (5C.7). We first normalize the variables by taking (see remark of Sect.5.5) the original variables to be

$$x = \ell X, \qquad t = \ell T/c_o, \qquad \eta = \varepsilon h \eta', \qquad \phi = \varepsilon c_o \ell \, \Phi_o.$$ (5C.9)

On substitution in (5C.6) we find that

$$\eta'_T + [(1 + \varepsilon \eta') \, \Phi_{oX} \,]_X - \frac{\delta^2}{6}(1+\varepsilon\eta')^3\Phi_{oXXXX} - \frac{\delta^2\varepsilon}{2}(1+\varepsilon\eta')^2\eta'_X\Phi_{oXXX} + \dots = 0,$$ (5C.10a)

$$\eta' + \Phi_{oT} + \frac{\varepsilon}{2}(\Phi_{oX})^2 - \frac{\delta^2}{2}(1+\varepsilon\eta')^2 [\, \Phi_{oTXX} + \varepsilon\Phi_{oX}\Phi_{oXXX} - \varepsilon(\Phi_{oXX})^2 \,] + \dots = 0.$$ (5C.10b)

Introducing the function $f = \Phi_{oX}$, retaining the terms in ε and δ^2, and differentiating equation (5C.10b) with respect to X yields

$$\eta'_T + [(1 + \varepsilon \eta') \, f \,]_X - \frac{\delta^2}{6} f_{XXX} = 0,$$ (5C.11a)

$$\eta'_X + f_T + \varepsilon f f_X - \frac{\delta^2}{2} f_{TXX} = 0.$$ (5C.11b)

When the terms in ε and δ^2 are neglected, equations (5C.11) reduce to

$$\eta'_T + f_X = 0, \qquad \eta'_X + f_T = 0.$$

These linear equations admit a simple solution

$$f = \eta', \qquad \eta_T = -\eta_X.$$

These relations mean that η' is a function of the variable $(X - T)$, i.e., it corresponds to a disturbance moving along the positive X direction. Note that for a wave moving along the negative X direction η' would be a function of $(X + T)$, corresponding to $\eta_T = \eta_X$. In order to determine an equation for h, valid to first order in ε and δ^2, we look for a solution of (5C.11) in the form

$$f = \eta' + \varepsilon F + \delta^2 G,$$ (5C.12)

where F and G are functions of η and its X derivatives. Then, inserting (5C.12) in (5C.11) yields

$$\eta'_T + \eta'_X + \varepsilon (F_X + 2\eta'\eta'_X) + \delta^2 (G_X - \frac{1}{6}\eta'_{XXX}) = 0,$$ (5C.13a)

$$\eta'_T + \eta' x + \varepsilon \left(F_T + \eta'\eta'_X \right) + \delta^2 \left(G_T - \frac{1}{2}\eta'_{XXT} \right) = 0. \tag{5C.13b}$$

Since η' is a function of the variable $(X - T)$, all T derivatives in (5C.13) may be replaced by minus X derivatives. To the order of accuracy being considered, after having subtracted (5C.13b) from (5C.13a), we obtain

$$\varepsilon \left(2F_X + \eta'\eta'_X \right) + \delta^2 \left(2G_X - \frac{2}{3}\eta'_{XXX} \right) = 0. \tag{5C.14}$$

Since the parameters ε and δ are independent, the coefficients of ε and δ^2 must vanish separately. Then (5C.13a) and (5C.13b) are consistent if

$$F = -\frac{\eta'^2}{4}, \qquad\qquad G = \frac{1}{3}\eta'_{XX} \tag{5C.15}$$

Inserting the above relations and (5C.12) for f into (5C.13a), we get the Korteweg– de Vries equation:

$$\eta'_T + \eta' x + \frac{3}{2}\varepsilon\, \eta'\eta'_X + \frac{\delta^2}{6}\eta'_{XXX} = 0. \tag{5C.16}$$

With the space and time variables restored (see (5C.9)) this equation becomes

$$\eta_t + c_o\eta_x + \frac{3}{2}\frac{c_o}{h}\eta\eta_x + c_o\frac{h^2}{6}\eta_{xxx} = 0. \tag{5C.17}$$

It can be cast in the form (5.36). Moreover, we can make a change of variables:

$$\tau = \frac{1}{6}\sqrt{\frac{g}{h}}\,t, \qquad \xi = \frac{x}{h}, \qquad \phi = \frac{3}{2}\frac{\eta}{h}. \tag{5C.18}$$

In term of these non dimensional variables (5C.17) becomes

$$\frac{\partial\phi}{\partial\tau} + 6\phi\frac{\partial\phi}{\partial\xi} + \frac{\partial^3\phi}{\partial\xi^3} = 0. \tag{5C.19}$$

Appendix 5D. Derivation of the nonlinear dispersion relation

The theory of progressive waves may be investigated in a very compact manner if we render the waves steady. We have to introduce (Rayleigh 1876) a current $(-c)$ equal and opposite to the wave velocity. With this artifice the complex potential (5B.12) for the current is $V = -cZ$, yielding $\phi = -cx$ and $\psi = -cz$. To illustrate the method we consider surface gravity waves. We can look for a series of the form

$$\frac{V}{c} = \frac{\phi + i\psi}{c} = -Z + ia\, e^{-ikZ} + ib\, e^{ikZ} \tag{5D.1}$$

or, by separating the real and imaginary parts

$$\frac{\phi}{c} = -x - (b\, e^{-kz} - a\, e^{kz})\sin kx, \qquad \frac{\psi}{c} = -z + (b\, e^{-kz} + a\, e^{kz})\cos kx. \tag{5D.2}$$

This represents a steady motion that is periodic along the x direction, superposed on a uniform current of velocity c. At the bottom defined by $z = -h$ we must have $\psi = C^{te}$; this requires that

$$b\, e^{kh} + a\, e^{-kh} = 0.$$

Thus, for deep water waves, that is, when the ratio of the depth h to the wavelength λ is sufficiently large (kh >>1), we have $b = 0$, and (5D.2) reduce to

$$\frac{\phi}{c} = -x + a\, e^{kz}\sin kx, \qquad \frac{\psi}{c} = -z + a\, e^{kz}\cos kx. \tag{5D.3}$$

The profile of the free surface must be a streamline; we choose it to be the line corresponding to $\psi = 0$. Consequently, from (5D.3), we get

$$z = a\, e^{kz}\cos kx. \tag{5D.4}$$

By expanding (5D.4) evaluated at $z = \eta$ we may write

$$z = a\,(1 + kz + \frac{1}{2}k^2 z^2 + \ldots\ldots)\cos kx \tag{5D.5}$$

and set

$$z = z_0 + \varepsilon z_1 + \varepsilon^2 z_2 + \ldots,$$

where $\varepsilon <<1$. Then replacing in (5D.5) to successive order of ε we determine z_0, z_1, and z_2 respectively. We finally obtain

$$z = \eta = \frac{1}{2}ka^2 + a\,(1 + \frac{9}{8}k^2 a^2)\cos kx + \frac{1}{2}ka^2\cos 2kx + \frac{3}{8}k^2 a^3\cos 3kx + \ldots. \tag{5D.6}$$

If we now set

$$a_m = a\,(1 + \frac{9}{8}k^2 a^2), \tag{5D.7}$$

140

Solving for a_m (5D.7) will give us a polynomial in a_m:

$$a = \delta_1 a_m + \delta_2 a_m^2 + \delta_3 a_m^3 + \ldots \qquad (5D.8)$$

to third order. Substituting in (5D.7), expanding, and retaining terms only in third order in a we obtain

$$\delta_1 = 1, \qquad \delta_2 = 0, \qquad \delta_3 = -(9/8)k^2.$$

Substituting (5D.8) and (5D.7) in (5D.6) yields

$$\eta = \frac{1}{2} k a_m^2 + a_m \cos kx + \frac{1}{2} k a_m^2 \cos 2kx + \frac{3}{8} k^2 a_m^3 \cos 3kx + \ldots \qquad (5D.9)$$

to third order. Thus, we have the surface wave profile, in a frame moving at velocity c, in terms of a power series in the amplitude a_m. The moving wave profile is obtained by replacing x by (x - ct) in (5D.9). Let us now calculate the wave speed c. In the steady regime we have $\phi_t = 0$ and the boundary condition (5B.19) becomes

$$\frac{p_a}{\rho} = -\frac{1}{2}(\phi_x^2 + \phi_z^2) - g\eta. \qquad (5D.10)$$

By using (5D.4) the latter equation can be written as

$$\frac{p_a}{\rho} = -\frac{c^2}{2}(1 - 2ka\, e^{k\eta} \cos kx + k^2 a^2 e^{2k\eta}) - g\eta. \qquad (5D.11)$$

The term $e^{2k\eta}$ can be expanded (for small amplitudes) and we may rewrite (5D.11) in the approximate form

$$\frac{p_a}{\rho} = -\frac{c^2}{2} + (kc^2 - g - k^3 c^2 a^2)\eta + \ldots. \qquad (5D.12)$$

Consequently, as the pressure must be uniform at the free surface, (5D.12) is satisfied if the coefficient of η is zero, that is, provided

$$c^2 = \frac{g}{k}(1 + k^2 a_m^2), \qquad (5D.13)$$

where we have replaced a with a_m, as given by equation (5D.7).

Appendix 5E. Details of the probes and the electronics

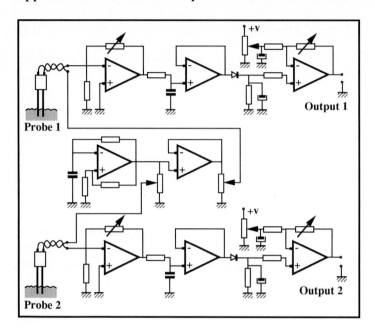

6 Mechanical Solitons

In the previous chapters we introduced solitons that are in fact *nontopological solitons*, because the system returns to its initial state after the passage of the wave. In the present Chap. we introduce a new class of solitons or *kink solitons* called *topological solitons*, because in some cases the structure of the system is modified after the passage of the wave .

We shall present an experimental mechanical transmisssion line, first studied by Scott in 1969, which is an efficient and pedagogical system for observing such solitons and for studying their remarkable properties (Nakajima et al. 1974). We introduce the large-amplitude *kink solitons* and discuss their remarkable properties for continuous and discrete systems. We describe simple experiments which allow one to study qualitatively their propagation and collisions.

6.1 An experimental mechanical transmission line

6.1.1 General description of the line

This system consists of a chain of pendulums, each pendulum being elastically connected to its neighbors by springs, as sketched in Fig.6.1. The equation of motion of the nth pendulum of the chain is given by

$$I \frac{d^2\theta_n}{dt^2} = \Gamma_{1n} + \Gamma_{2n},$$ (6.1a)

where

$$\Gamma_{1n} = - mgL\sin\theta_n, \quad \Gamma_{2n} = - \beta(\theta_n - \theta_{n+1}) - \beta(\theta_n - \theta_{n-1}),$$ (6.1b)

are respectively the gravitational restoring torque and the restoring torque owing to the coupling with the neighboring pendulums. Here, θ is the angle of rotation of the pendulums, I is the moment of inertia of a single pendulum of mass m and length L, g is the gravitation, and β is the torque constant of a section of spring between two pendulums. Setting

$$\omega_0^2 = \frac{mgL}{I}, \quad c_0^2 = \frac{a^2\beta}{I},$$ (6.2)

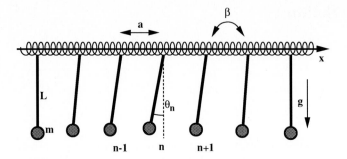

Fig.6.1. Sketch of the mechanical transmission line which consists of elastically coupled pendulums.

we rewrite (6.1) as

$$\frac{d^2\theta_n}{dt^2} + \omega_o^2\sin\theta_n = \frac{c_o^2}{a^2}(\theta_{n+1} + \theta_{n-1} - 2\theta_n) \quad (n = 1,2, \dots ,N). \qquad (6.3)$$

This system of N nonlinear differential equations, often called *the discrete Sine–Gordon equation*, cannot be solved analytically, and, as in the case of discrete nonlinear electrical transmission lines (Chap.3), we can use the continuum approximation

$$\theta_n(t) \to \theta(x,t), \qquad \theta_{n\pm1} = \theta \pm a\frac{\partial\theta}{\partial x} + \frac{a^2}{2}\frac{\partial^2\theta}{\partial x^2} \pm \frac{a^3}{3!}\frac{\partial^3\theta}{\partial x^3} + \frac{a^4}{4!}\frac{\partial^4\theta}{\partial x^4} + \dots$$

Note that this approximation is valid only if the angle of rotation θ varies slowly from one pendulum to the next which corresponds to the case where the amplitude of the restoring torque owing to the coupling is larger than the amplitude of the torque owing to the gravitation. In this case one has

$$\beta \gg mgL \text{ or } c_o^2/a^2 \gg \omega_o^2. \qquad (6.4a)$$

This condition can be rewritten as

$$c_o/\omega_o = d \gg a, \qquad (6.4b)$$

where d *is considered a discreteness parameter.* That is, if $d \cong a$ the angle of rotation varies abruptly from one pendulum to the next, and the continuum (long-wavelength) approximation cannot be used. Next, if condition (6.4) is assumed, (6.3) can be approximated by

$$\frac{\partial^2\theta}{\partial t^2} - c_o^2\frac{\partial^2\theta}{\partial x^2} + \omega_o^2\sin\theta = 0. \qquad (6.5)$$

Here, the first term corresponds to the inertial effects of the pendulums, the second term represents the restoring torque owing to the coupling between pendulums and the last term represents the gravitational torque. Equation (6.5) is called *the Sine-Gordon (SG) equation* (Rubinstein 1970, Lamb 1971). It has been used to model (Barone et al. 1971; Dodd et al. 1982) various phenomena in physics.

It is not obvious that (6.5) contains dispersion and nonlinearity. In fact, one can remark that $\omega_0^2 \sin\theta = \omega_0^2[\theta + (\sin\theta - \theta)]$, that is, dispersion is given by the linear part $\omega_0^2\theta$ and nonlinearity is given by $\omega_0^2(\sin\theta - \theta) = \omega_0^2[-\frac{\theta^3}{3!} + \frac{\theta^5}{5!} - \frac{\theta^7}{7!} + ...]$.

Before investigating solutions of (6.5) we consider its small amplitude limit, or linearization.

6.1.2 Construction of the line

Following Scott (1969), a simple mechanical transmission line, represented in Fig.6.2a, can be easily constructed. The pendulums, equidistant a length a, consist of screws of length L, which are fixed to brass cylinders connected by a steel spring and supported horizontally by a piano wire (Fig.6.2b). The stiffness of the springs is chosen in order to satisfy (6.4a). Under this condition the line can be modeled by (6.5), which describes waves that vary slightly over the distance a.

6.2 Mechanical kink solitons

If as an initial perturbation the angle of the first pendulum at the right end of the line makes only small excursions around the down position given by $\theta = 0$, then one can observe a small-amplitude wave packet, which propagates at group velocity, close to c_0, to the left and disperses. In practice, our measurements yield $c_0 \approx$ 60cm/s and $\omega_0 \approx 12.5$ rad s^{-1}; this leads to d $\approx$ 4a if one expresses c_0 in units of a.

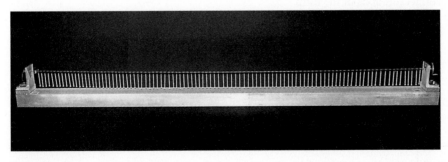

Fig.6.2a. The mechanical transmission line consists of a chain of N elastically coupled pendulums all spaced a distance a = 1.2cm apart. The length of the line is L = 146cm (Photo R. Belleville).

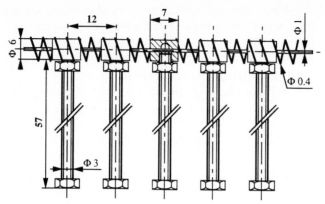

Fig.6.2b. Mechanical details: (i) Brass cylinder: diameter 6mm, length 7mm; (ii) spring: piano wire, diameter 0.4mm; (iii) screw diameter 3mm, length 57mm; (iv) Piano wire: diameter 1mm, length 1.5m.

If the initial disturbance is a complete rotation of the first pendulum, one observes that this rotation propagates collectively along the line, reflects at the opposite end, and travels again without any apparent modification of profile and velocity if one first ignores dissipative effects (see Sect.6.8). Before examining the properties of this localized wave, which is called a *kink soliton,* let us first consider the above small amplitude dispersive waves.

6.2.1 Linear waves in the low-amplitude limit

In this case, $\sin\theta \approx \theta$, and (6.5) reduces to its linear approximation

$$\frac{\partial^2\theta}{\partial t^2} - c_o^2\frac{\partial^2\theta}{\partial x^2} + \omega_o^2\theta = 0, \tag{6.6}$$

which is called *the Klein–Gordon (KG) equation.* This linear dispersive equation admits solutions of the form

$$\theta = \theta_o\cos(kx - \omega t). \tag{6.7}$$

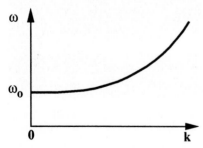

Fig.6.3. Representation of the dispersion relation (6.8) for k>0.

146

Replacing (6.7) in (6.6) yields the dispersion relation

$$\omega = \sqrt{\omega_0{}^2 + c_0{}^2 k^2},\qquad (6.8)$$

which is similar to the relation (2.27) obtained for linear electrical transmission lines. The frequency ω_0 is a cutoff frequency: the wave number k is imaginary for $\omega < \omega_0$; that is, waves with a frequency below ω_0 cannot propagate. The diagram $\omega = f(k)$ is represented in Fig.6.3 for $k \geq 0$.

6.2.2 Large amplitude waves: kink solitons

Let us seek traveling wave solutions which are not restricted by the previous small amplitude assumption. We first set

$$\omega_0 t = T, \quad \frac{\omega_0}{c_0} x = X,\qquad (6.9)$$

which corresponds to measuring distances in units of d (as given by (6.4b)) and time t in units of $1/\omega_0$. Thus, (6.5) reduces to the normalized form

$$\frac{\partial^2 \theta}{\partial T^2} - \frac{\partial^2 \theta}{\partial X^2} + \sin\theta = 0.\qquad (6.10)$$

We look for localized waves of permanent profile of the form

$$\theta = \theta(s) = \theta(X - uT),\qquad (6.11)$$

such as $\theta \to 0$ and $d\theta/ds \to 0$, when $s \to \pm \infty$, where s is a single independent variable depending on u which is an arbitrary velocity of propagation. For such solutions the calculations (see Appendix 6A) yield

$$\theta = 4 \arctan [\exp (\pm \frac{s - s_0}{\sqrt{1 - u^2}})].\qquad (6.12)$$

Relation (6.12) represents a localized solitary wave solution which can travel at any normalized velocity $-1 < u < 1$. Numerical investigations of collision events between such solitary waves indicate that they are solitons. The ($\pm$) signs correspond to localized soliton solutions which travel with opposite screw senses : they are respectively called a *kink soliton* and an *antikink soliton*. They are represented in Fig.6.4 as a function of s with $s_0 = 0$. When s increases from $-\infty$ to $+\infty$, the pendulums rotate from 0 to 2π for the kink and from 0 to -2π for the antikink.

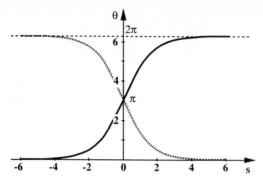

Fig.6.4. Representation of the kink and antikink waveform as a function of the variable s = X-uT.

At one end of the experimental line one can create a kink moving from the right to the left, as represented in Fig.6.5. After reflection at the opposite free end this kink becomes an anti kink, with opposite screw sense, which travels from the left to the right.

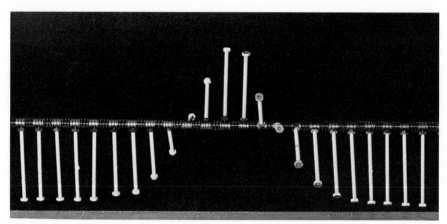

Fig.6.5. A kink traveling from the right to the left of the mechanical line (Photo R. Belleville).

The space derivative $\partial\theta/\partial x$, which is the gradient of the angle of rotation, can be calculated from (6.12). Setting

$$\gamma = \frac{1}{\sqrt{1 - u^2}} , \qquad (6.13)$$

we rewrite (6.12) as

$$\theta = 4\arctan\left[\exp\left(\pm\,\gamma s\right)\right]. \qquad (6.14)$$

Thus, one obtains

$$\frac{\partial\theta}{\partial X} = \frac{4\gamma e^{\gamma s}}{1 + e^{2\gamma s}} = 2\gamma \text{ sech } (\gamma s). \tag{6.15}$$

The time derivative $\partial\theta/\partial T$, which corresponds to the rotation velocity of the kink (antikink) is calculated as

$$\frac{\partial\theta}{\partial T} = -2u\gamma \text{ sech } (\gamma s). \tag{6.16}$$

Thus, the space or time derivatives of the kink soliton are localized pulse solitons. Their waveforms are represented as a function of s in Fig.6.6.

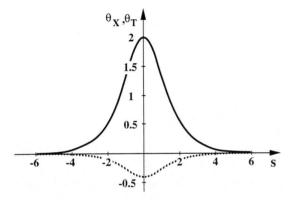

Fig.6.6. Representation of $\partial\theta/\partial X$ (<u>continuous line</u>) and $\partial\theta/\partial T$(<u>dotted line</u>) as a function of s = X - uT, for u = 0.1.

6.2.3 Lorentz contraction of the kink solitons

As the kink-soliton velocity approaches unity, its width gets narrower owing to the Lorentz contraction factor $\sqrt{1 - u^2}$ in the denominator of the exponent. This results from the form of the SG equation (6.10), which is invariant to the Lorentz transformation of the independent variables from (X,T) to (X', T'):

$$X \rightarrow X'= \gamma(X - uT) , \qquad T \rightarrow T'= \gamma(T - uX). \tag{6.17}$$

By a calculation similar to (6.11) we find that the relations (6.17) imply

$$\frac{\partial}{\partial X} \rightarrow \gamma(\frac{\partial}{\partial X'} - u\frac{\partial}{\partial T'}) , \qquad \frac{\partial}{\partial T} \rightarrow \gamma(-u\frac{\partial}{\partial X'} + \frac{\partial}{\partial T'}). \tag{6.18}$$

Substitution of (6.18) into the SG equation (6.10) leaves the form unchanged. Thus any solution of (6.10) for which the independent variables are transformed as (6.18) will remain a solution.

Now, taking account of the transformation (6.9) one gets the kink-soliton solution in laboratory coordinates (x, t):

$$\theta = 4 \arctan \left[\exp \left(\frac{\dfrac{\omega_0}{c_0} x - u\omega_0 t}{\sqrt{1 - u^2}} \right) \right]. \tag{6.19a}$$

By setting $v = uc_0$ and introducing the parameter d defined by relation (6.4b) we can transform the above solution into

$$\theta = 4 \arctan \left[\exp \left(\frac{x - vt}{d\sqrt{1 - \dfrac{v^2}{c_0^2}}} \right) \right]. \tag{6.19b}$$

As the soliton velocity v approaches the limiting value c_0, which is the velocity of linear waves, the soliton remains constant but its width gets narrower owing to the Lorentz contraction of its profile, given by $d\sqrt{1 - v^2/c_0^2}$, and its waveform approachs a step function. In other words, the soliton width decreases when v increases as represented in Fig.6.7. It is important to remark that the amplitude of the kink is independent of its velocity and when the velocity v= 0 (6.19b) reduces to

$$\theta = 4 \arctan \left[\exp (\pm x/d) \right].$$

Thus, by contrast to nontopological solitons the *kink soliton may be entirely static*, losing its wave character.

With the experimental line one can easily observe that the spatial extension of a static kink (v/c_0= 0) is larger than that of a rapidly moving kink ($v/c_0 \neq 0$). In practice one has $c_0 = 60$cm/s and one can launch kinks with velocities in the range $0 \leq v \leq 45$cm/s. This yields $0 \leq v/c_0 \leq 0.75$

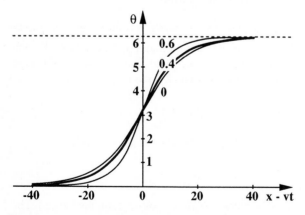

Fig.6.7. Lorentz contraction of the kink soliton represented for different values of its velocity, i.e., for $v/c_0 = 0$, 0.4, 0.6, and d = 1.

150

6.3 Particle properties of the kink solitons

To gain an insight into the corpuscular nature of these localized waves, let us calculate the energy of a kink. If one neglects dissipation effects, the total energy E, or Hamiltonian, of the chain of pendulums is the sum of
• the rotational kinetic energy,
• the potential energy owing to the coupling between pendulums,
• the potential energy owing to the gravitation.
It is written as

$$E = \sum_{n=1}^{N} [\frac{1}{2} I(\frac{d\theta_n}{dt})^2 + \frac{1}{2} \beta(\theta_{n+1} - \theta_n)^2 + mgL(1 - \cos\theta_n)]. \tag{6.20}$$

In the continuum approximation one has $|\partial\theta/\partial x| \approx |(\theta_{n\pm1} - \theta_n)/a|$ and the discrete sum can be replaced by an integral. Consequently (6.20) is approximated by

$$E = \frac{I}{a} \int_{-\infty}^{+\infty} [\frac{1}{2}(\frac{\partial\theta}{\partial t})^2 + \frac{1}{2}c_0^2(\frac{\partial\theta}{\partial x})^2 + \omega_0^2(1 - \cos\theta)] \, dx. \tag{6.21}$$

In terms of the variables X and T defined by the transformation (6.9), (6.21) becomes

$$E = \frac{I}{a} \omega_0 c_0 \int_{-\infty}^{+\infty} [\frac{1}{2}(\frac{\partial\theta}{\partial T})^2 + \frac{1}{2}(\frac{\partial\theta}{\partial X})^2 + (1 - \cos\theta)] \, dX, \tag{6.22}$$

where the factor $I\omega_0 c_0$ defines the energy scale. The energy E_K of the kink soliton (or antikink soliton) can now be calculated (see Appendix 6B) by substituting (6.12) in (6.22). One finds that

$$E_K = 4\frac{I}{a}\omega_0 c_0 \gamma (1 + u^2 + \frac{1}{\gamma^2}) = 8\gamma\frac{I}{a}\omega_0 c_0. \tag{6.23}$$

This can be written in the relativistic form

$$E_K = \frac{m_0 c_0^2}{\sqrt{1 - \frac{v^2}{c_0^2}}}, \tag{6.24}$$

where the soliton mass m_0 is given by

$$m_0 = 8\frac{I}{a}\frac{\omega_0}{c_0} = 8\frac{I}{a}\frac{1}{d}. \tag{6.25}$$

One can also introduce the relativistic momentum

$$p_K = \frac{m_0 v}{\sqrt{1 - \dfrac{v^2}{c_0{}^2}}} \ . \tag{6.26}$$

At rest ($v = 0$) one has p =0, and the static soliton energy is

$$E_{0K} = m_0 c_0{}^2. \tag{6.27}$$

Thus the kink soliton can be regarded, as a *relativistic particle* of energy E_K, mass m_0, and momentum p_K. The corpuscular nature of the kinks can be demonstrated on the mechanical line by observing their collisions, as presented in Sect.6.6.

The basic properties of the kink can be summarized as follows:
(i) *Its amplitude is independent of its velocity; it is constant and remains the same for zero velocity: the kink may be static.*
(ii) *Its width gets narrower as its velocity increases, owing to Lorentz contraction.*
(iii) *It has the properties of a relativistic particle.*
(iv) *The kink which has a different screw sense is called an antikink.*

The reader should compare these properties of the kink soliton, which is often called a *topological soliton* in the literature, to the properties of the KdV pulse soliton, described in Sect.3.3, which is called a *nontopological soliton* .

6.4 Kink–kink and kink–antikink collisions

From the kink-soliton solution (6.19) one can expect (Lamb 1971, 1980) that (6.10) admits more solutions of the form

$$\theta = 4 \arctan \left[\frac{F(X)}{G(T)} \right], \tag{6.28}$$

where F and G are arbitrary functions. Upon substitution of (6.28) into the SG equation (6.10) one can calculate (see Appendix6C) the following solutions:
• the kink–kink collision,
• *the kink –antikink collision solution*
• *the breather soliton*

The *kink–kink collision* solution has the form

$$\theta \, (X, T) = 4 \arctan \left[u \sinh \frac{X}{\sqrt{1 - u^2}} \ \mathrm{sech} \ \frac{uT}{\sqrt{1 - u^2}} \right]. \tag{6.29}$$

This solution first obtained by Perring and Skyrme (1962) describes the collision between two kinks with respective velocities +u and - u and approaching the origin from T → −∞ and moving away from it with velocities -u and +u respectively, for T → +∞ as represented in Fig.6.8.

The analytical solution also obtained by Perring and Skyrme to describe a collision process between a kink soliton and an antikink soliton in a center of mass coordinate system is

$$\theta (X, T) = 4\arctan [\frac{1}{u} \text{sech} (\gamma X). \sinh (\gamma uT)]. \tag{6.30}$$

Using (6.30), in Fig.6.9 we have represented a kink approaching the origin with velocity u from X = - ∞ and an antikink approaching the origin with velocity -u from X= +∞. Those collide and pass through each other, then at T = +∞ the kink is at X = +∞ and the antikink is at X = −∞

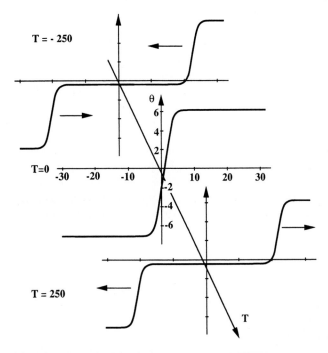

Fig.6.8. Representation of a kink-kink collision: at T=-250 the two kinks propagate with opposite velocities u = +0.4; and u = -0.4; at T = 0 they collide at the origin O; at T = +250 they move away from the origin with velocities u = -0.4 and u = +0.4.

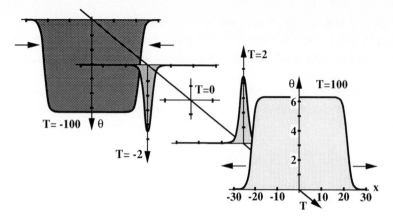

Fig.6.9. Representation of a kink–antikink collision: at T= -100 the kink and the antikink move with opposite velocities u = 0.2 and u =-0.2; the waveform which corresponds to their collision is represented at T=-2, 0 , 2; at T=100 they move away from the origin with respective velocities u= -0.2 and u = 0.2.

6.5 Breather solitons

The breather soliton solution has the form (see Appendix 6C) :

$$\theta_B (X, T) = 4\arctan [\frac{\sqrt{1 - \Omega^2}}{\Omega} \text{sech} (\sqrt{1 - \Omega^2} X) . \sin (\Omega T)], \qquad (6.31)$$

which represents a stationary-wave solution where a sech-shaped pulse envelope is modulated at frequency Ω . Owing to this structure represented in Fig.(6.10) the pulse looks like breathing and is therefore called a *breather mode or breather soliton* (Lamb 1980; Rajaraman 1982; Mankankov 1990). Sometimes it is also called a *bion* because it corresponds to a bound state of two solitons: a kink and an antikink, as observed in Fig.6.9 for -2 ≤ T≤2.

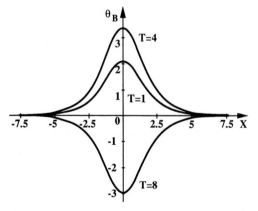

Fig.6.10. Representation of breather oscillations, as given by (6.31), for different times T=1,4, and 8.

From solution (6.31) one can get a propagating-wave solution since the SG equation is invariant under the Lorentz transformation of the variables (X,T) to (X', T'), as given by (6.17). Under this transformation the breather solution (6.31) is boosted into

$$\theta_B (X,T) = 4\arctan\left[\frac{\sqrt{1-\Omega^2}}{\Omega} \text{ sech } \left(\sqrt{1-\Omega^2}\frac{X-uT}{\sqrt{1-u^2}}\right)\sin\left(\Omega\frac{T-uX}{\sqrt{1-u^2}}\right)\right]. \quad (6.32)$$

Using relations : $u = v/c_0$, $d = c_0/\omega_0$, $\gamma = 1/\sqrt{1-u^2}$ and setting

$$\Omega = \omega_B/\omega_0, \quad (6.33)$$

one can express the breather solution (6.32) in laboratory coordinates (x, t) :

$$\theta_B (x, t) = 4\arctan\left\{\sqrt{\frac{\omega_0^2}{\omega_B^2}-1} \text{ sech } \left[\frac{\gamma}{d}\sqrt{1-\frac{\omega_B^2}{\omega_0^2}}(x-vt)\right] \sin\left[\gamma\omega_B(t-\frac{vx}{c_0^2})\right]\right\}. \quad (6.34)$$

Here ω_B, weighted by the Lorentz term γ, represents the frequency of internal oscillations of the breather soliton and one has $0 \le \omega_B < \omega_0$. Just as with the kink soliton, the breather soliton velocity is restricted to the range $0 \le |v| < c_0$ and the breather gets narrower owing to the Lorentz contraction of its profile, given by

$$d_B = \frac{d}{\gamma\sqrt{1-\frac{\omega_B^2}{\omega_0^2}}}. \quad (6.35)$$

The maximum amplitude of the breather is

$$A_B = 4 \arctan\left(\sqrt{\frac{\omega_0^2}{\omega_B^2}-1}\right). \quad (6.36)$$

One can calculate the breather energy E_B by substituting (6.32) in the energy expression (6.22) to obtain

$$E_B = 16\frac{I}{a}c_0\omega_0\gamma\sqrt{1-\frac{\omega_B^2}{\omega_0^2}}. \quad (6.37a)$$

Using (6.24) we can express the breather soliton energy in terms of the kink-soliton energy E_K and rewrite (6.37a) as

$$E_B = 2E_K\sqrt{1-\frac{\omega_B^2}{\omega_0^2}} \quad (6.37b)$$

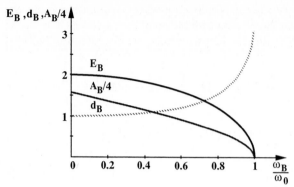

Fig.6.11. Variations of the energy E_B, amplitude $A_B/4$, and profile d_B of the breather soliton as a function of ω_B/ω_0, for $E_K = 1$, $d = 1$,

For $\omega_B \to 0$, the breather energy tends to twice the kink energy: $E_B \to 2E_K$; moreover, when $\omega_B \to 0$, $A_B \to \pi/2$ and $d_B \to d/\gamma$. One can interpret these results by considering that the breather is a bounded soliton–antisoliton pair. For $\omega_B \to \omega_0$, one has $E_B \to 0$, $A_B \to 0$, and $d \to \infty$, and the breather approaches a low– amplitude linear wave. From these results we see that the breather is a remarkable mode because depending on its internal frequency ω_B it interpolates between large amplitude nonlinear waves and weak–amplitude linear waves. Moreover, the breather can be considered a relativistic particle, with internal oscillations, of mass $m_B = 2m_0$ and momentum $p_B = 2p_K$. The quantities d_B, A_B, and E_B are represented as a function of ω_B/ω_0 in Fig.6.11, for $E_K = 1$.

6.6 Experiments on kinks and breathers

A kink-kink collision can be easily observed on the mechanical line. One creates a static kink (velocity u =0) and launches another kink with the same screw sense but with velocity u (Fig.6.12). When they collide, the moving pulse stops and the static pulse starts moving as if they were billiard balls, which shows the corpuscular nature of these solitons. The head-on collision between two kinks, as depicted in Fig.6.8, can be only approximately realized because it is difficult to create simultaneously two kinks with opposite velocities.

The collision without mutual destruction between a kink and an antikink as described by equation (6.30) and which travel with large velocities is also difficult to observe. In fact, owing to dissipation effects which inevitably occur for a real transmission line, the kink and antikink annihilate each other during the collision and radiate their energy onto the line.

However, if one launches on the mechanical line a kink and an antikink (with an opposite screw sense) with sufficiently low velocities, one can observe a bounded pair that is a breather soliton. Nevertheless, owing to dissipation effects present on the real line, only a few breathing oscillations can be observed.The oscillations then decrease with time and energy is radiated onto the line as shown in Fig.6.13.

156

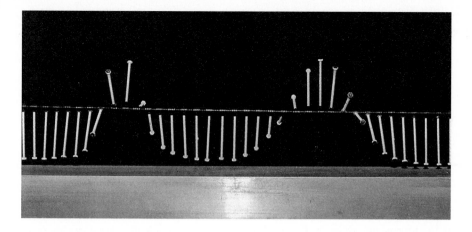

Fig.6.12. Collision between two kinks: a kink (originally static before the collision) moves from the right to the left and the other kink (originally traveling from the right to the left) becomes static (Photo R. Belleville).

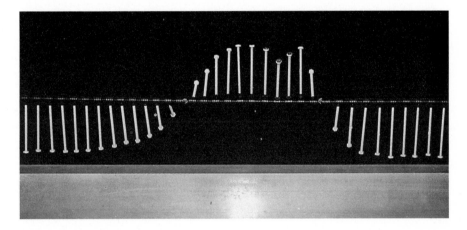

Fig.6.13. Observation of a breather soliton produced by a kink-antikink collision. Owing to dissipation, its lifetime is short (Photo R. Belleville).

6.7 Helical waves, or kink array

If one looks for nonlocalized solutions of the SG equation (6.10) for which θ and its first derivative do not tend to zero when $s \to \pm \infty$, one finds (see Appendix6D) periodic solutions of the form

$$\theta = 2 \arcsin \left[\pm \operatorname{sn} \left(\frac{s}{k\sqrt{1 - u^2}}, k \right) \right] + \pi, \qquad (6.38)$$

157

where sn is a Jacobian elliptic function (see Appendix 4B) of modulus k. The variations of θ as a function of s are represented in Fig.6.14. With respect to the variable s the spatial period or wavelength of the periodic function (6.55) is (see Appendix 4B)

$$\Lambda = 4K(k)k\sqrt{1 - u^2}. \tag{6.39}$$

This is sketched in Fig.6.15. The reader should verify that for a modulus k = 1, that is for $\Lambda \rightarrow \infty$, solution (6.38) reduces to

$$\theta = \theta_s = 4 \arctan [\exp (\pm \frac{s}{\sqrt{1 - u^2}})]. \tag{6.40}$$

One recovers (with $s_0 = 0$) the kink (antikink) soliton solution (6.12). Solution (6.38) represents an evenly spaced array of kinks (antikinks) or domains (see Fig.6.14), where θ changes by 2π, traveling uniformly for all X and for all T. In the mechanical model of pendulums this array of kinks represents in some sense a helical wave. It can be simply observed on the experimental mechanical line, as represented in Fig.6.16.

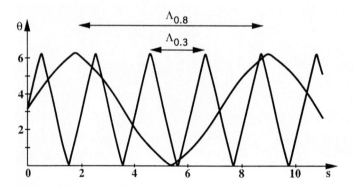

Fig 6.14. Representation of θ as a function of s = X-uT for k=0.3 and k=0.8, as given by (6.38).

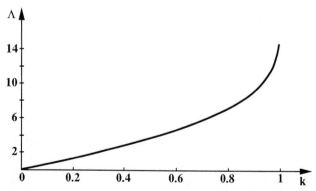

Fig 6.15. Wavelength Λ of the periodic wave versus the modulus k.

158

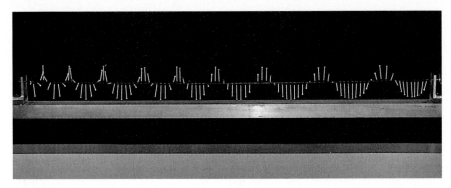

Fig.6.16. Observation of a helical wave or kink array on the mechanical line (Photo R. Belleville).

6.8 Dissipative effects

If one launches on the real mechanical transmission line a kink soliton with a given initial velocity, it gradually slows down owing to dissipative effects. This energy loss can be compensated for by applying an external constant torque to the kink. In this case the velocity u of the kink reaches a constant value, say u_∞, as time increases. Thus to account for this dissipation, that is, to model the real line, one has to replace (6.1) by

$$I \frac{d^2\theta_n}{dt^2} = \Gamma_{1n} + \Gamma_{2n} - \Gamma \frac{d\theta_n}{dt} - f, \qquad (6.41)$$

where $\Gamma d\theta_n/dt$ is a dissipative torque and f an external (positive or negative) constant torque which is assumed to be proportional to the rotation velocity. Both terms are chosen to be small, which is a reasonable assumption for the pendulums chain if one carefully selects the springs which ensure the coupling. Then, proceeding as in Sect.6.1 and using (6.8), we replace (6.10) by

$$\frac{\partial^2\theta}{\partial T^2} - \frac{\partial^2\theta}{\partial X^2} + \sin\theta = -F - \alpha \frac{\partial\theta}{\partial T}, \qquad (6.42)$$

where $F = f/I\omega_0^2$ and $\alpha = \Gamma/I\omega_0$. If we neglect these two perturbative terms, for any solution θ the total energy (6.22) of the system is constant in time,

$$dE/dT = 0. \qquad (6.43)$$

In this case, the energy of a single nonperturbed kink is constant; it is given by (6.23) and is equal to $E_K = 8\gamma$ when expressed in units of $(I/a)\omega_0 c_0$. On the other hand, if in (6.42) two perburbative terms are present, by multiplying this equation by $\partial\theta/\partial T$ and integrating with respect to X we obtain a power equation

$$\frac{dE}{dT} = - \int_{-\infty}^{+\infty} dX \ [\ \alpha(\frac{\partial\theta}{\partial T})^2 + F\frac{\partial\theta}{\partial T} \ ]. \tag{6.44}$$

Following Scott and Mc Laughlin (1978) we now assume that the predominant effect of the perturbative terms on a single kink is to modulate its velocity u; in other words, u -> u (T). Thus, we have

$$\frac{dE_K}{dT} = \frac{d}{dT} (\frac{8}{\sqrt{1 - u^2}}) = 8u \ (1 - u^2)^{-3/2} \frac{du}{dT} \ . \tag{6.45}$$

Making $E = E_K$ and inserting (6.16) into (6.44), after an integration similar to that of Sect.6.4, we get

$$\frac{dE_K}{dT} = 2\pi Fu - \frac{8\alpha u^2}{\sqrt{1 - u^2}} \ . \tag{6.46}$$

Finally, equating (6.45) and (6.46) yields

$$du/dT = (\pi F/4) \ (1 - u^2)^{3/2} - \alpha u \ (1 - u^2). \tag{6.47}$$

This differential equation describes the effect of dissipation and an external force on the velocity of the kink. When F = 0 and for small velocities, (6.47) can be approximated by

$$du/dT \approx - \alpha u.$$

In this case $u \approx \exp (- \alpha T)$; that is, the velocity of the kink decreases exponentially with time. When an external constant torque is present and as time increases the velocity of the kink tends asymptotically to a constant u_∞ . Consequently du/dT - > 0, and (6.47) yields

$$u_\infty = [1 + (4\alpha/\pi F)^2] - 1/2. \tag{6.48}$$

This simple result gives the velocity at which the power input to the kink is just balanced by its power loss to dissipation.

6.9 Envelope solitons

Let us ignore dissipation effects and confine ourselves (as in Sect.4.4) to those low-amplitude waves which consist of a slowly varying envelope modulating a carrier wave whose dispersion relation is that of a linear wave to order $\varepsilon \ll 1$. In the low amplitude limit $\theta \rightarrow \varepsilon\theta$, (6.5) is approximated by

$$\frac{\partial^2\theta}{\partial t^2} - c_o^2\frac{\partial^2\theta}{\partial x^2} + \omega_o^2(\theta - \varepsilon^2\frac{\theta^3}{6}) \approx 0. \tag{6.49}$$

We then assume that

$$\theta = \Theta\, e^{i(kx - \omega t)} + cc, \tag{6.50}$$

where cc denotes the complex conjugate. Here, as in Sects.4.4 and 5.7, the amplitude function $\Theta(\varepsilon x, \varepsilon t)$ is supposed to be constant with respect to the rapid variations in x and t of the carrier wave. Inserting (6.50) into (6.49) and neglecting the third-harmonic terms, to a first approximation we obtain a nonlinear dispersion relation

$$\omega^2 = \omega_o^2 + c_o^2 k^2 - \varepsilon^2\frac{\omega_o^2}{2}|\Theta|^2. \tag{6.51}$$

For very low amplitudes this relation, of the form $\omega = f(k, |\Theta|^2)$, reduces to the linear dispersion relation of the discrete pendulum chain,

$$\omega\ell^2 = \omega_o^2 + c_o^2 k^2. \tag{6.52}$$

Then, setting $X = \varepsilon x$ and $T = \varepsilon t$ and proceeding as in Sect.4.4 one can associate to (6.51) the equation

$$i\varepsilon[\frac{\partial\Theta}{\partial T} + v_g\frac{\partial\Theta}{\partial X}] + \varepsilon^2 P\frac{\partial^2\Theta}{\partial X^2} + \varepsilon^2 Q|\Theta|^2\Theta = 0, \tag{6.53}$$

which describes the evolution of the slow envelope of the wave. Here, the group velocity v_g, dispersion coefficient P, and the nonlinear coefficient Q are given (see (4.21)) by

$$v_g = \frac{\partial\omega}{\partial k}, \text{ at } k = k_c; \qquad P = \frac{1}{2}\frac{\partial^2\omega}{\partial k^2}, \text{ at } k = k_c; \qquad Q = -\frac{\partial\omega}{\partial|\Theta|^2}, \text{ at } \Theta_o = 0;$$

where ω_c and k_c are the frequency and wavenumber of the carrier wave. In a frame of reference moving with group velocity v_g, the substitution

$$\xi = X - v_g T, \qquad \tau = \varepsilon T \tag{6.54}$$

reduces (6.53) to the standard nonlinear Schrödinger equation

$$i \frac{\partial \Theta}{\partial \tau} + P \frac{\partial^2 \Theta}{\partial \xi^2} + Q|\Theta|^2 \Theta = 0. \tag{6.55}$$

In (6.55), the dispersion coefficient P and the nonlinear coefficient Q can be calculated from (6.51):

$$v_g = \frac{c_0^2 k_c}{\omega_c}, \qquad P = \frac{c_0^2 - v_g^2}{2\omega_c}, \qquad Q = \frac{\omega_0^2}{4\omega_c}, \tag{6.56}$$

We note that PQ > 0 for any k. Consequently in this case (6.55) admits envelope-soliton solutions (see (4.C15)) and the corresponding wavepackets obtained by combining the envelope waveform Θ in (6.50) and the sinusoidal carrier wave are

$$\theta(x, t) = 2\varepsilon \Theta_m \operatorname{sech} \left[\sqrt{\frac{Q}{2P}} \Theta_m \varepsilon (x - v_g t - \varepsilon c P t) \right] \cos \left[K_c x - \Omega_c t \right], \tag{6.57}$$

with

$$K_c = k_c + \varepsilon \frac{c}{2}, \qquad \Omega_c = \omega_c + \varepsilon v_g \frac{c}{2} - \varepsilon^2 \frac{|Q|}{2} \Theta_m^2 + \varepsilon^2 c^2 \frac{P}{4}.$$

Here, $2\Theta_m$ is the amplitude and c is a parameter defined in Appendix 4C. By setting

$$A_B \approx 4[(\omega_0/\omega_B)^2 - 1]^{1/2} = 2\varepsilon \Theta_m, \qquad v = \frac{c c_0^2}{2\omega_0}, \tag{6.58}$$

the reader may recognize that in the low-amplitude limit (Newell, 1978; Lamb 1980) and to order ε, the breather soliton (6.34) reduces to

$$\theta(x, t) = 2\varepsilon \Theta_m \operatorname{sech} \left[\frac{\omega_0}{2c_0} \Theta_m \varepsilon (x - vt) \right] \sin \left(\omega_0 t - \frac{\omega_0}{c_0^2} v x \right). \tag{6.59}$$

This corresponds to the NLS envelope soliton (6.57) for

$$k_c = 0, \qquad v_g = 0, \qquad P = c_0^2/2\omega_0, \qquad Q = \omega_0/4, \qquad \gamma = 1. \tag{6.60}$$

Such low-amplitude envelope solitons can be observed on the experimental line as a result of the decaying oscillations of an initial large-amplitude breather soliton (see Sect.6.6). Otherwise, they can be created directly by an initial small rotation ($\theta_{max} \approx \pi/3$) of the pendulums at one end of the line. Nevertheless, in both cases their lifetime is short owing to dissipation effects.

6.10 Lattice effects

6.10.1 Pocket version of the pendulum chain

One can construct a useful and versatile pocket version of the mechanical transmission line by using a rubber band and dressmakers' pins. For such a simple device, represented in Fig. 6.17a, the continuum approximation is valid if condition (6.4) is fulfilled and the pendulum chain can be modeled by the Sine–Gordon equation (6.5). This is the case in Fig.6.17a, where two kinks traveling from the right to the left are shown.

By increasing the mass m at the end of each pendulum one can also construct a chain for which condition (6.4) is not fulfilled. In this case the discretness parameter d is of the order of the lattice spacing a and the physics of the pendulum chain is modeled by the discrete Sine–Gordon equation (6.3)

$$\frac{d^2\theta_n}{dt^2} + \Gamma\frac{d\theta_n}{dt} + \omega_0^2\sin\theta_n = \frac{c_0^2}{a^2}(\theta_{n+1} + \theta_{n-1} - 2\theta_n), \qquad (6.61)$$

with a small dissipative term $\Gamma d\theta_n/dt$ (see (6.41)) that we can ignore to a first approximation.

Unlike the continuous Sine–Gordon equation (6.5), very little is known about the dynamics of (6.61). Numerical simulations of this discrete Sine–Gordon model have been performed by Currie et al (1977). They have shown that, when a kink with a width equal to two or three lattice spacings propagates in a one dimensional lattice, it changes its shape slightly by shrinking and radiating small-amplitude linear waves in its wake, which results in a spontaneous damping of its motion. This effect was interpreted theoretically by Peyrard and Kruskal (1984) who also found that in the absence of external driving, kinks propagate preferentially at a particular set of velocities. Thus, by using the discrete pendulum chain described above, one can create a kink and simply observe the small-amplitude wave it generates in its wake before pinning on the lattice, as represented in Fig. 6.17b. One can further increase the mass m in order to obtain a highly discrete kink that cannot propagate, as observed in Fig.6.17c where $(\theta_{n+1}-\theta_n)/(a)\approx(2\pi/3)/(1\text{cm})$ at the center of the kink.

With such a simple device, which can be modified very rapidly, one can do more. For example, by changing the mass (or length) of one or more pendulums one can observe the influence of defects or disorder on kink (continuous or discrete) propagation.

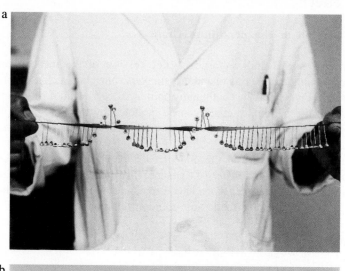

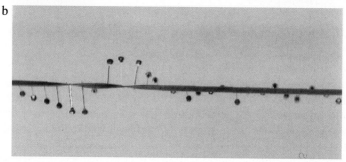

Fig. 6.17. (**a**)Two continuous kinks traveling form the right to the left of the mechanical line (pocket version). (**b**) A discrete kink (top view) traveling from the right to the left of the discrete pendulum chain radiates small oscillations in its wake. (**c**) A highly discrete kink, where $(\theta_{n+1}-\theta_n)/(a)=(2\pi/3)/(1\text{cm})$ at its center, that cannot move along the lattice. (Photos B. Michaux).

6.10.2 Pendulum chain with weak coupling

In order to make more precise observations, one can construct a chain of pendulums similar to the one described in Sect. 6.1.2. With such a lattice, strong discreteness effects are obtained by decreasing the stiffness of the springs or by increasing the moment of inertia I of each pendulum. Under these conditions the pendulum lattice can be modeled by the discrete Sine-Gordon equation (6.3) and the discreteness parameter d, defined in Sect. 6.1.1, is such that d<<a. Thus, one can observe the influence of lattice effects the mobility of kinks. For example, a highly discrete kink that is strongly pinned on the lattice and cannot travel is shown in Fig. 6.18a. In Fig 6.18b a higly discrete kink-antikink pair is shown. One can also observe sequence of kinks or antikinks or combination of them.

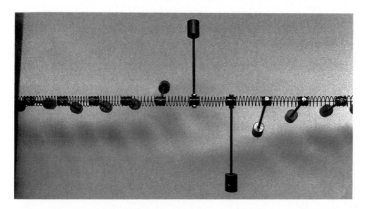

Fig. 6.18a. A discrete kink, with spatial extension 4a, as observed (top view) on the pendulum chain with very weak coupling. Starting from the right, the respective angles of the successive pendulums are : $\theta_1 = 10°$, $\theta_2 = 20°$, $\theta_3 = 90°$, $\theta_4 = 280°$, $\theta_5 = 350°$ (Photo R. Chaux)

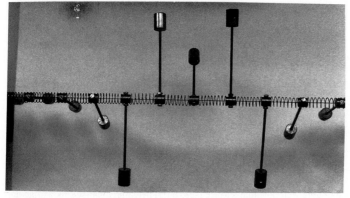

Fig. 6.18b. A discrete kink-antikink pair that cannot travel owing to strong lattice effects. Starting from the right, the respective angles of the successive pendulums are : $\theta_1 = 20°$, $\theta_2 = 80°$, $\theta_3 = 260°$, $\theta_4 = 300°$ (center of the pair), $\theta_{-3} = -260°$, $\theta_{-2} = -80°$, $\theta_{-1} = -20°$ (Photo R. Chaux)

6.11 A mechanical transmission line with two equilibrium states

6.11.1 Periodic and double-well substrate potentials

Analog mechanical systems, such as the experimental line we have considered in the above sections, play an important role in the study of kink solitons and their properties. In Sect 6.7 we have seen that this line can support kink arrays or sequences of kinks, because in (6.21) the potential energy owing to the gravitation, called *on-site* or *substrate potential* (see also (9.48)), is periodic: it presents an infinite number of equilibrium states (mod 2π) as represented in Fig 6.19a. Each kink joins two successive equilibrium states (potential minima).

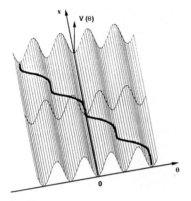

Fig 6.19a. Sketch of a sinusoidal substrate potential surface (see 6.22) of the form $(1-\cos\theta)$ and a sequence of kinks. Each kink connects two sucessive potential wells.

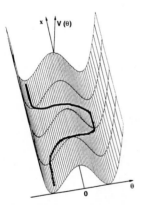

Fig 6.19b. Sketch of a double well potential surface (see (6.69) hereafter) with a kink and an antikink.

However, if the substrate potential has two equilibrium states only, (see Fig 6.19b) one cannot have successive kinks, but kinks and anti-kinks, alternatively, which connect the two potential wells.

166

6.11.2 General description of the mechanical chain

In order to observe kinks in systems with a double well potential (see Fig 6.19) we have constructed a mechanical analog which consists of an experimental chain of identical pendulums (Dusuel et al. 1998). Each basic unit , described in Appendix 6E, is similar to the pendulum recently studied by Peters (1995) : it can oscillate with a motion whose character is determined by the forces of torsion and gravity in opposition. Depending on the length d, which acts as a control parameter of each pendulum, the on-site potential $V(\theta_n)$, where $\theta_n(t)$ is the angular displacement as a function of time t of the nth pendulum, can present one or two minima. For the configuration presently considered it possesses two equilibrium positions (two wells). Each pendulum is connected to its neighbors by springs as sketched in Fig.6.20. When the dissipation is neglected and the difference between angular displacement of neighboring pendulums are small enough, the equation of motion of the nth chain unit is given (see Appendix 6F) by

$$I\frac{d^2\theta_n}{dt^2} = -K\theta_n + mgd\sin\theta_n + C_{o,\ell}\,(\theta_{n+1} + \theta_{n-1} - 2\theta_n)$$
$$-C_{o,n\ell}(\theta_n - \theta_{n+1})^3 - C_{o,n\ell}(\theta_n - \theta_{n-1})^3, \qquad (6.62)$$

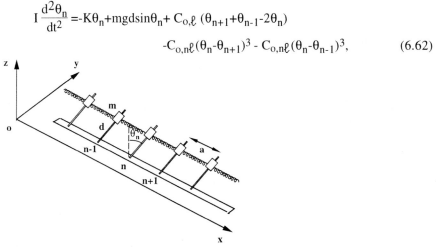

Fig. 6. 20. Sketch of the pendulum lattice apparatus. The pendulums are coupled to each other by springs and attached to the steel ribbon s (parallel to x axis). The angular diplacements of the nth pendulum is θ_n. The pendulums are at equilibrium in one of the two equivalent potential wells. The motion can occur in a plane perpendicular to the chain (x axis). See also Fig. 6. 21.

The terms on the right hand side of (6.62) represent the restoring torque owing to the torsion, the gravitational torque and the restoring torque owing to the coupling with the neighboring pendulums (see Appendix 6F). $I=md^2$ is the moment of inertia of a single pendulum of mass m and length d; g is the gravitation, K is the torsion constant. $C_{o,\ell}$ and $C_{o,n\ell}$ are the linear and nonlinear torque constants of a spring between two pendulums. They are given by

$$C_{o,\ell} = kd^2\,(1 - \frac{\ell_o}{\ell_1}), \qquad (6.63a)$$

$$C_{o,n\ell} = kd^2 \left(\frac{\ell_o d^2}{2\ell_1^3} - \frac{C_{o,\ell}}{6} \right), \qquad\qquad\qquad (6.63b)$$

where k is the spring stiffness, ℓ_o the natural length of a spring at rest and ℓ_1 the length of this spring when it is stretched between two adjacent pendulums at equilibrium (bottom of one well). Note that the nonlinear coupling term must be fully taken into account because the linear term is especially small when ℓ_1 is not very different from ℓ_o. Moreover, when $\ell_o = \ell_1$ we have have zero linear coupling: $C_{o,\ell} = 0$ and nonlinear coupling: $C_{o,n\ell} = kd^4/2K\ell_1^2$, as we will see in the following. Setting

$$\tau = \frac{K}{I} t, \qquad \Gamma = \frac{mgd}{K}, \qquad C_{1,\ell} = \frac{C_{o,\ell}}{K}, \qquad C_{n\ell} = \frac{C_{o,n\ell}}{K}, \qquad (6.64)$$

we transform (6.62) into

$$\frac{d^2\theta_n}{d\tau^2} + \theta_n - \Gamma\sin\theta_n + C_{1,\ell}\, (2\theta_n - \theta_{n+1} - \theta_{n-1}) + C_{n\ell}\, [(\theta_n - \theta_{n+1})^3 + (\theta_n - \theta_{n-1})^3] = 0.$$
$$(6.65)$$

In (6.65) the quantity $(-\theta_n + \Gamma\sin\theta_n)$ represents the *on-site* (zero coupling limit) torque. The interaction and on-site potential energy (compare to (6.20)), corresponding to (6.65) are

$$U(\theta_{n+1} - \theta_n) = \frac{C_{1,\ell}}{2} (\theta_{n+1} - \theta_n)^2 + \frac{C_{n\ell}}{4} (\theta_{n+1} - \theta_n)^4, \qquad\qquad (6.66)$$

and

$$V(\theta_n) = \frac{1}{2} (\theta_n^2 - \theta_m^2) + \Gamma\, (\cos\theta_n - \cos\theta_m). \qquad\qquad\qquad (6.67)$$

Here, the parameter Γ plays the role of a control parameter. For $\Gamma > 1$ it determines the depth and separation of the two wells (Peters 1995) and $\pm\theta_m$ correspond to the two equilibrium positions.

When the barrier height of the potential (6.67) is small compared to the coupling terms, θ_n varies slowly from one site to another and one can use, as in Sect 6.1.1, the standard continuum approximation : $\theta_n(t) \rightarrow \theta(x,t)$ and expand $\theta_{n\pm1}$. Under these conditions, setting $X = x/a$, we measure the distance x in units of lattice spacing a and (6.65) is reduced to

$$\frac{\partial^2\theta}{\partial\tau^2} - [C_{1,\ell} + 3C_{n\ell}\,(\frac{\partial\theta}{\partial X})^2]\frac{\partial^2\theta}{\partial X^2} - \frac{C_{1,\ell}}{12}\frac{\partial^4\theta}{\partial X^4} + \theta - \Gamma\sin\theta = 0. \qquad (6.68)$$

Note that C_ℓ represents the square of the velocity of linear waves in the chain. Equation (6.68) cannot be solved analytically. Nevertheless, in order to get some approximate solution one can replace the continuum approximation of potential (6.67) by the standard double-well potential called "Φ-four" potential (see Sect 9.7) given by

$$V(\Theta) = \frac{V_o}{2}\,(1-\Theta^2)^2, \qquad (6.69)$$

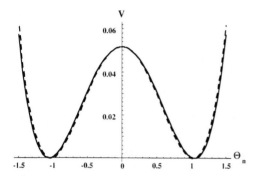

Fig. 6.21 Fittting of the double well potential (6.67) : dotted line, by a "Φ-four" potential (6.69): continuous line. For $-1.5<\Theta=\theta/\theta_m<1.5$ the two curves are practically superimposed and the approximation of on-site potential (6.67) of the real system by (6.69) is reasonable

where $\Theta= \theta/\theta_m$ and $V_o=-\theta_m^2+2\Gamma(1-\cos\theta_m)$. As depicted in Fig 6.21 the fit is good for $-1.5<\Theta<1.5$. Under these conditions eq (6.68) is approximated by

$$\frac{\partial^2\Theta}{\partial T^2} - [C_\ell+ 3C_{n\ell}(\frac{\partial\Theta}{\partial X})^2]\frac{\partial^2\Theta}{\partial X^2} - \frac{C_\ell}{12}\frac{\partial^4\Theta}{\partial X^4} - 2V_o'(\Theta-\Theta^3)= 0, \qquad (6.70)$$

where $T =\theta_m\tau$, $C_\ell = C_{1,\ell}/\theta_m^2$ and $V_o'= 2V_o/\theta_m^4$. We assume that

$$3C_{n\ell}\,(\frac{\partial\Theta}{\partial X})^2\frac{\partial^2\Theta}{\partial X^2} \gg \frac{C_\ell}{12}\frac{\partial^4\Theta}{\partial X^4}\,,$$

as it will be the case in the following.

6.11.3 Kink-soliton solutions

Qualitatively, from (6.70) we expect kink-solutions because the corresponding potential (6.69) is a double-well. As in Sect 6.2.2, we look for localized waves of permanent profile of the form $\theta(s)=\theta(X-uT)$, such that $\theta \to \pm 1$ and $d\theta/ds \to 0$, when $s \to \pm \infty$, where $s=X-uT$ is a single independent variable depending on u which is an arbitrary velocity of propagation. Integrating (6.70), taking into account the boundary conditions (see Appendix 6.A), and ignoring the small dispersive term $(C_\ell/12)\,(\partial^4\Theta/\partial X^4)$ we get

$$2(u^2-C_\ell)\Theta_s{}^2 - 3C_{n\ell}\Theta_s{}^4 + 2V_0{}'(1-\Theta^2)^2 =0. \tag{6.71}$$

Equation (6.71) cannot be integrated analytically, but numerically (Dusuel et al 1998). The kink (antikink) solution connect the two degenerate minima as sketched in Fig 6.19. First, let us consider the case $C_{n\ell} =0$. In this case (6.71) reduces to the standard continuous Φ-four model with linear coupling which, as discussed in Sect 9.7.2 and Appendix 9B, admits tanh-shaped kink solutions :

$$\Theta= \tanh\,[\frac{\Gamma(X - vT)}{\sqrt{2}D}], \tag{6.72}$$

where

$$D = \sqrt{C_\ell/V_0{}'}, \qquad v=u\sqrt{C_\ell}, \qquad \Gamma= (1 - v^2/C_\ell)^{-1/2}.$$

The kink waveform corresponding to (6.72) is represented in Fig 6.22.

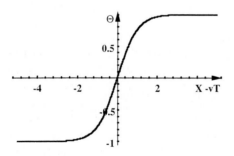

Fig. 6.22. Sketch of a thanh-shaped kink solution (6.72) for $\Gamma/D = \sqrt{2}$

6. 11.4 Compacton-like kinks or compactons

For $C_{n\ell} \neq 0$, (6.71) can be integrated if $u^2 - C_{\ell} = 0$, that is, for the two particular cases : $C_{\ell} = 0$ (zero linear coupling: linear waves cannnot exist) and $u=0$, or $u = \sqrt{C_{\ell}}$. Setting $\alpha = (2V_0'/3C_{n\ell})^{1/4}$, after integrating (6.71) , one obtains

$$\Theta_c (X) = \pm \sin [\alpha(X-X_0)], \tag{6.73}$$

when $\mid X-X_0 \mid < \frac{\pi}{2} \alpha$, and $\Theta = \pm 1$ otherwise. The constant of integration X_0 defines the position of the center of the kink (see Appendix 6A) .
For the second case we get

$$\Theta_c (X,t) = \pm\sin [(\alpha(s-s_0)], \tag{6.74}$$

when $\mid s-s_0 \mid = \mid X-\sqrt{C_{\ell}}\, T -X_0-\sqrt{C_{\ell}}\, T_0) \mid < \frac{\pi}{2} \alpha$ and $\Theta = \pm 1$ otherwise.

We remark that contrary to the kink solution (6.72) which has exponentially decreasing wings extending to infinity, solutions (6.73) and (6.74) are strictly localized : *they have no wings, that is, they have a compact shape.* Such solutions occur because in (6.71) and (6.70) the nonlinear coupling term, which corresponds to nonlinear dispersion, is preponderant: *the linear coupling term can be zero.* On the other hand we note that, unlike the kink-soliton solution (6.72) whose profile $(\sqrt{2D/\Gamma})$ is velocity dependent, the profile $1/\alpha$ of the solution (6.74) is independent of its velocity.

As a matter of fact, it was recently shown (Rosenau and Hyman 1993, Rosenau 1994, 1996, 1997) that solitary-wave solutions may compactify under the influence of nonlinear dispersion which is capable of causing deep qualitative changes in the nature of nonlinear phenomena. Such robust soliton-like solutions, characterized by the absence of an infinite tail and whose width is velocity independent , have been called *compactons* (Rosenau and Hyman 1993). They have been first obtained for a special class of KdV type equations with nonlinear dispersion, that is, for example, an equation similar to (3.21) where the linear dispersive term is replaced by a nonlinear term (see also Dey 1998; Dey and Khare 1998). In this context, it is interesting to note that a periodic wave ($\sim \cos^4$) solution, with large amplitude, was found theoretically by Nesterenko (1984) when studying the mechanical behavior of granular materials. Later, experimental evidence of such a wave was given (Lazaridi and Nesterenko 1985; Coste et al. 1997). In fact the nature of this periodic wave is critical : the half period of it (the solitary wave) can be wiewed as a compacton, that is, with the concept of compacton at hand, one has a better understanding of the work of these authors.

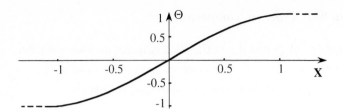

Fig. 6.23 A static compacton solution as given by (6.73) for $X_o=0$. The S shape (continuous line) corresponds to the compact part which connects the two constant parts ($\Theta = \pm 1$: horizontal dotted lines) of the solution

In this context, solution (6.73), which corresponds to zero linear coupling, represents a static *kink-compacton* (anti-compacton) or compacton for short. Its shape is represented in Fig 6.23 for $X_o=0$.

From the continuum approximation of (6.66) and (6.67) one can calculate the total potential energy E_{stat} localized in the static compacton

$$E_{stat} = \int_{-\pi/2\alpha}^{\pi/2\alpha} [\frac{1}{4} C_{n\ell}(\frac{\partial\Theta}{\partial X})^4 + \frac{1}{2}V_o'(1-\Theta^2)^2]\ dX. \tag{6.75}$$

Substituting (6.73) into (6.75) and integrating one finds

$$E_{stat} = aV_o'\frac{\pi}{4\alpha}, \tag{6.76}$$

which represents the mass (compare to Sect 6.3) of the static compacton.
Solution (6.74), which occurs when both the linear and nonlinear couplings are present, represents a dynamic compacton (anti-compacton) traveling at particular velocity $\sqrt{C\ell}$ (or $-\sqrt{C\ell}$). From (6.73) we obtain the field gradient

$$\frac{\partial\Theta_c}{\partial X} = \alpha \cos [\alpha (X - X_o)], \tag{6.77}$$

when $|X-X_o| < \frac{\pi}{2}\alpha$, and $\Theta=0$ otherwise. Thus, the potential energy density owing to the nonlinear coupling is $(C_{n\ell}/4) \alpha^4 \cos^4[\alpha (X - X_o)]$.

Remark

In the case of zero linear coupling ($C\ell = 0$) eq (6.70) reduces to

$$\frac{\partial^2 \Theta}{\partial T^2} - 3C_{n\ell}\left(\frac{\partial \Theta}{\partial X}\right)^2 \frac{\partial^2 \Theta}{\partial X^2} - 2V_o'(\Theta - \Theta^3) = 0.$$

In the small amplitude limit, when one considers oscillations, $\delta\Theta \ll 1$, near one of the two equilibrium positions : $\Theta = \pm 1 + \delta\Theta$ (see (9.64)), this equation can be linearized into

$$\frac{\partial^2 (\delta\Theta)}{\partial T^2} + 4V_o'\delta\Theta = 0.$$

Such an equation of motion, without any space dependent term, shows that the system oscillates as a whole (zero wavenumber) and that *linear waves cannot propagate in the system*. A situation somewhat similar was found by Nesterenko (1994) for the propagation of compression pulses in granular media. He has shown that when the linear interactions between neighboring elements are zero (no static external force) or small compared to the nonlinear interactions, no linear acoustic waves can propagate: the sound (linear waves) velocity is zero. Such a situation was called *sonic vacuum* by Nesterenko. He also showed that strongly nonlinear waves do propagate, even when the grains are just in contact, but in the absence of any static force.

6.11.5 Experiments

The experimental line is a lattice of 20 pendulums attached vertically to the center of an horizontal steel ribbon (2m long, 6mm wide and 0.1mm thick) supported by vertical metallic plates which are equidistant (a=10 cm) (see Figs. 3 and 5). A basic pendulum consists of a thin rod (diameter 3mm) along which a cylinder (mass m=67g) can be displaced and fixed. Depending on the vertical position d of the mass along the rod, the system can oscillate with a motion which depends on the potential shape and is determined, as mentioned earlier, by the forces of gravity and torsion in opposition. Here, with d=87mm and K= 0.03, the control parameter is Γ=1.2, thus the on-site potential is a symmetric double well potential.

Once its tension is adjusted, the ribbon is held tight on the top of each plate. With this precaution, the torsion constant is the same for each pendulum and the weak residual torsional coupling between pendulums can be neglected. Each pendulum (cylinder) is attached to its neighbors with a spring. Springs connecting pendulums that are at equilibrium, in the bottom of one of the potential wells, are horizontal as represented in Fig. 6.24.

Fig. 6.24. Experimental transmission line with 20 pendulums coupled by springs, all spaced at distance a=10 cm apart. Here, all the pendulums are at equilibrium in one of the two potential wells (Photo R. Chaux)

In the general case ($\ell_0 \neq \ell_1$, that is, $C_\ell \neq 0$ and $C_{n\ell} \neq 0$: see eqs (6.63)) with this mechanical line we can easily observe the dynamical properties of the kinks. For example, if one launches a moving kink with arbitrary initial velocity at one end of the line, after reflection at the opposite free end this kink becomes an anti-kink moving freely in the opposite direction and so on. Depending on its initial velocity a kink can reflect three or four times before gradually slowing down owing to dissipative effects which inevitably occur for a real mechanical line. With the above physical parameters no radiation of waves due to discreteness effects are observed. Thus, the continuum approximation is reasonnable.

With the following physical parameters: ($\ell_1 = \ell_0$= 68mm: C_ℓ=0), $C_{n\ell}$ =25 and k=120, a static compacton, as described by (6.73) can be observed as represented in Fig.6.25. The experimental shape fits approximatively (see Fig 6.26) the theoretical shape calculated from (6.73). Strictly speaking, for such a real system we can say that the experimental shape of the localized entity approaches a compact shape. In the context of phase transitions, discussed in Sect 9.7, a compacton may model a static domain wall. However, it is interesting to note that unlike a standard (tanh-shaped) kink which possesses (exponential) wings and can interact at distance with an anti-kink, a compacton and an anti-compacton will not interact unless they come into contact in a way similar to the contact between to hard spheres.

When $C_\ell \neq 0$ and $C_{n\ell} \neq 0$, solution (6.74) predicts a compacton moving at velocity $\sqrt{C_\ell}$ of the linear waves. In fact, in this case, we cannot conclude that the moving S-shaped entity we observe has a compact shape for the following reasons. First, we cannot control with sufficient precision the initial velocity of the kink. Second, even if we could launch a kink with exact velocity $\sqrt{C_1}$ it would gradually slow down owing to dissipative effects (that are important compared to small radiation effects predicted by numerical simulations (Dusuel et al 1998)); thus, its velocity becomes smaller than $\sqrt{C_\ell}$ and we can never observe a moving compacton.

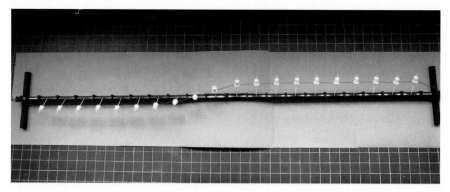

Fig. 6.25. Picture showing a static compacton, which interpolates between the two equilibrium staates, as observed on the experimental chain and predicted by (6.73). Note that the two transverse supports that are visible in Fig. 6.24, are now masked (Photo R. Chaux).

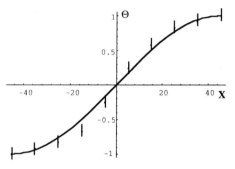

Fig. 6.26. Comparison of the experimental compacton shape depicted in Fig. 6.25, with the theoretical shape calculated from (6.73). Here, $\Theta = \theta/\theta_m = \pi/3$, and $X = x/a$ is dimensionless $(a = 10cm)$. The vertical lines represent the experimental precision.

Remark

The above mechanical line can be very useful to illustrate the problem of phase transitions in physical systems . The kink represents a transition from one potential minimum to the other one. The structural character of the system is changed by the passage of the kink which can be considered as a *domain wall* through which the transition takes place (see Sect 9.7.2 and Charru 1997).

6.12 Solitons, compactons and nanopterons

In this Sect. we compare to the soliton two different kinds of nonlinear localized waves: the compacton and the *nanopteron* that may be generated in a continuum system, depending on the interplay between nonlinear and dispersive phenomena.

175

(a) As discussed throughout this book, we recall that :

A solitary wave ϕ_{sw} (s=x-vt) is a localized traveling wave, whose transition from one constant asymptotic state as s $\rightarrow -\infty$ to another as x $\rightarrow +\infty$ is essentially localized in s. A soliton ϕ_s (s=x-vt) is a solitary wave which asymptotically preserves its shape and velocity upon collision with other solitary waves.

Two types of solitons: the pulse-soliton and the kink-soliton, are sketched in Fig.6.27a.

(b) The above working definition of the soliton shows that it is not a strictly localized entity, it possesses asymptotic wings. However, observed patterns in nature whether stationary or traveling are of finite extent. In this regard, we have seen (see Sects. 6.11.4 and 6.11.5) that :

The compacton is a robust and strictly localized nonlinear wave, which, unlike the soliton, is characterized by the absence of infinitesimal wings extending to infinity . and has a profile or spatial width that is independent of its velocity.

The essential ingredient for the exitence of compactons is nonlinear dispersion (nonlinear coupling: Sect 6.11.4). A pulse-compacton (see Rosenau and Hyman 1993) and a kink-compacton are sketched in Fig. 6.27b.

(c) A third kind of nonlinear localized,wave, that we would like to mention, which, to our knowledge, has not yet been observed in the real world but looks interesting, is *the nanopteron or weakly nonlocal solitary wave*. The nanopteron, which may occur when linear dispersion is large compared to nonlinearity, was introduced by Boyd (1989, 1990a). It has the following characteristics (Boyd 1990b):

(i)*The nanopteron is a permanent and steadily translating wave at a constant velocity v.*

(ii) *It satisfies all the other requirements of a solitary wave except that it asymptotes not to zero but to a small amplitude oscillation when s $\rightarrow \pm \infty$.*

(iii) *The amplitude of the far field oscillations is an exponentially small function of the maximum amplitude (core) of the wave.*

The profile of a pulsed-shaped nanopteron is depicted in Fig. 6.27c. As one can see from this figure, the nanopteron motion occurs on the background of a periodic wave. This wave (oscillating background) may be nonlinear or linear as recently discussed (Zolotaryuk 1998) for the propagation of kink-shaped nanopterons in a Klein-Gordon lattice. A kink-shaped nanopteron is also sketched in Fig. 6.27c.

Physically, a nanopteron can be considered as the bound state resulting from the nonlinear interaction between the core (pulse or kink) and the periodic wave. In this context it has been shown (Buryak 1995) that for some generalized nonlinear Schrödinger models, stationary soliton bound states can exist in resonance with small amplitude linear waves.

176

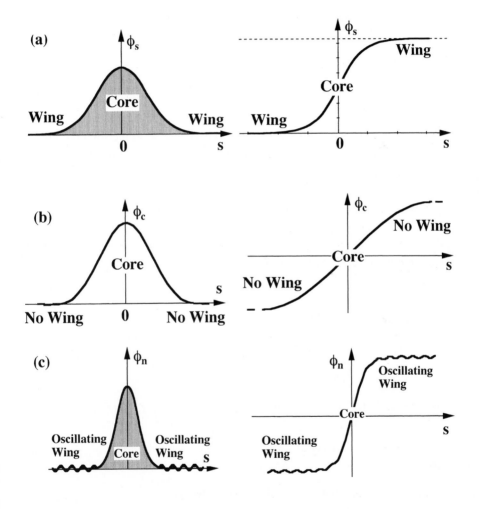

Fig. 6.27a-c. Sketch of the profiles of : (a) a pulse-soliton and a kink-soliton. (b) A pulse-compacton and a kink-compacton. (c) A pulse-shaped nanopteron and a kink-shaped nanopteron.

Appendix 6A. Kink soliton and antikink soliton solutions

From (6.10), identifying the two expressions of $d\theta$ in terms of the old variables (x,t) and the new variable s,

$$d\theta = \frac{\partial\theta}{\partial X}dX + \frac{\partial\theta}{\partial T}\,dT = \frac{\partial\theta}{\partial s}dX - u\frac{\partial\theta}{\partial s}\,dT, \qquad (6A.1)$$

yields

$$\partial/\partial X = \partial/\partial s \, , \, \partial/\partial T = -u\,\partial/\partial s.$$

Thus, (6.10) reduces to an ordinary differential equation,

$$(1 - u^2)\, d^2\theta/ds^2 = \sin\theta. \qquad (6A.2)$$

Note that if one replaces θ by $(\pi + \theta)$, (6A.2) is analogous to the simple pendulum equation. Multiplying both sides of (6A.2) by $d\theta/ds$ and integrating yields

$$\frac{d\theta}{ds} = \pm\sqrt{\frac{2(C - \cos\theta)}{1 - u^2}}\,, \qquad (6A.3)$$

where C is an arbitrary constant of integration. If one looks for localized solutions for which $\theta \to 0$ and $d\theta/ds \to 0$, when $s \to \pm\infty$, one gets $C = 1$. Thus, solutions to (6A.3) can be written as

$$\pm\frac{s - s_0}{\sqrt{1 - u^2}} = \int_{\theta(s_0)}^{\theta(s)} \frac{d\theta}{\sqrt{2(1 - \cos\theta)}}\,,$$

with $s_0 = X_0 - uT_0$, where X_0 is the wave position at time T_0. Integrating gives

$$\pm\frac{s - s_0}{\sqrt{1 - u^2}} = Ln\,(\mathrm{tg}\frac{\theta}{4}), \qquad (6A.4)$$

with $\theta(s_0) = \pi$. The solution (6A.4) is then written as

$$\theta = 4\arctan\left[\exp\left(\pm\frac{s - s_0}{\sqrt{1 - u^2}}\right)\right]. \qquad (6A.5)$$

Appendix 6B. Calculation of the energy and the mass of a kink soliton

Equation (6.22) can be written as

$$E = \frac{I}{a} \omega_0 c_0 \, (E_1 + E_2 + E_3), \tag{6B.1}$$

and E_1 , E_2 , E_3 are the three energy terms, respectively. The first term or kinetic-energy term is

$$E_1 = \frac{1}{2} \int\limits_{-\infty}^{+\infty} [4\gamma^2 u^2 sech^2\gamma(X - uT)] \, dX, \tag{6B.2}$$

which after integration results in

$$E_1 = 2\gamma u^2 \, [\, \tanh (X - uT) \,] \, _{-\infty}^{+\infty} = 4\gamma u^2. \tag{6B.3}$$

The second term is also easily calculated as

$$E_2 = \frac{1}{2} \int\limits_{-\infty}^{+\infty} [4\gamma^2 sech^2\gamma(X - uT)] \, dX = 4\gamma. \tag{6B.4}$$

The third term becomes

$$E_3 = \int\limits_{-\infty}^{+\infty} 2 \, sin^2\frac{\theta}{2} \, dX. \tag{6B.5}$$

Taking into account (6.14) one has

$$sin(\theta/2) = sin \, [2 \arctan e^{\gamma s}] = sin \, 2y,$$

with $y = \tan^{-1} e^{\gamma s}$. Noting that

$$sin \, 2y = \frac{2\tan y}{1 + \tan^2 y} = 2\frac{e^{\gamma s}}{1 + e^{2\gamma s}} = sech \, \gamma s,$$

the E_3 term is then simply written as

$$E_3 = 2 \int\limits_{-\infty}^{+\infty} [sech^2\gamma(X-uT)] \, dX = \frac{4}{\gamma}. \tag{6B.6}$$

Finally the kink-soliton energy is

$$E_K = 4 \frac{I}{a} \omega_0 c_0 \, \gamma \, (1 + u^2 + \frac{1}{\gamma^2}) = 8\gamma \frac{I}{a} \omega_0 c_0. \tag{6B.7}$$

Appendix 6C. Solutions for kink–kink and kink–antikink collisions, and breathers

We look for functions F and G satisfying (6.28) and (6.10). Thus, setting

$$\theta/4 = \phi, \quad \tan\phi = F/G = y \qquad (6C.1)$$

and using the identity

$$\sin 4\phi = 4\tan\phi \, \frac{1 - \tan^2\phi}{(1 + \tan^2\phi)^2}, \qquad (6C.2)$$

(6.10) is transformed into

$$\frac{\partial^2\phi}{\partial T^2} - \frac{\partial^2\phi}{\partial X^2} = y \, \frac{(1 - y^2)}{(1 + y^2)^2}. \qquad (6C.3)$$

Upon substitution of (6C.1) and (6C.2) into (6C.3) we get

$$\frac{2y}{(1 + y^2)^2}(y_X)^2 - \frac{y_{XX}}{(1 + y^2)} - \frac{2y}{(1 + y^2)^2}(y_T)^2 + \frac{y_{TT}}{(1 + y^2)} + y\frac{(1 - y^2)}{(1 + y^2)^2} = 0, \qquad (6C.4)$$

which simplifies to

$$\frac{1 + y^2}{y}(y_{XX} - y_{TT}) + 2\left[(y_T)^2 - (y_X)^2\right] = (1 - y^2). \qquad (6C.5)$$

Here the subscripts X and T denote the derivatives. By using (6C.1) we now express y in terms of F and G:

$$y_X = \frac{F_X}{G}, \quad y_{XX} = \frac{F_{XX}}{G}, \quad y_T = -\frac{FG_T}{G^2}, \quad y_{TT} = -\frac{FG_{TT}}{G^2} + \frac{2FG_T^2}{G^3}. \qquad (6C.6)$$

Substituting relations (6C.6) in (6C.5) yields

$$(G^2 + F^2)\left(\frac{F_{XX}}{F} + \frac{G_{TT}}{G}\right) - 2\left[(F_X)^2 + (G_T)^2\right] + F^2 - G^2 = 0. \qquad (6C.7)$$

Deriving (6C.7) with respect to X and with respect to T one obtains successively

$$G^2\left(\frac{F_{XX}}{F}\right)_X + (FF_{XX})_X + (F^2)_X\frac{G_{TT}}{G} - \left[2(F_X)^2 - F^2\right]_X = 0, \qquad (6C.8a)$$

$$(G^2)_T\left(\frac{F_{XX}}{F}\right)_X + (F^2)_X\left(\frac{G_{TT}}{G}\right)_T = 0. \qquad (6C.8b)$$

Separating the variables in (6C.8b) gives

$$\frac{1}{(F^2)_X} \left(\frac{F_{XX}}{F} \right) X = - \frac{1}{(G^2)_T} \left(\frac{G_{TT}}{G} \right) T = A, \tag{6C.9}$$

where A is a constant. Equations (6C.9) are integrated to give

$$F_{XX} = AF^3 + B_1 F, \tag{6C.10a}$$

$$G_{TT} = - AG^3 + B_2 G, \tag{6C.10b}$$

where B_1 and B_2 are constants. Multiplying (6C.10a) and (6C.10b) respectively by $2F_X$ and $2G_T$ and integrating we get

$$(F_X)^2 = \frac{A}{2} F^4 + B_1 F^2 + C_1, \tag{6C.11a}$$

$$G_T)^2 = - \frac{A}{2} G^4 + B_2 G^2 + C_2, \tag{6C.11b}$$

where C_1 and C_2 are two new integration constants. Substituting (6C.11) in (6C.7) implies that

$$B_1 - B_2 = 1, \quad C_1 - C_2 = 0. \tag{6C.12ab}$$

Setting now

$$\frac{A}{2} = - q^2, \quad B_1 = b^2, \quad C_1 = n^2, \tag{6C.13}$$

eqs (6C.11) finally become

$$\left(\frac{dF}{dX} \right)^2 = -q^2 F^4 + b^2 F^2 + n^2, \tag{6C.14a}$$

$$\left(\frac{dG}{dT} \right)^2 = q^2 G^4 + (b^2 - 1) G^2 - n^2, \tag{6C.14b}$$

where q, b and n are arbitrary constants whose values will determine the kinds of solution of (6.30). Let us examine some particularly interesting cases:
the kink–soliton solution for q =0, b>0, and n = 0;
the kink–kink collision solution for q =0, $b^2 > 1$, and n ≠ 0;
the breather–soliton solution for q ≠ 0, $b^2 < 1$, and n = 0;
the kink–antikink collision solution for q ≠ 0, b>1, and n =0.

6C.1 Kink solutions: $q = 0$, $b > 1$, and $n = 0$

In this case (6C.14) reduce to

$$\frac{dF}{F} = \pm b\, dX, \qquad \frac{dG}{G} = \pm \sqrt{b^2-1}\; dT. \qquad (6C.15)$$

These equations can be easily integrated to give

$$F = K_1 \exp(\pm bX), \qquad G = K_2 \exp(\pm \sqrt{b^2-1}\; T), \qquad (6C.16)$$

where K_1 and K_2 are integration constants. Replacing relations (6C.16) in (6C.1), we find that

$$\theta(X, T) = 4 \arctan\left[\frac{K_1}{K_2} \exp\left(\pm \frac{X \pm uT}{\sqrt{1-u^2}}\right)\right], \qquad (6C.17)$$

where $u = \sqrt{b^2-1}/b$. Setting $K_1/K_2 = 1$ and choosing the adequate signs one recovers the kink-soliton (antisoliton) solution (6.12).

6C.2 Kink–kink collisions: $q = 0$, $b^2 > 1$, and $n \neq 0$.

We assume that $n = b$, and solutions of (6C.14) are

$$F = \sinh bX, \qquad G = \frac{b}{\sqrt{b^2-1}} \cosh \sqrt{b^2-1}\; T. \qquad (6C.18)$$

We then set $u = \sqrt{b^2-1}/b$. From (6C.1) and (6C.18) we obtain the solution

$$\theta(X, T) = 4\arctan\left[u \sinh \frac{X}{\sqrt{1-u^2}} \; \text{sech} \frac{uT}{\sqrt{1-u^2}} \right]. \qquad (6C.19)$$

Let us show that this solution now describes the collision between two kinks.

For $X \to -\infty$, $\theta \to 4 \arctan\left[\frac{-ue^{-\gamma x}}{2\cosh\gamma uT} \right]$,

which reduces to

$$\theta \to \theta_{k1} = 4 \arctan\left[-ue^{-\gamma(x - uT)} \right] \text{ for } T \to -\infty, \qquad (6C.20a)$$

$$\theta \to \theta_{k2} = 4 \arctan\left[-ue^{-\gamma(x + uT)} \right] \text{ for } T \to +\infty. \qquad (6C.20b)$$

Using the relation

$$\arctan y = -\arctan \left[\frac{1}{y}\right] + \frac{\pi}{2} ,$$

we can transform (6C.20) into

$$\theta_{k1} = 4 \arctan \left[\frac{1}{u} e^{\gamma(x-uT)}\right] + 2\pi \text{ for } T \to -\infty, \qquad (6C.21a)$$

$$\theta_{k2} = 4 \arctan \left[\frac{1}{u} e^{\gamma(x+uT)}\right] + 2\pi, \text{ for } T \to +\infty. \qquad (6C.21b)$$

For $X \to \infty$, $\theta \to 4 \arctan \dfrac{u e^{\gamma x}}{2\cosh\gamma uT}$,

which reduces to

$$\theta \to \theta_{k1} = 4 \arctan u e^{\gamma(x+uT)} \text{ for } T \to -\infty, \qquad (6C.22a)$$

$$\theta \to \theta_{k2} = 4 \arctan u e^{\gamma(x-uT)} \text{ for } T \to +\infty. \qquad (6C.22b)$$

Relations (6C.20) and (6C.22) represent a kink k1 and a kink k2 with respective velocities +u and -u approaching the origin from $T \to -\infty$ and moving away from it with velocities -u and +u respectively for $T \to +\infty$.

6C.3 Breather solitons: $q \neq 0$, $b^2 < 1$, and $n = 0$.

Here (6C.14) reduce to

$$\frac{dF}{F\sqrt{b'^2 - F^2}} = \pm qdX , \qquad \frac{dG}{G\sqrt{-b''^2 + G^2}} = \pm qdT, \qquad (6C.23)$$

where $b'^2 = b^2/q^2$ and $b'' = \sqrt{1-b^2}/q$. The solutions of (6C23) are calculated as follows.

Tables of integrals give

$$\int \frac{dx}{x\sqrt{R}} = -\frac{1}{\sqrt{A}} Ln \left[\frac{\sqrt{R} + \sqrt{A}}{x} + \frac{B}{2\sqrt{A}}\right], \text{ for } A > 0;$$

$$\int \frac{dx}{x\sqrt{R}} = \frac{1}{\sqrt{-A}} \sin^{-1} \left[\frac{Bx + 2A}{x\sqrt{B^2 - 4AC}}\right], \text{ for } A < 0;$$

where $R = A + Bx + Cx^2$. Using these relations with $B = 0$, from (6C.23) we obtain

$$-\frac{1}{b'} \text{Ln} \left[\frac{\sqrt{b'^2 - F^2} + b'}{F} \right] = \pm qX + C_1, \qquad \frac{1}{b''} \sin^{-1}\left[\frac{b''}{G}\right] = \pm qT + C_2,$$

where the constants C_1 and C_2, which correspond to initial space and time coordinates, can be set equal to zero. From these relations we get

$$F = b'\text{sech } (b'qX), \qquad G = \pm \frac{b''}{\sin(b''qT)}. \qquad (6C.24)$$

Substituting these solutions of (6C.23) in (6.28) and taking account of (6C.24) we find the breather-soliton solution

$$\theta_B (X, T) = 4 \arctan \left[\frac{\sqrt{1 - \Omega^2}}{\Omega} \text{ sech } (\sqrt{1-\Omega^2}\, X) \sin (\Omega\, T) \right]. \qquad (6C.25)$$

6C.4 Kink–antikink collision: $q \neq 0$, $b > 1$, and $n = 0$

One can write

$$\sqrt{1 - b^2} = i \sqrt{b^2 - 1} = \Omega. \qquad (6C.26)$$

Substituting this relation in (6C.25) one immediately obtains

$$\theta (X, T) = 4 \arctan \left[\frac{b}{\sqrt{b^2 - 1}} \text{ sech } (bX) \sinh (\sqrt{b^2-1}\, T) \right]. \qquad (6C.27)$$

We then set $u = \sqrt{b^2-1}/b$ and rewrite expression (6C.27) as

$$\theta (X, T) = 4 \arctan \left[\frac{1}{u} \text{ sech } (\gamma X) \sinh (\gamma\, uT) \right]. \qquad (6C.28)$$

This analytical solution describes a collision process between a kink soliton and an antikink soliton in a center-of-mass coordinate system. In fact, using (6C.28) we have the following asymptotic results:

$$\text{for } T \to -\infty, \quad \theta \to 4 \arctan \frac{-e^{-\gamma uT}}{u2\cosh\gamma X},$$

which reduces to

$$\theta \to \theta_k = -4 \arctan \frac{1}{u} e^{\gamma(x-uT)} \text{ for } X \to -\infty, \tag{6C.29a}$$

$$\theta \to \theta_{-k} = -4 \arctan \frac{1}{u} e^{-\gamma(x+uT)} \text{ for } X \to +\infty. \tag{6C.29b}$$

On the other hand

$$\text{for } T \to +\infty, \ \theta \to 4 \arctan \frac{e^{\gamma u T}}{2u \cosh\gamma X}.$$

This reduces to

$$\theta \to \theta_{-k} = 4 \arctan \frac{1}{u} e^{\gamma(x+uT)}, \text{ for } X \to -\infty, \tag{6C.30a}$$

$$\theta \to \theta_k = 4 \arctan \frac{1}{u} e^{-\gamma(x-uT)}, \text{ for } X \to +\infty. \tag{6C.30b}$$

Except for the unimportant factors u or 1/u, which can be absorbed in the phase, at time $T = -\infty$, (6C.29a) describes a kink approaching the origin with velocity u from $X = -\infty$, and (6C.29b) represents an antikink approaching the origin with velocity -u from $X = +\infty$. These collide and pass through each other; then at $T = +\infty$ the kink is at $X = +\infty$ and the antikink is at $X = -\infty$ as described by relations (6C.30b) and (6C.30a), respectively.

Appendix 6D. Solutions for helical waves

If the integration constant C in (6A.3) is not equal to unity, solutions can be written as elliptic integrals:

$$\pm \frac{s}{\sqrt{1 - u^2}} = \int_{\pi}^{\theta_p} \frac{d\theta}{\sqrt{2(C - \cos\theta)}}. \tag{6D.1}$$

If $C > 1$ and $u < 1$, (6D.1) can be written as

$$\pm \frac{2s}{\sqrt{1-u^2}} = \int_{\pi}^{\theta_p} \frac{d\theta}{\sqrt{(D+\sin^2\frac{\theta}{2})}}, \qquad (6D.2)$$

where $D = (C-1)/2$ is a constant that we transform into $D = (1-k^2)/k^2$ by introducing a new parameter k which varies in the interval 0 to 1. Therefore, (6D.2) becomes

$$\pm 2U = \int_{\pi}^{\theta} \frac{d\theta}{\sqrt{(1-k^2\cos^2\frac{\theta}{2})}}, \qquad (6D.3)$$

with $U = s/(k\sqrt{1-u^2})$. We now change to the new variable $\sin\phi = \cos\frac{\theta}{2}$ and transform the above equation into

$$\pm U = \int_{0}^{\phi} \frac{d\phi}{\sqrt{(1-k^2\sin^2\phi)}}. \qquad (6D.4)$$

From the definition (see Appendix 4B) of the Jacobian elliptic function sn (U,k) with modulus k, we rewrite (6D.4) as follows:

$$\sin\phi = \pm \text{sn}(U,k),$$

or

$$\cos\frac{\theta}{2} = -\sin(\frac{\theta}{2}-\frac{\pi}{2}) = \pm \text{sn}(U,k), \qquad (6D.5)$$

where the period of sn (U,k) is $4K(k)$, with $K(k)$ the complete elliptic integral of the first kind (see Appendix 4B). Replacing U by its expression we finally have

$$\theta = 2\sin^{-1}[\pm \text{sn}(\frac{s}{k\sqrt{1-u^2}}, k)] + \pi. \qquad (6D.6)$$

Appendix 6E. Pendulum with torsion and gravity

A basic pendulum consists of a thin rod along which a cylinder of mass m can be displaced and fixed (see Fig 6E.1). Depending on the position d of the mass along the rod the system can oscillate with a motion which is determined by the torque $mgd\sin\theta$ owing to gravity and the torque $-K\theta$ owing to torsion, which act in opposition. The equation of motion of the pendulum is

$$I\frac{d^2\theta}{dt^2} = -K\theta + mgd\sin\theta, \tag{5E.1}$$

where I is the moment of inertia ($I \approx md^2$, if the mass of the rod is small compared to the mass of the cylinder).

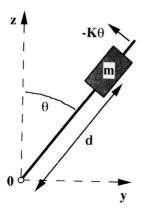

Fig. 6E.1 Sketch of one pendulum where torsion and gravity act in opposition.

Appendix 6F. Model equation for the pendulum chain

In this appendix we derive eq (6.62). The general equation of motion of the nth pendulum of the chain represented in Fig 6.20 is

$$I\frac{d^2\theta_n}{dt^2} = -K\theta_n + mgd\sin\theta_n + M_{n-1,n} - M_{n,n+1}, \tag{6F.1}$$

where $M_{n-1,n}$, $M_{n,n+1}$ are the torque exerted by pendulum n-1 on pendulum n and pendulum n on pendulum n+1, respectively.
In terms of the components $y_n = -d\sin\theta_n$ and $z_n = d\cos\theta_n$ of the displacement, the elongation of the spring (see Fig 6.20) between pendulums n and n+1 is

$$\Delta\ell = \sqrt{\ell_1{}^2 + (y_{n+1} - y_n)^2 + (z_{n+1} - z_n)^2} - \ell_o \tag{6F.2}$$

where ℓ_o is the length of the spring at rest, ℓ_1 the minimal length of the stretched spring between two pendulums. Thus, we have

$$M_{n,n+1} = \frac{k\Delta\ell}{V}(y_{n+1}z_n - y_n z_{n+1}), \tag{6F.3}$$

where

$$(y_{n+1}z_n - y_n z_{n+1}) = d^2 \sin(\theta_n - \theta_{n+1}), \tag{6F.4}$$

and

$$V = \sqrt{\ell_1{}^2 + 4d^2\sin^2\frac{(\theta_n - \theta_{n+1})}{2}}. \tag{6F.5}$$

One obtains

$$M_{n,n+1} = kd^2\left[1 - \frac{\ell_o}{\ell_1\sqrt{1 + \frac{4d^2}{\ell_1{}^2}\sin^2\frac{(\theta_n - \theta_{n+1})}{2}}}\right]\sin(\theta_n - \theta_{n+1}) \tag{6F.6}$$

$M_{n-1,n}$ is obtained by replacing n by n-1 in the above expression. When the difference $|\theta_n - \theta_{n+1}|$ between the angular displacement of neighboring pendulums is small enough (weakly discrete limit), the torques can be replaced by their expansion in terms of these angular differences

$$M_{n,n+1} \approx kd^2\left\{(1 - \frac{\ell_o}{\ell_1})(\theta_n - \theta_{n+1}) + \left[\frac{\ell_o d^2}{2\ell_1{}^3} + \frac{1}{6}(\frac{\ell_o}{\ell_1} - 1)\right](\theta_n - \theta_{n+1})^3\right\}. \tag{6F.7}$$

Thus inserting (6F.7) and a similar expression for $M_{n-1,n}$ in (6F.1), we obtain (6.62).

7 Fluxons in Josephson Transmission Lines

The superconducting Josephson junctions (Barone and Paterno 1982; Parmentier 1978; Lomdahl 1985; Likharev 1986; Pedersen 1986) have proven to be one of the most successful testing grounds for nonlinear wave theory; their use for information processing and storage is quite attractive. In the long Josephson junction or transmission line, the physical quantity of interest is a quantum of magnetic flux, or a fluxon, which has a soliton behavior. It is a remarkably robust and stable object, which can be easily manipulated at high speed and stored electronically. Consequently it should be used as a basic bit in information processing systems (for a recent review see Ustinov (1998)).

In this chapter we consider successively the localized short junction, and the transmission line, which are at present well characterized in the context of low-temperature superconductors and might be modified in the near future with the new high temperature superconductors. We then discuss the fluxon properties, which present remarkable similarities with the mechanical kink-solitons, considered in the previous chapter, although they correspond to a completely different physical phenomenon.

7.1 The Josephson effect in a short junction

Before considering the Josephson transmission line, it is appropriate to say a few words about the superconducting tunneling effect predicted by B. Josephson in 1962. When one applies an external potential difference to a system (Fig.7.1) constituted by a thin insulating layer sandwiched between two simple conductors, the insulating layer acts as a potential barrier which stops the flow of electrons between the two conductors; the conductivity of this junction is practically infinite. However if the insulator is thin enough, say 20–10 Å, some electrons are able to penetrate the potential barrier and give ohmic conductivity. This is the quantum tunneling effect. Josephson has investigated the tunneling effect that occurs when the two conductors are replaced by superconductors. The theory of low temperature conductivity tells us that a superconductor is a system where, owing to a particular interaction of the free electrons with the surrounding lattice, a fraction of the conduction electrons forms pairs called Cooper pairs. In these pairs the two electrons have opposite momentum and spin. These pairs are able to condense in the same quantum state so that the superconductor can be described by a single *macroscopic wave function* of the form

$$\psi = \sqrt{\rho}\ e^{i\phi}. \tag{7.1}$$

In this picture a single wave function is associated with a macroscopic number of electrons, which are assumed to condense in the same "macroscopic quantum state"; ρ represents the pair density and ϕ is the quantum phase common to all the pairs.

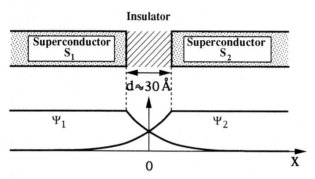

Fig.7.1. Scheme of a Josephson system

Now, we consider a junction where two superconductors S_1 and S_2 are separated by a macroscopic distance d. If d is large enough the wavefunctions ψ_1 and ψ_2 and the phases ϕ_1 and ϕ_2 of the two superconductors are independent. If d is reduced to about 100 Å, single electrons but not pairs can flow by tunneling from one superconductor to the other. If d is further reduced down to about 30 Å, Cooper pairs can flow from one superconductor to the other; this is so-called *Josephson tunneling*. In this situation the two phases are no longer independent.

7.1.1 The small Josephson junction

As a result of this lack of phase independence, we consider a small Josephson junction (Josephson, 1962, 1964), as shown in Fig. 7.2a, which consists of two layers of superconducting metal separated by a thin dielectric barrier layer, which is small enough to permit tunneling of superconducting electron pairs or, equivalently, coupling of the wave functions of the two superconductors. To understand the essential features of the Josephson effect and derive the basic Josephson equations, we can use a simple approach (see Appendix 7A) due to Feynman (1960). One finds that

$$J = J_o \sin \phi, \qquad \frac{d\phi}{dt} = \frac{2e}{\hbar} V. \tag{7.2}$$

Here J is the superconducting current, J_0 is defined by (7A.5); $\hbar = h/2\pi$, where h is the Planck constant. We note that a Josephson junction has the electrical behavior of a nonlinear inductor (Scott 1969) in parallel with a linear capacitance and a resistance (Fig.7.2b), which represents the small dissipative effects that we ignore in this section and the next. If the magnetic flux is defined by

$$\frac{\partial \Phi}{\partial t} = V, \tag{7.3}$$

from (7A.7) in Appendix 7A we obtain

$$\phi = 2\pi \frac{\Phi}{\Phi_0}, \tag{7.4}$$

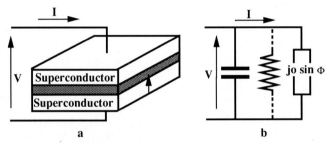

Fig.7.2a,b. Schematic representation of a small junction

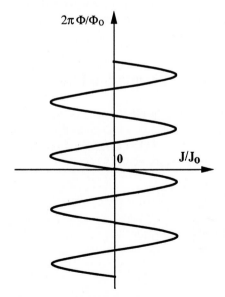

Fig.7.3. Representation of the nonlinear relation between normalized magnetic flux and current density as given by (7.5)

where $\Phi_0 = h/2e = 2,064 \ 10^{-15}$ Wb is the *flux quantum*. Thus, the first of relations (7.2) may be rewritten as

$$\Phi = (\Phi_0/2\pi) \sin^{-1} (J/J_0). \tag{7.5}$$

In practice, this nonlinear flux–current relation can be thought of as representing a nonlinear inductance; it is plotted in Fig.7.3. The Josephson equations (7.2) lead to remarkable properties: the DC and AC Josephson effects.

If the voltage across the junction is zero $V = 0$, from (7.3) $\phi =$ constant but not necessarily zero. Consequently, (7.2) implies that *a finite current density J can exist even with zero external voltage.* The existence of this DC superconducting current was predicted by Josephson in 1962 and observed experimentally by Anderson and Rowell in 1963.

Now, if a constant voltage $V = V_0$ is applied to the junction, it follows by integration of (7.3) that $\phi = \phi_0 + (2e/h)V_0 t$. Consequently the phase varies with time and an alternating current density

$$J = J_0 \sin (\phi_0 + \frac{2e}{h} V_0 t), \tag{7.6}$$

appears in the junction: this is the AC Josephson effect. The corresponding frequency is $F_J = \omega_J/2\pi = 2eV_0/h$, which gives $F_J/V_0 = 483.6$ MHz / μV.

In summary, the Josephson junction is a remarkable system where the *quantum phase difference between the wavefunctions of the two superconductors has a macroscopic meaning and is directly observable* via the current flowing into the junction and the voltage across it.

7.2 The long Josephson junction as a transmission line

Let us consider a long Josephson junction, sketched in Fig.7.4, which consists of two identical superconducting strips of width a separated by a thin dielectric layer of thickness d. This long tunneling junction can be regarded as a transmission line as far its electromagnetic behavior is concerned (see Barone et al. 1971).

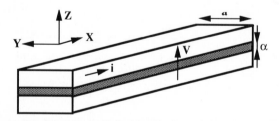

Fig.7.4. Sketch of the long junction or Josephson transmission line.

In other words, a shunt capacitance per unit length and a series inductance per unit length can be defined. The capacitance

$$\mathcal{C} = \varepsilon_r \varepsilon_0 a/d, \qquad (7.7)$$

where ε_r the relative dielectric constant and $\varepsilon_0 = 1/36\ \pi\ 10^{-9}$ F/m, is due to the storage of the electric field energy, which is confined within the dielectric layer. The inductance

$$\ell = \mu_0 (2\lambda_L + d)/a, \qquad (7.8)$$

where $\mu_0 = 4\pi\ 10^{-7}$ m^{-1}, represents the magnetic field energy stored in a region of thickness $d + 2\lambda_L$. Here, λ_L is the London penetration depth, which gives a measure of the field penetration into the superconducting electrodes, and the quantity a represents the width of the superconducting strips. We neglect the dissipative effects and consider a long junction, with small enough width to treat the electromagnetic fields in the insulating barrier as uniform in one space coordinate : the y direction is very small compared to the x direction. Under these conditions, the equivalent electrical transmission line corresponding to the Josephson transmission line is sketched in Fig.7.5. A unit section contains a linear inductance and an effective nonlinear inductance in parallel with a linear capacitance. The reader should compare such an electrical transmission line model with the classical ones considered in previous chapters.

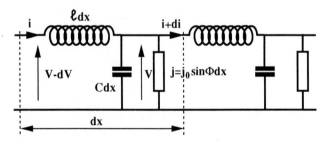

Fig.7.5. Schematic representation of the nondissipative Josephson transmission line and its electrical equivalent

From Kirchhoff's law applied to one unit section of length dx we obtain the corresponding set of nonlinear partial differential equations governing the electrodynamics of the junction:

$$\frac{\partial V}{\partial x} = -\ell\,\frac{\partial i}{\partial t}, \qquad (7.9a)$$

$$\frac{\partial \Phi}{\partial t} = V,$$

(7.9c)

where i represents the series current. Equations (7.9) combine into the following partial differential equation for the flux:

$$\frac{\partial^2 \Phi}{\partial x^2} - \ell\mathcal{C}\frac{\partial^2 \Phi}{\partial t^2} = \ell J_0 \sin\frac{2\pi\Phi}{\Phi_0}.$$

(7.10)

We introduce the characteristic velocity of linear waves, known as the Swihart velocity (Swihart 1961), and the Josephson plasma frequency,

$$c_J = (\ell\mathcal{C})^{-1/2}, \qquad \omega_J = (2\pi\, J_0/\Phi_0\, \mathcal{C})^{1/2}.$$

(7.11)

In practice one has $c_J \approx c/20$ where c is the velocity of light. Using (7.4) and (7.11), (7.10) is rewritten as

$$\frac{\partial^2 \phi}{\partial t^2} - c_J{}^2\frac{\partial^2 \phi}{\partial x^2} + \omega_J{}^2 \sin\phi = 0.$$

(7.12)

This describes the evolution of the quantum phase difference or, equivalently, the evolution of the flux in units of $2\pi\Phi/\Phi_0$. The reader may note that (7.12) is a Sine–Gordon equation, similar to (6.5) which describes the angular motion of pendulums in a mechanical transmission line. Indeed, the mechanical line was originally constructed by Scott (1969) to serve as a mechanical analog to the Josephson transmission line. In the small-amplitude static limit, (7.12) reduces to

$$\frac{\partial^2 \phi}{\partial x^2} = \frac{1}{\lambda_J{}^2}\, \phi,$$

(7.13)

where $\lambda_J = c_J / \omega_J$. The physical solution of (7.13) has the form

$$\phi(x) \sim \exp(-x / \lambda_J).$$

Thus, λ_J gives a measure of the typical distance over which the phase (or magnetic flux) changes; it is called the Josephson penetration depth. This quantity allows one to define precisely a *small* and a *long* junction. A junction is said to be small if its geometric dimensions are small compared with λ_J. On the contrary a junction is said to be long if its geometric dimensions are large compared with λ_J. Next, introducing the transformations

$$X = x / \lambda_J, \qquad T = t\, \omega_J,$$

(7.14)

$$X = x / \lambda_J, \qquad T = t \, \omega_J, \qquad\qquad (7.14)$$

yields the two dimensional Sine–Gordon equation in normalized form,

$$\frac{\partial^2 \phi}{\partial T^2} - \frac{\partial^2 \phi}{\partial X^2} + \sin \phi = 0. \qquad\qquad (7.15)$$

Equation (7.15) is identical to (6.10). Consequently, all the nonlinear solutions obtained in Chap.6 can be interpreted directly on the Josephson transmission line. That is, from solutions (6.15), (6.12), and (6.16) and using laboratory coordinates we obtain

$$\Phi = 4 \frac{\Phi_0}{2\pi} \tan^{-1} \left[\exp \left(\frac{x - vt}{\lambda_J \sqrt{1 - \frac{v^2}{c_J^2}}} \right) \right], \qquad\qquad (7.16a)$$

$$V = \frac{\Phi_0}{2\pi} \frac{\partial \phi}{\partial t} = -\frac{\Phi_0 \omega_J}{2\pi} \frac{2v}{\sqrt{1 - \frac{v^2}{c_J^2}}} \operatorname{sech} \left[\frac{x - vt}{\lambda_J \sqrt{1 - \frac{v^2}{c_J^2}}} \right], \qquad\qquad (7.16b)$$

and

$$i = -\frac{\Phi_0}{2\pi\ell} \frac{\partial \phi}{\partial x} = -\frac{\Phi_0}{2\pi\ell\lambda_J} \frac{2}{\sqrt{1 - \frac{v^2}{c_J^2}}} \operatorname{sech} \left[\frac{x - vt}{\lambda_J \sqrt{1 - \frac{v^2}{c_J^2}}} \right], \qquad\qquad (7.16c)$$

The kink (antikink) soliton solution (7.16a), moving with velocity v, corresponds to a $\pm 2\pi$ jump in phase difference ϕ across the insulating barrier separating the two superconductors, as sketched in Fig.7.6. In other words it is a flux quantum vortex, where a current loop consists of a surface current and a tunneling supercurrent, connecting the two surface layers via the barrier. This current loop supports one quantum Φ_0 of magnetic flux and the soliton (antisoliton) is therefore often called a *fluxon* (*antifluxon*). This moving fluxon is accompanied by a negative voltage pulse (7.16b) and a negative current pulse (7.16c), which correspond to the space and time derivatives of the fluxon (antifluxon) solution, as shown in Fig.7.7.

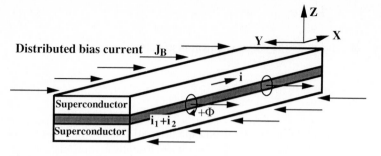

Fig.7.6. Sketch of the fluxons that can move along the transmission line.

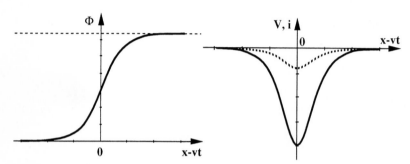

Fig.7.7. Ilustration of a fluxon (kink) and the corresponding voltage V(<u>dotted line</u>) and current pulses i (<u>continuous line</u>) as a function of x (v=0.1), in arbitrary units

7.3 Dissipative effects

Up to now, to the lowest-order nonlinear approximation, the Sine–Gordon equation was used to describe the dynamics of physical quantities, flux, voltage, and current. However, for the short junction (see Fig.7.2), in dealing with real transmission lines we must take into account losses, bias, and junction irregularities, which influence the motion of real fluxons. On the equivalent transmission line model, dissipative effects can be represented (Fig.7.8) by the following elements:

• a series resistance R, which represents the scattering of quasiparticles in the surface layers of the two superconductors (Scott 1964)
• a parallel conductance G, which is referred to as the quasiparticle tunneling current (Josephson 1964).

In addition to the losses, one also considers a uniformly distributed bias current J_B per unit length. Applying Kirchhoff's laws to the model of Fig.7.8 gives

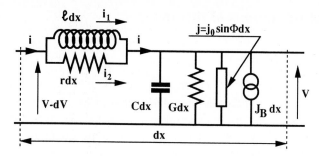

Fig.7.8. Equivalent electrical transmission line of the dissipative long Josephson junction

$$\frac{\partial V}{\partial x} = -\ell \frac{\partial I_1}{\partial t}, \tag{7.17a}$$

$$\frac{\partial V}{\partial x} = -RI_2, \tag{7.17b}$$

$$\frac{\partial}{\partial x}(I_1 + I_2) = -\mathcal{C}\frac{\partial V}{\partial t} - GV - J_o \sin 2\pi \frac{\Phi}{\Phi_o} + J_B, \tag{7.17c}$$

$$\frac{\partial \Phi}{\partial t} = V. \tag{7.17d}$$

Combining relations (7.17) yields a modified SG equation,

$$\phi_{XX} - \phi_{TT} - \alpha\phi_T + \beta\,\phi_{XXT} = \sin\phi - \gamma, \tag{7.18}$$

expressed in terms of the normalized coordinates (7.14). The constants α, β, and γ are defined by

$$\alpha = G/\omega_J\mathcal{C}, \ \beta = \omega_J\ell/R, \ \gamma = J_B/J_o. \tag{7.19}$$

For b =0, the power balance analysis described in Sect.6.8 can be used to calculate the limiting velocity of the fluxon. When α and $\beta \neq 0$ are sufficiently small, solutions of (7.19) can be calculated using the perturbation analysis (McLaughlin et al. 1978), considered in Sect.6.9, which has proven quite successful in describing experimental results in the limit of a small bias current (Davidson 1986). The effect of the damping terms α and β was analyzed numerically (Lomdahl et al. 1982); when they are important, they can lead to interesting properties of fluxons for digital applications (Nitta et al. 1984).

7.4 Experimental observations of fluxons

7.4.1 Indirect observation

The modified Sine–Gordon equation (7.18) can be used to analyze the d.c current singularities which occur in the voltage–current characteristics of a long one-dimensional Josephson junction and are an experimental manifestation of fluxon motion in the line. From (7.16b) we know that to the propagating fluxon there corresponds a voltage pulse which can be detected at either end of the transmission line. Current singularities in the current–voltage characteristics were observed in the absence of an external applied magnetic field; an explanation of their existence was proposed by Fulton and Dynes (1973) on the basis of fluxon motion inside the junction. These singularities, known as zero-field steps, which are essentially a series of current spikes at approximately equidistant intervals of d.c. voltage, were first observed by Chen et al. (1971). The fact that they occur at nonzero voltages suggests an explanation in terms of fluxon motion. Consider, in the presence of damping and a bias current, a single fluxon moving with velocity v along a transmission line of length L. Starting from one end, it is reflected at the opposite end as an antifluxon, comes back, and is next reflected, and so on. The period of the motion is $\tau = 2L/v$, and for one period the corresponding phase change is $\Delta\phi = 4\pi$. For N fluxons moving on the line the phase change will be $\Delta\phi_N = 4N\pi$. The mean voltage over one period τ dc voltage is calculated from (7.17d):

$$\frac{2\pi}{\Phi_0} V_{DC} = \frac{1}{\tau} \int_0^\tau \frac{\partial \phi}{\partial t} dt = \frac{\Delta\phi_N}{\tau}. \tag{7.20}$$

Replacing τ by its value, one gets the voltage,

$$V_{DC} = \Phi_0(Nv_\infty/L), \tag{7.21}$$

at which the so-called zero field steps occur. Here v_∞ is the limiting velocity of the fluxon in the presence of damping α and bias. For a given junction geometry, if we assume $\beta = 0$ and if α and γ are treated as small perturbations, this steady-state fluxon velocity is similar to that calculated for a mechanical kink soliton. From (6.48), one gets

$$v_\infty = [1 + (4\alpha/\pi\gamma)^2]^{1/2}. \tag{7.22}$$

If the junction length is very large, the details at the boundaries play a minor role, and the voltage V_{dc} at the first step is given by (7.21)· For N = 1, one has the first

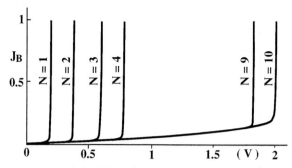

Fig.7.9. Schematic representation of the zero-field steps in the voltage–current characteristic of a Josephson transmission line

zero field step; for N = 2 one has the second one; and so on (Parmentier 1978; Pedersen et al. 1984), as shown in Fig.7.9.

It has also been proved possible to calculate (from the model equation (7.18)) the shape of the zero-field steps in transmission lines with different geometries: for instance, an inline junction, where the bias current is fed in the direction of propagating fluxons, and an overlap junction, where the bias current is fed at right angles to the fluxon velocity. The experimental findings agree very well with the results of this modified Sine–Gordon model (Levring et al. 1982; Pedersen et al. 1984).

7.4.2 Direct observation

Direct-observation experiments have been developed to study the propagation properties of the fluxons (Matsuda et al. 1982; Nitta et al. 1984). In these experiments an input pulse is fed into one end of the transmission line and the output voltage which appears across the terminal resistance at the output end is detected. A low-noise and fast-rise-time detector together with minicomputer signal processing are used to improve the signal/noise ratio and observe the fluxon. The recorded waveforms show that the number of pulses varies according to changes in input pulse height.

The experimental behavior of a fluxon–antifluxon collision was also investigated using an improved direct-measurement system (Matsuda 1986) based on a signal processing technique. The experiment showed an increase in propagation delay time following a collision. This result was qualitatively explained as the effect of dissipation (see (7.22)) using numerical simulation and perturbation theory, where the origin of the increase in the delay time is explained in terms of a power balance equation. In the presence of dissipation one expects (compare to Sect.6.6) that a fluxon–antifluxon collision can result in a breather decay mode, as sketched in Fig.7.10. A direct experimental observation of a fluxon–antifluxon

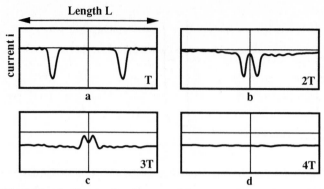

Fig.7.10a–d. Sketch of a fluxon–antifluxon (current pulses) collision as can occur in an experiment. Owing to dissipation effects it results in a breather mode which decays as oscillation period T increases

collision was reported on a discrete Josephson transmission line, i.e., a line composed of 31 discrete junctions (Fujimaki et al, 1987). By precise control of the collision of a fluxon and an antifluxon in the line, the annihilation process into a breather decay mode was observed in time but also in space. This destructive collision can be well described in terms of the modified (dissipative) Sine–Gordon equation.

A spatially-resolved technique such as the low temperature scanning electron microscopy (LTSEM) has been proved to be a powerful tool for studying fluxon dynamics in large Josephson junctions. It consists in irradiating the sample with an electron beam and at the same time, measuring the voltage. Thus, even if the measured quantity, the average voltage across the whole junction, is global, the perturbation is local and therefore information can be retrieved about the local behavior of the junction (for a review see Doderer 1997). Another thechnique is the low temperature scanning laser microscope (LTSLM) which allows to perform spatially-resolved imaging of superconducting properties of a sample by using local heating of a sample area of about one micron in diameter.

Remarks

(i) Although the solutions of (7.18) provide a deep insight into nonlinear wave propagation, for their existence, they require transmission lines of infinite length. In Josephson transmission lines, a more realistic approach of real finite junctions implies the imposition of finite boundary conditions. For example, open circuit termination at the two ends of the line corresponds to setting the following boundary conditions: $\phi_X (0, t) = \phi_X (L, t) = 0$. In this case plasma oscillations, fluxon oscillations, and breather oscillations are exactly calculated in terms of elliptic functions (Parmentier 1978; Costabile et al. 1978).

(ii) In the junction geometries mentioned above, the boundary effects and fluxon–antifluxon interactions are present and are expected to be of importance. However, a new experimental approach was to choose a geometry that has no boundaries at all: an annular or ring-shaped junction which consists of an overlap junction folded back into itself (Davidson et al. 1986). This geometry not only eliminates boundary effects, but also, owing to flux quantization, makes it possible to trap a single fluxon and study its undisturbed motion, as well as fluxon–antifluxon collisions. Moreover, these are the closest experimental conditions under which the perturbation theory on its original form is expected to apply.

7.4.3 Lattice effects

For a discrete Josephson transmission line, that is, a lattice of Josephson junctions, the physics is governed by a discrete Sine–Gordon equation

$$\frac{d^2\phi_n}{dT^2} + \alpha\frac{\partial\phi_n}{\partial T} + \sin\phi_n = \frac{\Lambda_J^2}{a^2}(\phi_{n+1} + \phi_{n-1} - 2\phi_n) + \gamma_n, \qquad (7.23)$$

with n = 1,2, ... ,N. Here T is the normalized time coordinate (7.14), α and γ_n are the damping term and the bias current as defined in (7.19), and a is the lattice spacing. As we have seen in Sect. 6.10, (7.23) may be viewed as the equation of motion (6.61) of N pendulums, each of which is damped, driven by a constant torque, and coupled to its neighbors by torsional springs. Recently, Van der Zandt et al. (1995) have measured the current–voltage characteristics of discrete rings of Josephson junctions. Their results show that, in agreement with a theoretical prediction of Ustinov et al. (1993), the resonant steps that appear in these characteristics are due to phase locking between a kink or fluxon propagating along the ring and the linear waves radiated in its wake. For information about lattices of Josephson junctions the reader is referred to the book edited by Parmentier and Pedersen (1994) and references therein.

A recent study (Van der Zant et al. 1998) on the dynamics of one-dimensional arrays of Josephson junctions has shown that these arrays are nice model systems for (7.23), an excellent agreement between theory and experiment was found.

Appendix 7A. Josephson equations

In this approach the wavefunctions of the two superconductors forming the Josephson junction satisfy the following system of linearly coupled Schrödinger equations:

$$i\hbar\frac{\partial\psi_1}{\partial t} = E_1\psi_1 + K\psi_2, \qquad i\hbar\frac{\partial\psi_2}{\partial t} = E_2\psi_2 + K\psi_1. \qquad (7A.1)$$

Here, E_1 and E_2 are the ground state energies of the superconductors. K is a real coefficient which describes the coupling interaction between the two superconductors and depends on the specific junction structure; $\hbar = h/2\pi$ is Planck's constant. When the separation distance between S_1 and S_2 is significant

$K = 0$ and (7A.1) reduce to independent Schrödinger equations. When d is small enough, the wavefunctions overlap, as represented in Fig.7.1. If a d.c. potential difference V is applied across the junction, the energy difference is shifted to E_1 - $E_2 = 2eV$, where 2e is the charge of a pair, twice the electron charge e. To simplify, the zero of the energy level can be redefined halfway between the two values E_1 and E_2 such that $E_1 = eV$ and $E_2 = - eV$. Under these conditions (7A.1) become

$$i\hbar\frac{\partial\psi_1}{\partial t} = + eV\psi_1 + K\psi_2, \qquad i\hbar\frac{\partial\psi_2}{\partial t} = - eV\psi_2 + K\psi_1. \tag{7A.2}$$

Using an expression similar to (7.1) for each wavefunction and separating the real and imaginary parts in (7A.2) we obtain

$$\hbar\frac{\partial\rho_1}{\partial t} = - 2K \sqrt{\rho_1\rho_2} \sin\phi, \qquad \hbar\frac{\partial\rho_2}{\partial t} = + 2K \sqrt{\rho_1\rho_2} \sin\phi; \tag{7A.3}$$

$$- \hbar\frac{\partial\psi_1}{\partial t} = K \sqrt{\frac{\rho_2}{\rho_1}} \cos\phi + eV, \quad - \hbar\frac{\partial\psi_2}{\partial t} = K \sqrt{\frac{\rho_1}{\rho_2}} \cos\phi - eV. \tag{7A.4}$$

Here, $\phi = \phi_2 - \phi_1$ is the phase difference between the macroscopic quantum wavefunctions. Since ρ_1 and ρ_2 represent the density of pairs in each super-conductor, the quantities $\partial\rho_1/\partial t = J_1$ and $\partial\rho_2/\partial t = J_2$ are the current densities. From (7A.4) we can write $-J_1 = J_2 = J$. In practice, the actual values of ρ_1 and ρ_2, being the excess charges supplied by the external voltage source, remain constant. We then set

$$\rho_1 = \rho_2 \approx \rho_o , \qquad \frac{2K}{\hbar} \sqrt{\rho_1\rho_2} \approx \frac{2K\rho_o}{\hbar} = J_o \tag{7A.5}$$

Therefore from (7A.3) and (7A.4), it follows that

$$J = J_o \sin\phi. \tag{7A.6}$$

Equations (7A.5) can be combined to give

$$\frac{d\phi}{dt} = \frac{2e}{\hbar} V.. \tag{7A.7}$$

Equations (7A.6) and (7A.7) are the two basic equations which describe the Josephson effect.

8 Solitons in Optical Fibers

In the near future nonlinear optics should probably revolutionize the world of telecommunications and computer technologies. With lasers producing high-intensity and short-duration optical pulses, it is now possible to probe the interesting, and potentially useful, nonlinear effects in optical systems and waveguides. Among the guiding structures, the optical fiber is an interesting (Gloge 1979) and important device (Mollenauer and Stolen 1982; Doran and Blow 1983). In an optical transmission system using linear pulses, the bit rate of transmission is limited by the dispersive character of the material, which causes the pulse to spread out an eventually overlap to such an extend that all the information is lost. To overcome this limitation Hasegawa and Tappert (1973) proposed to compensate the dispersive effect by the nonlinear change (Kerr effect) of the refractive index, or dielectric response, of the fiber material. When the frequency shift due to the Kerr effect is balanced with that due to the dispersion, the initial optical pulse may tend to form a nonlinear stable pulse called "an optical soliton" which in fact is an envelope soliton. This prediction and subsequent observation of optical-solitons propagation (Mollenauer et al. 1980) in a single fiber has stimulated theoretical and experimental studies on nonlinear guided waves. The exploitation of the typical nonlinear effects in optical devices has led to the demonstration of various (quasi) soliton properties. The consequences for future long-distance high-bit-rate communication systems are the subject of much study (Mollenauer and Smith 1988). In recent years, a considerable body of literature have been published on the subject and the reader is refered to the books by Hasegawa (1989), Agrawal (1989), Newell and Moloney (1992), Taylor (1992), Abdullaev et al. (1993), Hasegawa and Kodama (1995) Akmediev and Ankiewicz (1997) and references therein.

In this chapter we introduce the reader to the modeling of optical nonlinear fibers and to the typical properties of optical solitons.

8.1 Optical-fiber characteristics

Generally an optical fiber is a cladded cylindrical waveguide which consists of a highly transparent (high-index) dielectric core surrounded by a second layer of a second dielectric (Fig.8.1) with a lower optical index. Optical fibers guide light by total internal reflection: this is achieved by a refractive index $n(x,y)$ (or dielectric function $\varepsilon(x,y)$) variation decreasing in the radial direction from the center to the periphery (Fig.8.1).

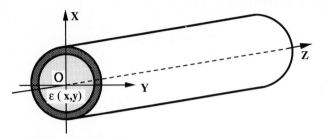

Fig.8.1. Sketch of an optical fiber with an optical index n(x, y), or a dielectric function ε (x,y) which varies in the transverse section

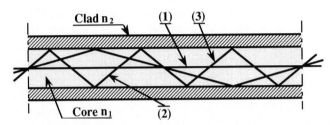

Fig.8.2. Representation of an optical fiber with different modes of propagation: (1) lowest-order mode; (2) middle-order mode; (3) high-order mode

In multimode fibers certain rays or modes can propagate (Fig.8.2) at different velocities; these are in fact solutions of the wave equation for the cylindrical dielectric waveguide. One can design the fiber in such a way that it transmits only one mode and attenuates all other modes by absorption or leakage. This is done by modifying the index profile at a sufficiently small index difference so that only one state is supported: one then has a monomode fiber. While multimode fiber is cheaper and easier to produce and can be used with incoherent sources, single-mode fibers offer much greater bit capacity than multimode fibers and are of primary interest for optical communication systems. Nowadays they are easy to produce and are becoming cheaper and cheaper. *In the following the term optical fibers refers to monomode fibers*, unless specified otherwise.

8.1.1 Linear dispersive effects

Optical fibers are dispersive. There are three main types of dispersion:
• the group-velocity dispersion (GVD), which results from the propagation velocity of the signal being related to the light wavelength (Fig.8.3);
• the modal dispersion, which is inherent to multimode fibers;
• the waveguide or geometrical effects.
Let us focus on the GVD, which plays a critical role in the propagation of optical pulses and manifests through the frequency dependence of the linear part $n_0(\omega)$ of the refractive index. Specifically, as will be shown in the following (see eq(8.11))

204

$$n_0(\omega) = \frac{ck(\omega)}{\omega} \ , \tag{8.1}$$

where c is the velocity of light. Generally, this dispersion relation, even in the linear approximation, is not known explicitly, contrary to the electrical transmission line studied in Chap.2. However, the wavenumber k can be expanded (compare to Chap.4) as a function of ω in a Taylor series about the center wavenumber k_0:

$$k\ (\omega) = k_0 + (\frac{\partial k}{\partial \omega})(\omega - \omega_0) + \frac{1}{2}(\frac{\partial^2 k}{\partial \omega^2})\ (\omega - \omega_0)^2 + ..., \tag{8.2}$$

where the derivatives are evaluated at $k = k_0$. Here, one can write

$$(\frac{\partial k}{\partial \omega}) = \frac{1}{v_g} = \frac{\partial}{\partial \omega}(\frac{\omega\, n_0(\omega)}{c}) = \frac{1}{c}\,[\,n_0(\omega) + \omega\frac{\partial n_0(\omega)}{\partial \omega}\,], \tag{8.3a}$$

and

$$(\frac{\partial^2 k}{\partial \omega^2}) = \beta = \frac{\partial}{\partial \omega}(\frac{1}{v_g}) = -\frac{1}{v_g^2}\frac{\partial v_g}{\partial \omega} = \frac{1}{c}\,[\,2\frac{\partial n_0(\omega)}{\partial \omega} + \omega\frac{\partial^2 n_0(\omega)}{\partial \omega^2}\,]; \tag{8.3b}$$

v_g is the group velocity (as defined in Chap.2) and β corresponds to the group-velocity dispersion (GVD), which depends not only on the property of the glass material, but also on the waveguide property of the optical fiber. In practice, the above quantities are expressed in terms of the wavelength $\lambda = 2\pi c/\omega$. In this case one has

$$\frac{\partial n_0}{\partial \omega} = \frac{\partial n_0}{\partial \lambda}\frac{\partial \lambda}{\partial \omega} = -\frac{\lambda^2}{2\pi c}\frac{\partial n_0}{\partial \lambda}\ .$$

Relations (8.3) thus become

$$\frac{1}{v_g} = \frac{1}{c}\,[\,n_0(\lambda) - \lambda\frac{\partial n_0(\lambda)}{\partial \lambda}\,], \tag{8.4a}$$

and

$$\beta = (\frac{\partial^2 k}{\partial \omega^2}) = -\frac{\lambda^2}{2\pi c}\frac{\partial}{\partial \lambda}(\frac{1}{v_g}) = \frac{\lambda^2}{2\pi c}\,D, \tag{8.4b}$$

where D is a dispersion parameter

$$D = \frac{\lambda}{c}\frac{\partial^2 n_0(\lambda)}{\partial \lambda^2}\ . \tag{8.5}$$

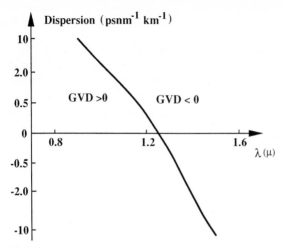

Fig.8.3. Dispersion versus wavelength in glass

In Fig.8.3 the measured variation of the dispersion parameter D, expressed in units of ps nm^{-1}km^{-1}, is represented as a function of the wavelength λ. Note that D vanishes for $\lambda = \lambda_D \approx 1.3$ µm. Consequently, for $\lambda < \lambda_D$ the GVD given by (8.3b) is positive and the fiber is said to exhibit *normal dispersion*. In the opposite case when $\lambda > \lambda_D$, the GVD is negative: this is the *anomalous dispersion regime*. If we choose $\lambda = 1.5$µm, from Fig.8.3 one gets D = -10 ps nm^{-1}km^{-1}. Using (8.4b) one obtains $\beta = -8$ ps^{2}km^{-1}. This result indicates that an optical pulse with a width of a few picoseconds would deform as it travels along the fiber over a distance of 1 km.

Note that the GVD defined by (8.3b) is related to the dispersion coefficient 2P introduced by (2.43) by $\beta v_g^3 = -2P$.

8.1.2 Nonlinear effects

The response of any dielectric medium to an intense electromagnetic field becomes nonlinear. In optical fibers the nonlinearity originates from the dependence of the induced electrical polarization to the electric field. This effect results in an intensity dependence of the refractive index of the form

$$n(\omega, |E|^2) = n_0(\omega) + n_2|E|^2. \qquad (8.6)$$

Here, n_2, often called the Kerr coefficient is the nonlinear part of the refractive index. It has a very small value of the order of 10^{-22}(m/V)^2. The Kerr effect has a short response time of the order of 10^{-15} s because it is due to the response of the electron orbits in SiO$_2$ molecules to the external electric field. The sign of the nonlinear coefficient is such as to increase the refractive index with intensity. Its frequency dependence is generally ignored since it changes little over the spectral width of the wave pulse for pulses with widths of a few picoseconds or more.

8.1.3 Effect of losses

Although the quality of optical fibers has dramatically improved over recent years, the fiber does exhibit some attenuation, resulting in an attenuation of the optical-pulse transmission, which can no longer be neglected for long-distance telecommunications. The power loss rate Γ_p expressed in km^{-1}, or Γ_{dB} expressed in dBkm^{-1}, which measures the power loss during transmission along a fiber of length L = 1km, is defined as

$$\Gamma_p L = - \log_e (P_t/P_i), \qquad \Gamma_{dB} L = -10 \log_{10} (P_t/P_i). \qquad (8.7a)$$

Here, P_i and P_t represent the input and output power, measured respectively at distance z =0 and z = L . From (8.7a) an attenuation length L_p for the power or L_a for the amplitude, which corresponds to the square root of the power, can be defined as

$$L_p = \frac{1}{\Gamma_p} = \frac{1}{\Gamma_{dB}} \frac{10}{\log_e(10)} , \qquad L_a = \frac{1}{\Gamma} = \frac{1}{\Gamma_{dB}} \frac{20}{\log_e(10)} . \qquad (8.7b)$$

There are special wavelengths or spectral windows where the attenuation is particularly low as shown in Fig.8.4. For instance, in the linear regime of a commercially available monomode optical fiber the loss rate is about or less than 0.2dB /km at the wavelength $\lambda \approx 1.55\mu$m. For shorter wavelengths, i.e., for visible light, the loss rate is more important, reaching the values 1 to 10 dB/km.

For example, from Fig.8.4 at $\lambda \approx 1.55\mu$m we have approximately $\Gamma_{dB} = 0.20$dB km^{-1}, and from (8.7b), $L_a = 2L_p \approx 44$ km. In this case the power and the amplitude of a linear wavetrain will decay by factors (e^{-1})/2 and e^{-1} over distances of 22 km and 44 km, respectively.

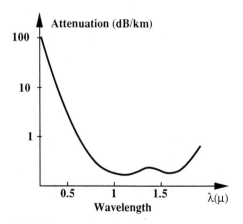

Fig.8.4. Attenuation (log scale) versus wavelength

8.2 Wave-envelope propagation

Let us now consider the propagation of the wave envelope. Starting from the Maxwell equations one can show (see Appendix 8A) that the wave propagation through a medium is governed by the equation

$$\nabla^2 E - \frac{1}{c^2}\frac{\partial^2 E}{\partial t^2} = \mu_0 \frac{\partial^2 P}{\partial t^2} , \tag{8.8}$$

where the polarization is made up of a linear and a nonlinear part. Here, for the sake of simplicity we have assumed a linearly polarized wave traveling in the z direction; that is, we have ignored the transverse (x, y) dependence of **E**. For a detailed calculation which accounts for the transverse effects, the reader is refered to the books by Hasegawa (1989), Agrawal (1989), and references therein. Hence, we now consider a monochromatic field amplitude of the form

$$E(z,t) = \frac{1}{2} [A(Z,T) e^{i(kz-\omega t)} + A^*(Z,T) e^{-i(kz-\omega t)}], \tag{8.9}$$

with the z axis as the direction of propagation. Here, $A(Z,T)$ is the slowly varying transverse component of the electric field, which depends on the slow space $Z = \varepsilon z$ and the slow time $T = \varepsilon t$ (see Chap.4); A^* is its complex conjugate, k the wavenumber, and ω the frequency. Inserting (8.9) into (8.8), and ignoring the rapid third-harmonics terms, to a first approximation one gets a nonlinear dispersion relation

$$\frac{c^2 k^2}{\omega^2} = 1 + \chi_L(\omega) + 3\chi_{NL}|A|^2 = n^2(\omega), \tag{8.10}$$

where χ_L and χ_{NL} represent respectively the linear and nonlinear (Kerr effect) dielectric susceptibilities (see Appendix 8A). By using (8.6) and (8.9) the index of refraction is written

$$n(\omega) = n_0(\omega) + n_2 |A|^2. \tag{8.11}$$

Then, (8.11) and (8.10) yield

$$n_0^2(\omega) \approx 1 + \chi_L(\omega), \qquad 2n_0 n_2 \approx 3\chi_{NL}. \tag{8.12}$$

The nonlinear dispersion (8.10) relation for the refractive index can be formally written as

$$k = k(\omega, |A|^2). \tag{8.13}$$

208

We now follow the procedure (Karpman et al. 1973; Hasegawa et al. 1980) used for the electrical transmission lines (see Chap.4) to construct the equation which governs the wave-envelope propagation in the z direction. To the Taylor expansion (8.2) we add a nonlinear term $(\partial k/\partial |A|^2)$ which is evaluated at amplitude $|A_0|^2 = 0$. We then have

$$k - k_0 = \frac{\partial k}{\partial \omega}(\omega - \omega_0) + \frac{1}{2}\frac{\partial^2 k}{\partial \omega^2}(\omega - \omega_0)^2 + \frac{\partial k}{\partial |A|^2}(|A|^2) + \ldots \qquad (8.14)$$

As in Sect. 4.4 we replace $(k-k_0) = K$ by a spatial operator $-i\varepsilon\partial/\partial Z$, and $(\omega - \omega_0) = \Omega$ by a temporal operator $i\varepsilon\partial/\partial T$, and let (8.14) operate on the complex amplitude function A. We get

$$\varepsilon i \left(\frac{\partial A}{\partial Z} + \frac{1}{v_g}\frac{\partial A}{\partial T}\right) - \frac{1}{2}\varepsilon^2 \beta \frac{\partial^2 A}{\partial T^2} + \varepsilon^2 \frac{\partial k}{\partial |A|^2}|A|^2 A = 0, \qquad (8.15)$$

where we have assumed that the small nonlinear term is $O(\varepsilon^2)$. Here, $1/v_g$ and the group-velocity dispersion β are defined by (8.3). In the absence of group dispersion β and nonlinearity $\partial k/\partial |A|^2$ (8.15) gives $A = A(T-Z/v_g)$. Hence, it is appropriate (see Sects.2.5 and 4.4) to describe the dynamical evolution of the electric field amplitude A by using a coordinate system that moves at the group velocity :

$$s = \varepsilon Z = \varepsilon^2 z, \qquad t' = T - Z/v_g = \varepsilon(t - z/v_g). \qquad (8.16)$$

Equation (8.15) then reduces to the nonlinear Schrödinger equation,

$$i\frac{\partial A}{\partial s} - \frac{1}{2}\beta\frac{\partial^2 A}{\partial t'^2} + Q|A|^2 A = 0, \qquad (8.17)$$

where $Q = (\partial k/\partial |A|^2)$. In practice (Hasegawa 1989; Agrawal 1989), the nonlinear coefficient is related to n_2 by the relation

$$Q = n_2 k_0/A_e, \qquad (8.18)$$

where A_e, the effective core area, takes into account the fact that the light intensity varies in the cross section of the fiber. Its value depends on the mode distribution. For $Q = 0$, (8.17) becomes a linear Schrödinger equation similar to (2.48) if one exchanges the space and time variables. Consequently, the spatial width of a Gaussian pulse given by (2.53) is a temporal width. It is given by

$$T_L = T_0 \sqrt{1 + \left(\frac{\beta L}{T_0^2}\right)^2}. \qquad (8.19)$$

209

This relation shows that the width which was initially T_0 at distance s=0 becomes T_L at distance s=L. One defines the dispersion length

$$L_d = T_0^2/|\beta|, \qquad\qquad (8.20)$$

which provides a length scale over which the dispersive effects become important. For example, if $|\beta| = 10$ ps^{2}km^{-1} and $T_0 = 10$ ps , the dispersive effects are negligible for L<$L_d = 10$ km. For $T_0 = 3$ps the dispersive effects are negligible if L< $L_d = 0.9$km.

Now, in the absence of dispersion ($\beta =0$), (8.17) reduces to

$$i\frac{\partial A}{\partial s} + Q|A|^2A = 0. \qquad\qquad (8.21)$$

This nonlinear equation can be solved for an arbitrary initial condition $A_0(t')$:

$$A(s, t') = A_0(t') \exp [\, i\, Q\, |A_0|^2s]. \qquad\qquad (8.22)$$

The nonlinear phase

$$\phi = Q\, |A_0|^2s, \qquad\qquad (8.23)$$

acquired during propagation depends on the amplitude squared, that is, on the light intensity or initial power. Consequently, the frequency which is the derivative of the pulse is *chirped*. This effect is called *self-phase modulation* (SPM). Noting that the coefficient of s has the dimension of length one, by setting $|A_0|^2 = P_i$, where P_i is the input (peak) power, one can define a nonlinear length

$$L_{nl} = 1/\, QP_i. \qquad\qquad (8.24)$$

This provides the length scale over which the nonlinear effects become important. With the dispersion length, it is one of the two important scales which need to be considered in optical-fiber propagation. In the visible range one can choose the typical value $Q = 20$W^{-1}km^{-1}. Thus, for $P_i = 0.1$W and $P_i = 1$W, the nonlinear effects are negligible when L<$L_{nl} = 500$m and L<$L_{nl} = 50$m, respectively.

8.3 Bright and dark solitons

Having discussed the linear and nonlinear parts of (8.17) we now return to the full equation and define a dimensionless parameter which measures the ratio of dispersion to nonlinearity, such as

$$N^2 =L_d/\, L_{nl} \qquad\qquad (8.25)$$

As we saw in Sect.4.5 and Appendix 4C the soliton solutions of (8.17) depend on the relative signs of the dispersive term -β and nonlinear term Q. For optical fibers Q is always positive, but β can be negative or positive as illustrated in Fig.8.3.

8.3.1 Bright solitons

For $-\beta Q > 0$, that is, in the regime of anomalous dispersion, a direct calculation (Appendix 4C) shows that (8.17) admits an envelope-soliton solution of amplitude A_0:

$$A = A_0 \, \text{sech} \left[\sqrt{\frac{Q}{-\beta}} \, A_0 \varepsilon (t - \frac{z}{v_g}) \right] \exp \left\{ i \, \frac{Q}{2} A_0^2 \varepsilon^2 z \right\}, \qquad (8.26)$$

which is expressed in terms of the original space and time variables defined by (8.16). The parameter $\varepsilon \ll 1$, which reminds us that the envelope wave varies slowly in space and time can be dropped out.

The total electrical field E(z, t), which takes account of the carrier (light) wave, whose inverse frequency is in the femtosecond (10^{-15} s) range, is obtained by replacing (8.26) in (8.9). In the context of optics the envelope soliton is called a *bright soliton* because it corresponds to a pulse of light (Fig.8.5). The broadening effects of anomalous dispersion are exactly balanced by the narrowing effects of focusing nonlinearity and the pulse never changes (in the limit of zero loss) its shape when propagating along the fiber. Thus, one can write N =1 and the power to launch a single soliton is obtained from (8.25). It is given by

$$P_i = \frac{-\beta}{Q T_0^2} . \qquad (8.27)$$

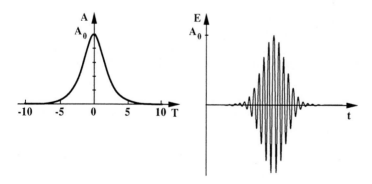

Fig.8.5. Representation of the envelope function A(Z,T) and the corresponding electric field E(z,t) of a bright soliton in its moving frame.

For example, if $|\beta| = 10$ ps^{2}km^{-1} and $Q = 5$W^{-1}km^{-1} near 1.55μm, for $T_0 = 10$ ps or 2 ps the required power is about $P_i = 20$mW or 500 mW.

In the regime of anomalous dispersion, (8.17) admits a continuous wave train solution (see (4.31) and (8.22)) with a phase which depends on the square of the amplitude. In terms of the laboratory coordinate z it has the form

$$A\,(z, t) = A_0 \exp\,(iQA_0^2 \varepsilon^2 z).$$

This solution is unstable to a small perturbation: this is another example of the Benjamin–Feir or modulational instability, which was discussed in chapters 1,4 and 5. The solution (8.26), which can be obtained by a simple calculation (Appendix 4C), represents a single or one-soliton solution. This solution can also be obtained by more sophisticated mathematical techniques such as the inverse scattering technique (IST) (Zhakarov and Shabat 1972) and the Hirota method (Hirota 1973), which are described in Chap.10. Moreover, these techniques allow one to obtain two-soliton solutions ((see (10.67)) and N-soliton solutions, which represent combinations between solitons. The IST also allows (see (10.47)) to determine the soliton solutions which arise from an initial disturbance (Satsuma and Yajima (1984). For example, the input pulse shape, generated from a laser, can be approximated by

$$q(s', 0) = N\,\mathrm{sech}\,(\tau), \tag{8.28}$$

where

$$q = \sqrt{QL_d}\,A, \qquad s' = s/L_d, \qquad \tau = t'/T_0, \tag{8.29}$$

are normalized variables and N is an integer which represents the soliton order. The solutions generated by this pulse have interesting features. As seen previously, for $N = 1$ we have a pulse with just the right amplitude of one soliton which remains forever like that at the input, as sketched in Fig.8.5. An input pulse with $N = 2$, that is with twice the amplitude (or four times the power), produces a two-soliton solution. For higher N an input pulse generates N interacting solitons. In contrast to the one-soliton solution, which has a permanent profile, these N-soliton solutions undergo oscillations in shape, with period $s'_0 = \pi/2$, when propagating (see figures 8.6 and 10.7). In fact, the solitons travel with the same velocity and the resulting waveform oscillates owing to the phase interference between the solitons. Using (8.16) and (8.20) we have

$$z_0 = \frac{\pi}{2}\,L_d = \frac{\pi}{2}\,\frac{T_0^2}{|\beta|}\,. \tag{8.30}$$

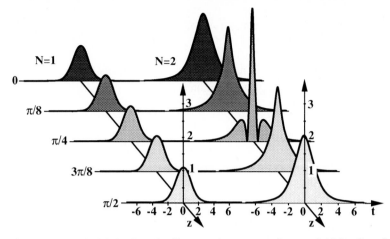

Fig.8.6. Theoretical behavior of solitons as they emerge from an initial pulse of the form N sech(0, t) and propagate along the fiber. For N = 1, the shape of the one-soliton solution remains unchanged. For N = 2, the two-soliton solution (calculated from (10.67) with x→ t , t → z) pulsates with period s' = $\pi/2$.

With the numerical values considered above one gets z_0 = 15.70km for T_0=10ps and z_0= 2.51 km for T_0 = 2ps. Generally, for pulse widths between about 1 and 30ps and a typical range of GVD for silica-core fibers the soliton period can range from less than 1 to 100 km. It can be substantially increased by lengthening the input pulse and reducing the core fiber diameter.

Remark

Although the IST mentioned above, is a very powerful and useful method to find soliton solutions, it was also used by Leon and coworkers (Claude and Leon 1995; Claude et al. 1995; Leon and Mikhailov 1998) in the study of Stimulated Raman Scattering, to show that some solutions are non solitonic (see remark p 276).

8.3.2 Dark solitons

For -βQ < 0, that is, for positive dispersion, bright pulses cannot propagate: the combination of GVD and nonlinearity leads to temporal broadening. However, in this case (8.17) admits a hole-soliton (see section 4.5) or *dark-soliton* solution in the context of optics (for a recent review see Kivshar and Luther-Davis 1998) . In terms of the original laboratory coordinates this solution has the form

$$\psi = A\ (Z,T)\ e^{i\phi(z,t)}. \tag{8.31}$$

Here, the amplitude function A(z,t) is given by

$$A = A_0\{ 1 - m^2 \operatorname{sech}^2 [\sqrt{\frac{Q}{\beta}} mA_0 \, \varepsilon(t - \frac{z}{v_g})] \}^{1/2}, \qquad (8.32)$$

where A_0 is the amplitude and m is a parameter which controls the depth of the modulation of the amplitude. The phase function $\theta(z,t)$ is (see Appendix 4C):

$$\phi = \sqrt{\frac{Q}{\beta}} [\{ A_0(1-m^2)^{1/2} \varepsilon(t - \frac{z}{v_g}) + \arctan\{ \frac{m}{\sqrt{1-m^2}} \tanh [\sqrt{\frac{Q}{\beta}} mA_0 \varepsilon(t - \frac{z}{v_g})]\}]$$

$$- \varepsilon^2 \frac{Q}{2} (3-m^2)A_0{}^2 z. \qquad (8.33)$$

The total electrical field E(z, t) which takes account of the carrier (light) wave, is obtained by replacing (8.31) in (8.9). In optics the hole soliton corresponds to a hole in the continuous light (carrier) wave, i.e., the portion where light is absent (see Fig.8.7): it is called a *dark soliton* (see Weiner 1992). Thus, the parameter m controls the darkness of the soliton.

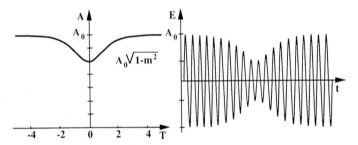

Fig.8.7. Representation of the envelope function A(Z,T) of a dark soliton and the corresponding electric field E(z,t) in its moving frame

Unlike a bright soliton, no higher-order dark solitons exist. The dark soliton broadens at lower powers or narrows at higher powers, as predicted by numerical simulations.

8.4 Experiments on optical solitons

Hasegawa and Tappert (1973) were the first to model the propagation of a guided mode in a perfect nonlinear monomode fiber by the NLS equation (8.17). They found that the optical pulse is an envelope or bright soliton and they predicted the stationary transmission of the pulse in the anomalous dispersion ($\beta<0$) regime. This prediction was then successfully verified by the experiments of Mollenauer et al. (1980). Using a monomode fiber they demonstrated a dispersionless transmission of an optical pulse (7ps width) with a peak power of about $P = 1W$ at a wavelength $\lambda = 1.45\mu m$ for a distance $d \cong 700$ m.

214

Modulational instability was predicted (A. Hasegawa and Brinkman 1980; Hasegawa 1984) and observed in optical fibers (Tai et al. 1986). 100ps pulses emitted from a laser operating at 1.32 μm providing a quasi-continuous environment for the observation of modulational instability were transmitted

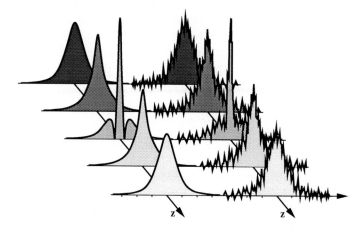

Fig.8.8. Predicted theoretical pulse evolution: narrowing and splitting, and illustration of the real shapes of the pulse as they can be observed experimentally in the presence of noise and parasitic effects

The pulse-shape narrowing and the splitting of pulses, with prediction based on the NLS equation, were observed experimentally by Mollenauer et al. (1980). The 7 ps duration pulses generated from a mode-locked color-center laser were observed by autocorrelation on a monomode silica-glass fiber at negative group-velocity dispersion (Stolen et al. 1983). In the autocorrelation technique the beam is divided in two roughly equal parts, and, after propagating along separated paths, is brought together at a product detector. The autocorrelation function measured in the experiments is then

$$C(\tau) = \int_{-\infty}^{\infty} E(t)E(t+\tau)dt,$$

where τ represents a time shift. In Fig.8.8 we have sketched the behaviors of predicted theoretical autocorrelation functions and experimental autocorrelation traces as they could be observed when an input pulse (N =2) is launched into the fiber.

Experiments on dark solitons in optical fibers have been performed in recent years (see Weiner 1992 and references therein). The experimental results obtained are consistent with the formation of dark solitons, as predicted by the nonlinear Schrödinger equation (8.17).

8.5 Perturbations and soliton communications

8.5.1 Effects of losses

In a real fiber, for propagation over a long distance, such as for communications applications, the optical soliton will deform progressively as a result of fiber loss. The attenuation is small and can be treated as a perturbation for relatively small distances. That is, linear absorptive or scattering loss simply adds a small term $i\Gamma A$ to (8.17) where Γ is the attenuation coefficient given by (8.7b). In terms of the variables defined by the transformation (8.29), (8.17) becomes

$$i\frac{\partial q}{\partial s'} + \frac{1}{2}\frac{\partial^2 q}{\partial \tau^2} + |q|^2 q + i\gamma q = 0, \tag{8.34}$$

where $\gamma = \Gamma L_d$. This term is small and can be treated as a perturbation. Following the perturbation method, developed by Kodama and Ablowitz (1980) and Hasegawa and Kodama (1981), the damped bright pulse solution can be calculated:

$$q(\tau,s') \cong q_0 \exp(-2\gamma s') \operatorname{sech}[\tau q_0 \exp(-2\gamma s')] \exp(i\sigma) + O(\gamma^2), \tag{8.35}$$

with

$$\sigma = (q^2_0/8\gamma)[1-\exp(4\gamma s')]. \tag{8.36}$$

Equation (8.35) describes a pulse which spreads exponentially: its amplitude decays as $(\exp-2\gamma s')$ and its width increases as $\exp(2\gamma s)$. From (8.29) and (8.16) one gets $\gamma s' = \varepsilon^2(z/L_d)(\Gamma L_d)$; thus, one can define an attenuation length for the soliton in the form

$$L_s = (1/2\Gamma) = (L_a/2), \tag{8.37}$$

which is half the linear attenuation length L_a of a linear wavetrain defined by (8.7b). According to the example of Sect.8.2, for $\Gamma_{dB} = 0.2$dB we have $L_a = 44$ km; this yields $L_s = 22$ km. We have seen that if the losses are ignored, we should have a single soliton if $L_{nl} \approx L_d$. In the presence of losses we can see that the nonlinear length L_{nl} ($\approx L_d$) must be smaller than L_s in order for a light wave having its electric field become a single soliton. In fact, since fibers that are commercially available have a loss rate Γ_{dB} of less than 0.2 dB at $\lambda = 1.5$ μm, the above condition can be easily satisfied.

For pulse widths of several picoseconds or more, a description of pulses in optical fibers in term of the NLS equation perturbed by losses seems to fit the experimental findings reasonably well. However, experiment in the high-power and the ultrashort-pulse regimes, as seen from the studies of pulse compression and frequency shift, requires further study to determine the effects of the higher-order terms on the NLS model. For information on these effects the reader is referred to the literature cited in the introduction to this chapter.

8.5.2 Soliton communications

In an optical communications system (Haus and Wong 1996), the information is coded in binary pulses which modulate the light carrier wave (as in Fig.8.5). The performance of a high- speed fiber communications system is limited by dispersion, losses, and parasitic effects induced by electronic repeaters. The use of a single soliton as a bit of information is interesting and solves the first problem, since the dispersion broadening effects are compensated for by the nonlinear focusing effects. The attenuation of solitons due to fiber loss can be compensated for by amplifying the solitons periodically, in order that they recover their original width and power. In an all optical communications system, which does not use electronic repeaters, this amplification can be achieved by using (Hasegawa 1983; Mollenauer 1985) stimulated Raman scattering.

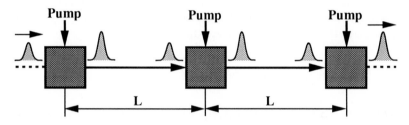

Fig.8.9. Illustration of a communications system using bright solitons

In this set-up, continuous laser power is pumped into the fiber along which the solitons are propagating at specific amplification periods L, as illustrated in Fig.8.9. Pump lasers inject continuous light which interacts with the incident light, in the fiber. From the pumping frequency, a small amount of energy is scattered to another frequency which is that of the soliton carrier, resulting in an amplification effect. The fiber losses are then compensated for and the soliton can continue on its way over the amplification length L. The same process then continues periodically. With such soliton-based communication systems one expects to transmit information over long distances, that is, for over more than 5000 km at bit rate of 100 G bit /s.

8.6 Modulational instability of coupled waves

In this section we consider briefly only the aspects important from a modulational instability standpoint since recent books (Agrawal 1989, Newell and Moloney 1992, Hasegawa and Kodama 1995) describe these phenomena in detail. As mentioned in Sect.4.4, spontaneous modulational instability of a single light wave in a fiber, in the anomalous dispersion regime ($-\beta Q > 0$: see Sect. 8.3.1), was first observed by Tai et al. (1986a). Then induced modulational instability was observed and used (Tai et al. 1986b) to generate subpicosecond soliton-like optical pulses at a 0.3 THz repetition rate.

However, when there are two waves with different frequencies but with the same polarization propagating in a fiber, they can interact with each other (see Sect.4.8) through the Kerr nonlinearity. In this case, following the procedure of Sect.4.8, one can obtain (Agrawal 1989) two coupled NLS equations similar to (4.56) for electrical transmission lines. Let us discuss the case where the two optical waves correspond to the two polarization components of the same optical field. This is the case of a birefringent nonlinear medium, where the vector nature of the optical field should be taken into account since the two orthogonally polarized components, at frequency $\omega = \omega_0$, travel at different velocities because of different refractive indices associated with them. The polarization effect is included by considering (see (8.9)) the electric field E in the form

$$\mathbf{E}\,(z,t) = \frac{1}{2}\,[A_x\,(Z,T)\,\exp[i(k_x z - \omega_0 t) + A_x^*\,(Z,T)\,\exp\,[-\,i(k_x z - \omega_0 t)]\,\mathbf{x}$$

$$+ \frac{1}{2}\,[A_y\,(Z,T)\,\exp[i(k_y z - \omega_0 t) + A_y^*\,(Z,T)\,\exp\,[-\,i(k_y z - \omega_0 t)]\,\mathbf{y}. \quad (8.38)$$

Here, A_x and A_y represent the slow envelope components of the electric field in the transverse x and y directions. Proceeding as in Sect. 4.8, that is, inserting (8.38) in (8.8) yields nonlinear dispersion relations for k_x and k_y. These relations can be used (Menyuk 1987) to derive the two coupled NLS equations

$$i[\frac{\partial A_x}{\partial Z} + \frac{1}{v_{g,x}}\frac{\partial A_x}{\partial X}] + \beta_x \frac{\partial^2 A_x}{\partial T^2} + \chi_{NL}(|A_x|^2 + \frac{2}{3}|A_y|^2)A_x$$

$$+ \frac{\chi_{NL}}{6}\,A_y^2 A_x^* \exp[-2i(k_x - k_y)\,z] = 0, \quad (8.39a)$$

$$i[\frac{\partial A_y}{\partial Z} + \frac{1}{v_{g,y}}\frac{\partial A_y}{\partial X}] + \beta_y \frac{\partial^2 A_y}{\partial T^2} + \chi_{NL}(|A_y|^2 + \frac{2}{3}|A_x|^2)A_y = 0,$$

$$+ \frac{\chi_{NL}}{6}\,A_x^2 A_y^* \exp[2i(k_x - k_y)\,z] = 0, \quad (8.39b)$$

where $v_{g,x}$, $v_{g,y}$, β_x and β_y are the group velocities and dispersion coefficients (see

(8.3)), evaluated at ω_0, of each component wave. In the final terms of (8.39) the wave-vector mismatch $(k_x-k_y)=\Delta k$ appears due to linear birefringence of the fiber. Equations (8.39) are quite complicated and can be solved in only an approximate way for certain specific cases of interest.

For strongly birefringent (polarization-preserving) fibers, Δk is large enough (large phase mismatch) for the final terms to be neglected because they are rapidly varying. In this case, as we have seen in Sect.4.8 for equations similar to (8.39), the modulational instability conditions are extremely rich. For example, Rothenberg (1990) has observed that modulational instability can occur for normal dispersion when orthogonally polarized visible waves are coupled in a strongly birefringent fiber. The spectrum of each wave shows the exponential buildup from threshold of a single sideband shifted by the modulation frequency, in contrast to the symmetric spectra (see Fig.4.10) observed (Tai et al. 1986a) in single-wave modulational instability. For weakly birefringent fibers, Δk is small and the last terms in (8.39) cannot be neglected. However, one can ignore the difference $\Delta v_g = v_{g,x} - v_{g,y}$ in group velocities of the two linearly polarized components, as this is assumed to be a higher-order effect. Modulational instability generated by the coherent interaction of two polarization modes has been observed (Murdoch et al. 1995) in such fibers.

In the general case both Δk and Δv_g must be taken into account and the situation is more complicated. A theoretical and numerical investigation (Soto-Crespo et al. 1995) of stationary soliton-like pulses represents a first step in trying to solve this problem.

Remarks

(i) It has been observed (Millot et al. 1997) that induced modulational instability in birefringent fibers can lead to femtosecond pulse-trains at repetition rate of 220-300 GHz in the normal group velocity dispersion regime. Starting from Maxwell's equations the propagation of polarized light in these fibers was modeled by two coupled nonlinear Schrödinger equations and numerical simulations have revealed that these pulse-trains have the structure of dark solitons or "domain walls".
(ii) A polarization domain wall soliton (compare to the domain wall of Sects. 6.11.3 and 9.7.2) was also recently observed (Seve et al. 1998) by mixing two intense counterpropagating laser beams in a nonlinear isotropic dielectric (a fiber). The soliton represents the switching of the state of polarization of light between two domains where both waves are circularly polarized and corotating.

8.7 A look at quantum optical solitons

In this chapter we have focused on optical fiber solitons because of their potential applications (Hasegawa and Kodama 1995) in high-performance optical communication systems. Such robust pulses can propagate inside optical fibers as a result of the balance of dispersion and nonlinearity. They are often called *temporal solitons* because they are solutions of a classical NLS equation (see (8.17)) which is both time and space dependent. In fact, such an equation is an approximate wave-envelope equation, which ignores quantum effects. From the correspondence principle of quantum mechanics it is natural to expect that a quantum version of the NLS equation can also serve (see Abdullaev et al. 1993) as a quantum model for pulse propagation in nonlinear optical fibers. Such a quantum NLS equation (QNLS equation for short), where field operators for the photon replace the envelope function of (8.17), has been obtained. At this point, the reader should wonder why a nonlinear evolution equation, formulated in the Heisenberg picture, is called a QNLS equation, a connotation which looks somewhat confusing. In fact, it is important to point out that the QNLS equation can be formulated in the Schrödinger picture. In this case, the quantum problem is described by a linear Schrödinger equation (Lai and Haus 1989a) which governs the time evolution of the state of the system. Thus, the superposition principle of quantum mechanics can be used.

The simplest *quantum soliton* is an eigenstate of the QNLS equation consisting of a cluster of photons with a well-defined energy and photon number. Nevertheless, the classical result of a soliton as a phase-shifted, undistorted pulse never occurs in quantum theory. Quantum states consisting of linear superpositions of such solitons will also spread out, as sketched in Fig. 8.10, because each component soliton of the wave packet has a different phase and group velocity (Lai and Haus 1989b). Thus, such entities are no longer solitons in the strict sense.

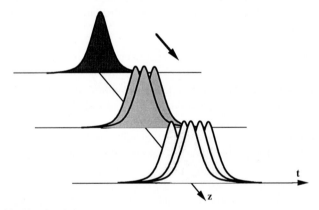

Fig.8.10. Sketch of the quantum spreading of a light pulse: due to quantum noise, each component soliton of the pulse has a different phase and group velocity

Even for a quantum soliton with a well-defined photon number, the center-of-mass position obeys the Heisenberg uncertainty principle, which implies the presence of quantum noise even when no photons are present. Position and momentum are conjugate observables and thus not simultaneously measurable. As a result, a soliton with a given velocity cannot have a well-defined position at all times; this is the quantum diffusion phenomenon for position. Other pairs of conjugate variables, such as the number of photons (quanta) and phase, obey an uncertainty relation. Thus, phase diffusion will occur because the photon number, like momentum, is invariant during propagation. As a result of this phenomenon, the distribution of the amplitudes narrows, or squeezes, for certain phases.

220

The uncertainty principle allows the reduction or squeezing of the photon-number noise. Quantum solitons and phase diffusion provide a promising way for generating squeezed states of light (Walls 1983) and reducing quantum noise in signal transmission (see Drummond et al. 1993, for a review). In this regard, a method for realizing macroscopic quantum soliton solutions has been proposed recently (Werner 1998): simultaneous photon number and momentum squeezing have been predicted using soliton propagation in an interferometer.

8.8 Some other kinds of optical solitons : spatial solitons

Up to now we have focused on temporal optical solitons in fibers, that is, robust nonlinear pulses which result from the balance between the nonlinear Kerr effect and the linear group velocity dispersion. In fact, in the early days of nonlinear optics it was proposed (Chiao et al. 1964) that light beams can trap themselves by creating their own waveguide through the nonlinear Kerr effect. Such self-trapping in dielectric waveguide modes appeared to be possible in intense laser beams.

The connection between self-trapping and soliton theory was first made by Zakharov and Shabat (1972) who showed that there are two kinds of optical solitons, spatial and temporal, where the nonlinearity balances dispersion and diffraction, respectively. For spatial pulses, each plane wave (spatial Fourier) component of the beam acquires a different phase and the beam broadens or diffracts. This diverging effect can be balanced by nonlinear effects, that is, material properties change in the presence of light and can counteract diffraction by what is called *light-induced lensing*. The simplest *spatial soliton* is a solution of a nonlinear Schrödinger equation, similar to (8.17), but where a second space variable replaces time. There are several kinds of spatial optical solitons such as: Kerr-type solitons, photorefractive solitons, quadratic solitons and incoherent solitons, which exhibit a variety of rich phenomena and whose study will lead to a deeper understanding of nonlinear phenomena (for a recent overview see Segev and Stegeman 1998).

It has been proposed that localized entities similar to spatial solitons in continuous media, considered above, should occur in an infinite array of identical coupled waveguides. In such a device, when low intensity light is injected into one, or a few neighboring waveguides, it will couple to more and more waveguides as it propagates, thereby broadening its spatial distribution. This spatial diverging effect is analoguous to diffraction in continuous media. Through the nonlinear Kerr effect an important intensity can change the refractive index of the input waveguides and decouple them of the rest of the array. It has been shown that certain light distributions propagate while keeping a fixed spatial profile among a limited number of waveguides, in an analogous way to spatial solitons. Such spatially localized waves are termed *discrete spatial optical solitons*. They are solutions of a discrete nonlinear Schrödinger equation which models (Christodoulides and Joseph 1988; Aceves et al. 1994) the evolution of the eletrical field in one waveguide of the array.

Recently, the observation of discrete spatial solitons in an array of 41 waveguides was reported (Eisenberg et al. 1998). Light, provided by a pulsed source, was coupled to the central waveguide and interesting results were obtained. At low power, the propagating electrical field spreads as it couples to more waveguides; when sufficient power is injected, the field is localized close to the input waveguides and its distribution agrees with the theoretical predictions based on the discrete nonlinear Schrödinger equation.

Appendix 8A. Electromagnetic equations in a nonlinear medium

Let us start with the Maxwell equations, which govern all electrical phenomena. In MKS units, we have

$$\nabla \times \mathbf{H} = \mathbf{J} + \frac{\partial \mathbf{D}}{\partial t}, \qquad \mathbf{D} = \varepsilon_0 \mathbf{E} + \mathbf{P}, \qquad (8.A.1)$$

$$\nabla \times \mathbf{E} = -\frac{\partial \mathbf{B}}{\partial t}, \qquad \mathbf{B} = \mu_0 \mathbf{H} + \mathbf{M}, \qquad (8.A.2)$$

$$\nabla \cdot \mathbf{D} = \rho, \qquad \nabla \cdot \mathbf{B} = 0, \qquad (8A.3)$$

where $\mathbf{E}(\mathbf{r},t)$ is the electric field, $\mathbf{H}(\mathbf{r},t)$ the magnetic field, $\mathbf{D}(\mathbf{r},t)$ the electric displacement, $\mathbf{P}(\mathbf{r},t)$ the electric polarization, $\mathbf{M}(\mathbf{r},t)$ the magnetic polarization, and $\mathbf{B}(\mathbf{r},t)$ the magnetic induction at point $\mathbf{r}$ (x, y, z) and time t. ρ is the free charges density, $\mathbf{J}$ the current density vector, ε_0 is the the vacuum permittivity and μ_0 the vacuum permeability.

After taking the curl of both sides of (8A.2), replacing $\nabla \times \mathbf{H}$ by (8A.1) and using the relation $\nabla \times \nabla \times \mathbf{E} = \nabla (\nabla \cdot \mathbf{E}) - \nabla^2 \mathbf{E}$, we get

$$\nabla^2 \mathbf{E} - \nabla(\nabla \cdot \mathbf{E}) = \mu_0 \frac{\partial^2 \mathbf{D}}{\partial t^2} . \qquad (8A.4)$$

In an isotropic dielectric medium, which is nonconductive and nonmagnetic, as is the case for a fiber material, ρ, $\mathbf{J}$ and the magnetic polarization $\mathbf{M}$ are zero. As given by (8A.1) the dielectric displacement or the total response $\mathbf{D}(\mathbf{r},t)$ of an isotropic nonlinear dielectric medium to an arbitrary electric field $\mathbf{E}(\mathbf{r},t)$ depends on the polarization

$$\mathbf{P}(\mathbf{r},t) = \mathbf{P}_L(\mathbf{r},t) + \mathbf{P}_{NL}(\mathbf{r},t),$$

which is made up of a linear part $\mathbf{P}_L(\mathbf{r},t)$ and a nonlinear part $\mathbf{P}_{NL}(\mathbf{r},t)$. In the general case one considers the spatial and temporal dispersion i.e., if one takes into account the fact that both the linear and nonlinear responses of the medium at the point $(\mathbf{r},t)$ are determined not only by the field at this point, but also by the fields (Akmanov et al. 1968) at points $(\mathbf{r}_1, t_1)$, one can write

$$P_L(\mathbf{r},t) = \varepsilon_0 \int_{-\infty}^{t} dt_1 \int d\mathbf{r}_1 \; \chi_L(t-t_1, \mathbf{r}-\mathbf{r}_1) \; E(t_1, \mathbf{r}_1), \qquad (8A.5a)$$

and

$$P_{NL}(\mathbf{r},t) = \varepsilon_0 \int\int_{-\infty}^{t}\int dt_1 \, dt_2 \, dt_3 \iiint d\mathbf{r}_1, d\mathbf{r}_2, d\mathbf{r}_3 \; \chi_{NL}(t_1, t_2, t_3, \mathbf{r}_1, \mathbf{r}_2, \mathbf{r}_3)$$

$$\text{x } \mathbf{E}(t - t_1, \mathbf{r} - \mathbf{r}_1) \, \mathbf{E}(t - t_1 - t_2, \mathbf{r} - \mathbf{r}_1 - \mathbf{r}_2) \, \mathbf{E}(t - t_1 - t_2 - t_3, \mathbf{r} - \mathbf{r}_1 - \mathbf{r}_2 - \mathbf{r}_3),$$

$$(8A.5b)$$

where χ_L and χ_{NL} represent respectively the linear and nonlinear (Kerr effect) dielectric susceptibilities. Equation (8A.5a) represents the local linear response of the medium. In the text we ignore the spatial dispersion, that is, χ_L and χ_{NL} are assumed to be independent of $\mathbf{r}$.

Note that (8A.5b) does not account for second-order nonlinear effects, since the molecule SiO_2 has a center of symmetry. In other terms the lowest order nonlinear effects in optical fibers originate from the third order susceptibility. On a microscopic level, the polarization is related to the motion of the electrons under the influence of the electric field and, for a medium lacking an inversion symmetry, one has to add a second-order susceptibility term to (8A.5b).

If in a first approximation we ignore the spatial dispersion of the fiber material and express the linear part (8A.5a) as a temporal convolution product only,

$$P_L(t) = \varepsilon_0 \, \chi_L(t) * E(t), \qquad (8A.6)$$

Equation (8A.5b) represents the local nonlinear response. In a first approximation we ignore the spatiotemporal dispersion and rewrite (8A.5) as

$$P_{NL} = \varepsilon_0 \, \chi_{NL}|E|^2 \, E. \qquad (8A.7)$$

Hence, we may rewrite the (8A.1) for $\mathbf{D}$ symbolically as

$$\mathbf{D} = \varepsilon * \mathbf{E}, \qquad (8A.8)$$

where the dielectric response function is given by

$$\varepsilon(t) = \varepsilon_0[1 + \chi_L(t) + \chi_{NL} \, E^2]. \qquad (8A.9)$$

From Maxwell's equations (8A.1) and (8A.3) we have

$$\nabla D = \nabla.(\varepsilon * \mathbf{E}) = \varepsilon * (\nabla.\mathbf{E}) + (\nabla\varepsilon*). \, \mathbf{E} = 0, \qquad (8A.10)$$

which means that $\nabla.\mathbf{E}$ is zero because we have ignored the spatial dispersion ($\nabla\varepsilon = 0$).

223

Thus, when the transverse inhomogeneity is small, the electric field can be approximated by a divergence free field with only transverse components and, by using (8A.10), (8A.4) is reduced to

$$\nabla^2 \mathbf{E} - \frac{1}{c^2} \frac{\partial^2 \mathbf{E}}{\partial t^2} = \mu_o \frac{\partial^2 \mathbf{P}}{\partial t^2} . \tag{8A.11}$$

9 The Soliton Concept in Lattice Dynamics

In previous chapters we have considered nonlinear waves in the macroworld. We have examined different systems which provide the simplest examples of one-dimensional systems or devices, where the localized waves or pulses called solitons can be simply and coherently created, easily observed, and manipulated on a macroscopic scale. At the microscopic level the localized nonlinear wave modes have a spatial extension ranging from less than a few microns to a few angströms. These excitations, which correspond to large-amplitude atomic or molecular motions, are mainly created by thermal processes, sometimes by some external stimulus; their experimental manifestation is indirect; their observation is more subtle than for the nonlinear macrowaves. This should explain why the number of experimental investigations is rather small compared to the considerable body of literature devoted to theoretical and numerical studies.

In this chapter we consider nonlinear lattice dynamics, that is, the one-dimensional propagation of waves in lattices modeling discrete microscopic structures. Such lattice models, in spite of their relative simplicity, are associated with rather important problems in physics (Toda 1978). First, they have played an important role in the discovery of the soliton (see Chap.1) and are of a certain conceptual value (Jackson 1990). Second, they allow us to clarify many features of molecular chains studied in physics, chemistry, and biology (Davydov 1985; Collins 1983). In the first sections we examine soliton excitations which may occur in one-dimensional lattices with nonlinear atomic interactions. The next sections are devoted to one-dimensional lattices with a nonlinear *substrate* or *"on site" potential*. In the last sections we consider energy localization in lattices and present new analog experiments allowing to observe nonlinear localized modes.

9.1 The one-dimensional lattice in the continuum approximation

The model consists of a one-dimensional chain of N atoms. Each one of mass m, interacts through a nearest neighbor interatomic potential U. In the equilibrium position, the particles are uniformly spaced each at distance a from its nearest neighbor, as represented in Fig.9.1. If $u_n(t)$ is the longitudinal displacement of the nth atom from its equilibrium position, the relative displacement between atoms n and n+1 is

$$r_n = u_{n+1} - u_n. \tag{9.1}$$

225

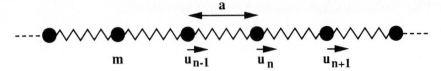

Fig.9.1. Monoatomic chain with longitudinal displacements

We denote by $U(r_n)$ the potential which describes the interaction between atom n and its nearest neighbor n+1. Thus, the force that atom n+1 exerts on atom n is

$$F = dU/dr_n. \tag{9.2}$$

In the general case, the interaction potential U may be chosen to have the form of a standard atomic-potential model like the Morse or Lennard-Jones potential. One can also choose the Toda potential (see Sect.3.1), which is efficient and physically realistic in modeling electrical lattices. However, although this potential yields forces which have the nature of physical atomic forces, it is not so realistic for modeling atomic interactions because it is unbounded as the difference between atomic displacements goes to infinity.

From Newton's law we get the equation governing the motion of the nth particle

$$m\frac{d^2u_n}{dt^2} = F(u_{n+1}-u_n) - F(u_n - u_{n-1}), \tag{9.3}$$

where the last term represents the force that atom n -1 exerts on atom n. Like Fermi, Pasta and Ulam (1955), we consider a lattice model where the interaction potential has the simple form

$$U(r) = \frac{1}{2}Gr^2 + \frac{1}{3}GAr^3. \tag{9.4a}$$

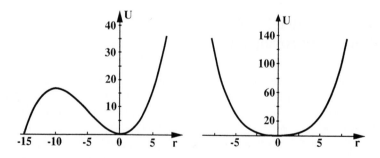

Fig. 9.2. Representation of U(r) given by (9.4 a, b), G=1, A=0.1, B=0.1.

Here, G represents the coefficient of linear interaction (Hooke's law); A is a small coefficient compared to G, it represents *weak nonlinear interactions*. Note that this asymmetric potential should model (in one dimension) a real crystal without a center of symmetry. Otherwise, if a center of symmetry exists, one may choose a symmetric potential of the form

$$U(r) = \frac{1}{2} Gr^2 + \frac{1}{4} GBr^4. \qquad (9.4b)$$

The linear and simple nonlinear (anharmonic) force corresponding to potential (9.4a) is

$$F(r) = G r + GAr^2. \qquad (9.5)$$

For a complete study, the boundary conditions, that is, the motion of the first and last atom in the chain, must be specified. For example, one may have periodic boundary conditions: $u_{N+n} = u_n$, fixed ends : $u_1 = u_N = 0$, or an infinite lattice: $N \to \infty$, for which the end effects are irrelevant. In the following, we consider the infinite lattice case.

Next, replacing (9.5) in the equation of motion (9.3) we obtain

$$m\frac{d^2u_n}{dt^2} = G[(u_{n+1} - u_n) - (u_n - u_{n-1})] + GA[(u_{n+1} - u_n)^2 - (u_n - u_{n-1})^2]. \qquad (9.6)$$

Note that (9.6) can be obtained from the Hamiltonian

$$H = \sum_{n=1}^{N} [\frac{1}{2} m(\frac{du_n}{dt})^2 + \frac{G}{2} (u_{n+1} - u_n)^2 + \frac{1}{3}GA(u_{n+1} - u_n)^3] \qquad (9.7a)$$

and

$$m\frac{d^2u_n}{dt^2} = -\frac{\partial H}{\partial u_n}, \quad n = 1,2,....N. \qquad (9.7b)$$

System (9.6) with N nonlinear differential equations cannot be solved analytically and, as for previous chapters (see Sects.2.3 and 4.1), we use the continuum approximation

$$u_{n\pm1} = u \pm a\frac{\partial u}{\partial x} + \frac{a^2}{2} \frac{\partial^2 u}{\partial x^2} \pm \frac{a^3}{3!} \frac{\partial^3 u}{\partial x^3} + \frac{a^4}{4!} \frac{\partial^4 u}{\partial x^4} +....$$

Substituting in (9.6) and omitting the subscript n we obtain

$$u_{tt} - c_0^2 u_{xx} = c_0^2 (2aAu_x u_{xx} + \frac{a^2}{12} u_{xxxx}), \qquad (9.8)$$

227

where $c_0 = a\sqrt{G/m}$ is the velocity of linear (sound) waves in the chain. Weak nonlinearity and weak dispersion, which appear on the right hand side of (9.8), will balance if A is of order a. Note that the linear approximation (A=0) of (9.8) is similar to (2.36), which models electrical networks. The linear dispersion relation corresponding to (9.8) is

$$\omega = c_0 k \sqrt{1 - k^2 a^2/12} \ .$$ (9.9)

It is valid for small k (see Sect.2.3). By differentiating (9.8) with respect to x and setting

$$u_x = w,$$ (9.10)

we obtain

$$w_{tt} - c_0^2 w_{xx} = p(w^2)_{xx} + h w_{xxxx},$$ (9.11)

where

$$p = aAc_0^2, \qquad h = c_0^2 a^2/12.$$ (9.12)

Equation (9.11), which describes waves propagating to the left and the right, is called the *Boussinesq equation* (see Sect.3.4). We then look for a solution with permanent profile propagating with constant velocity v. That is, we assume that

$$w(x - vt) = w(s), \qquad \partial/\partial x = \partial/\partial s, \qquad \partial/\partial t = -v \, \partial/\partial s,$$

and (9.11) becomes

$$(v^2 - c_0^2)w_{ss} = p(w^2)_{ss} + h w_{ssss}.$$

Two integrations with respect to s yield

$$(v^2 - c_0^2)w = p(w^2) + h w_{ss} + K_1 w + K_2.$$ (9.13)

We look for localized solutions; this implies that $K_1 = 0$ and $K_2 = 0$ in (9.13), which is integrated (see Appendix 3B) to give a solitary-wave solution,

$$w(x,t) = \frac{3}{2aA} \left(\frac{v^2}{c_0^2} - 1\right) \operatorname{sech}^2 \left[\frac{1}{a}\sqrt{3\left(\frac{v^2}{c_0^2} - 1\right)} \, (x - vt)\right],$$ (9.14)

which represents the gradient of atomic displacements. Then, taking into account (9.10) and using (9.11) we get the atomic displacement

$$u(x, t) = \frac{1}{2A}\sqrt{3\left(\frac{v^2}{c_0^2} - 1\right)} \tanh \left[\frac{1}{a}\sqrt{3\left(\frac{v^2}{c_0^2} - 1\right)} \, (x - vt)\right],$$ (9.15)

228

which has the form of a kink. The solitary waves can propagate in both directions because we can have $v>0$ or $v<0$, and they are supersonic because we must have $|v| > c_0$.

We can reduce (9.8) to the KdV equation, which describes waves propagating in one direction only. By using the reductive perturbation technique, we make the transformations

$$\xi = \varepsilon^{1/2}(x - c_0 t), \qquad \tau = \varepsilon^{3/2} c_0 t,$$

and express the displacement u in a perturbation series

$$u = \varepsilon u_1 + \varepsilon^2 u_2 + ...$$

After we have set $\partial u / \partial \xi = y$, calculations similar to those of Appendix 3C yield

$$y_\tau + Aayy_\xi + (a^2/24)y_{\xi\xi\xi} = 0. \tag{9.16}$$

The soliton solution traveling at velocity c has the form

$$y(\xi,\tau) = \frac{3c}{Aa} \operatorname{sech}^2 [\frac{1}{a} \sqrt{6c} \, (\xi - c\tau)]. \tag{9.17}$$

In terms of the original variables (x, t) one gets the atomic displacement

$$u(x, t) = \frac{1}{2A}\sqrt{6v/c_0 - 1} \tanh [\frac{1}{a}\sqrt{6v/c_0 - 1} \, (x - vt)], \tag{9.18}$$

where $v = c_0(1+c)$. We note that v must be positive, only. The shape of kink (9.15) approaches the shape of kink (9.18) for small supersonic velocities; that is, for small amplitudes, $v \to c_0$, $3(v^2/c_0^2 -1) = 3(v/c_0 -1)(v/c_0 +1) \to 6(v/c_0 -1)$. The two profiles are compared in Fig.9.3: for $v/c_0 =1.05$ they are similar. This means that in a real experiment one could not detect a difference between the two models.

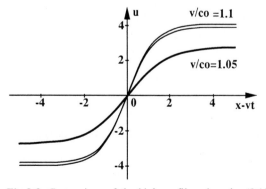

Fig.9.3. Comparison of the kink profiles given by (9.15) and (9.18) respectively: for $v/c_0 = 1.05$ they are similar; for $v/c_0 = 1.1$ there is a slight difference

The reader may repeat the above calulations for symmetric potential (9.4b) and find sech-shaped solitary waves for w.

9.2 The quasi-continuum approximation for the monatomic lattice

In the previous section we saw that the lattice equations can be reduced to a partial-differential equation which can be solved exactly only if the interactions between atoms are approximated by a simple form: a low-order polynomial. However, as already mentioned in Chap.2, this continuum approximation is not accurate enough if the interactions are strongly nonlinear, and it must be improved. This problem has been considered by Collins (1981), who introduced a "quasi-continuum approximation". By using (9.1) and (9.2) let us rewrite (9.3) as

$$m\frac{d^2r_n}{dt^2} = U'(r_{n+1}) + U'(r_{n-1}) - 2 \cdot U'(r_n),\qquad(9.19)$$

where $U'(r_n)$ is the derivative of a general interaction potential U, as defined by (9.2). If the solitary waveform does not change very rapidly with n, we can regard n as a continuous variable rather than an integer, and we can write

$$U'(n+1) + U'(n-1) - 2U'(n) = [\exp(d/dn) + \exp(-d/dn) - 2]\, U'(n),$$

which yields

$$m\frac{d^2r_n}{dt^2} = 2[\cosh(\frac{d}{dn}) - 1]U'(r_n).\qquad(9.20)$$

The continuum approximation used in the previous section consists in expanding the cosh operator in a Taylor series. The quasi-continuum approximation additionally takes into account that the force $U'(n)$ may vary quite rapidly, even if r_n itself varies slowly. At this point, we do not use the procedure of Collins, who inverted the cosh operator in (9.20), but follow the iteration procedure used by Hochstrasser et al. (1988). Thus, with $x = na$ and $r_n(t) \to r(x,t)$, (9.2) is expanded as

$$m\frac{\partial^2 r}{\partial t^2} = (\,a^2\frac{\partial^2}{\partial x^2} + \frac{a^4}{12}\frac{\partial^4}{\partial x^4} + \,....)U'(r),\qquad(9.21)$$

where the lattice spacing a is regarded as an expansion parameter. Then, setting

$$\alpha(x,t) = a^2\frac{\partial^2}{\partial x^2}\,U'(r),\qquad(9.22)$$

230

we have

$$O(0) : m\frac{\partial^2 r}{\partial t^2} = \alpha, \qquad O(a^2) : m\frac{d^2 r}{dt^2} = \alpha + \frac{a^2}{12}\frac{\partial^2}{\partial x^2}\alpha.$$

These two equations and (9.22) yield

$$m\frac{\partial^2 r}{\partial t^2} = a^2\frac{\partial^2}{\partial x^2}U'(r) + m\frac{a^2}{12}\frac{\partial^2}{\partial x^2}\left(\frac{\partial^2 r}{\partial t^2}\right). \tag{9.23}$$

This equation was also obtained by Rosenau (1987). First, let us compare this equation with that obtained with the continuum approximation. For this, we insert in (9.23) the simple potential (9.4a) and get

$$\frac{\partial^2 r}{\partial t^2} = c_o^2\frac{\partial^2 r}{\partial x^2} + Ac_o^2\frac{\partial^2 r^2}{\partial x^2} + \frac{a^2}{12}\frac{\partial^4 r}{\partial x^2\partial t^2}.$$

Noting that $r_n = u_{n+1} - u_n \rightarrow a\partial u/\partial x$ in the quasi-continuum approximation and using (9.10), we can transform this equation into

$$w_{tt} - c_o^2 w_{xx} = p(w^2)_{xx} + (h/c_o^2)u_{xxtt}. \tag{9.24}$$

Comparing (9.24) with (9.11) shows that only the last term, which represents dispersion, is different. Therefore, the calculation of the solitary-wave solutions of (9.24) is exactly the same as for (9.11) if one simply replaces h by $h(v/c_o)^2$ in (9.13). Under this condition, one obtains

$$w(x,t) = \frac{3}{2aA}\left(\frac{v^2}{c_o^2} - 1\right)\operatorname{sech}^2\left[\frac{1}{a}\sqrt{3\left(1 - \frac{c_o^2}{v^2}\right)}(x - vt)\right]. \tag{9.25}$$

The solitary-wave solutions (9.25) and (9.14) have the same amplitude but their profiles are different. In Fig.9.4 we have represented the shape of these solutions. We can note that for $A = 0.1G$, which represents a strong nonlinearity in practical cases, the waveforms are slightly different. The difference will increase with A, and the quasi-continuum approximation will become more and more accurate.

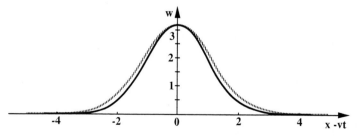

Fig. 9.4. Comparison of the soliton profiles given by (9.14) (continuous line) and (9.25) (dotted line) for a=1, G=1, A=0.1, v/c_o=1.1

Remark

For atomic motions with very large amplitude it is more realistic to consider the full nonlinear potential U(r) rather than its polynomial expansion. In this case, we look for localized solutions with permanent profile: $r(x,t) = r(x-vt) = r(s)$. The integration of (9.23) yields

$$(\frac{dr}{ds})^2 = \frac{24}{v^2} [\frac{mv^2}{2a^2} r^2 - U(r)].$$ (9.26)

This equation will have solitary-wave solutions if its right-hand side is positive. However, it cannot be integrated analytically for most models of potential such as the Lennard-Jones potential or Morse potential.

9.3 The Toda lattice

The discrete equations of motion (9.3), or equivalently (9.19), are generally intractable. However, in 1967 Toda proposed a nonlinear lattice (see Sect.4.1) with a particular potential function U(r). This lattice model allows the integration of (9.19) and has wide applicability. The nonlinear potential has the form

$$U(r) = \frac{A}{B} (e^{-Br} - 1) + Ar.$$ (9.27)

For A,B > 0, it presents strong repulsion and much weaker attraction, as sketched in Fig.9.5. For small |B| or for small amplitudes an expansion of (9.27) yields

$$U(r) = \frac{AB}{2} (r^2 - \frac{B}{3} r^3 + ...).$$

Consequently, for a motion with sufficiently small amplitude the lattice behaves like a harmonic lattice with spring constant G=AB. The force, as defined by (9.2), corresponding to (9.27) is

$$F(r) = A(e^{-Br} - 1).$$ (9.28)

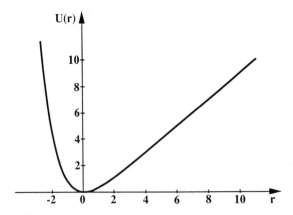

Fig. 9.5. Sketch of the Toda potential for A=B=1

With (9.27) the equation of motion (9.19) takes the form

$$m\frac{d^2r_n}{dt^2} = A[\ 2\exp(-Br_n) - \exp(-Br_{n+1}) - \exp(-Br_{n-1})]. \tag{9.29}$$

Using (9.28), we can rewrite (9.29) as

$$\frac{d^2}{dt^2} \text{Ln}\ (1 + \frac{F_n}{A}) = \frac{B}{m}\ (\ F_{n+1} + F_{n-1} - 2F_n\ ). \tag{9.30}$$

Note that this equation is similar to the Toda equation (see Sect.4.1) for the electrical network where the voltage plays the role of a force. For detailed solutions of these equations the reader may consult the book by Toda (1978). Here, we just consider the discrete soliton solution. We look for a sech-shaped localized solution:

$$F_n = (m\beta^2/AB)\text{sech}^2(Pna \pm \beta t). \tag{9.31}$$

Substituting in (9.30) one can verify that (9.31) is an exact solution of (9.30) if

$$\beta = \sqrt{AB/m}\ \sinh P.$$

Here, P is a parameter with dimensions equal to the inverse of a length. The lattice solitary wave has the velocity

$$v = \beta/P = \sqrt{AB/m}\ (\sinh P/P).$$

It travels faster than linear waves with long wavelength, which propagate with velocity $v_0 = \sqrt{G/m}$. In fact, this lattice-solitary wave is very stable and robust: it remains unchanged through collisions with other solitary waves, as shown by numerical simulations. Moreover, the Toda-lattice equations have been solved by the inverse-scattering technique described in Chap.10. Thus, it is a soliton in the strict sense.

9.4 Envelope solitons and localized modes

As for the electrical networks (see Chap.4) we now examine modulated waves or wavepacket solutions. We consider a nonlinear lattice in which the interaction potential between nearest neighbor particles is symmetric and has the form (9.4b) with A, B and G >0. With this potential, (9.19) becomes

$$m\frac{d^2r_n}{dt^2} = G\ (r_{n+1} + r_{n-1} - 2\ r_n) + GB\ (\ r^3_{n+1} + r^3_{n-1} - 2\ r^3_n\ ). \tag{9.32}$$

233

In the weak-amplitude limit we assume that the longitudinal atomic displacement has the form

$$r_n = \varepsilon R \, e^{i\,(kna\,-\,\omega t)} + \varepsilon R^* \, e^{-\,i\,(kna\,-\,\omega t)}, \tag{9.33}$$

where $\varepsilon \ll 1$ and the amplitude $R(\varepsilon x, \varepsilon t)$ is constant with respect to the rapid variables $x = na$ and t. Then, inserting this expression into (9.32) and ignoring the third harmonic terms, to a first approximation we obtain

$$\omega = 2\sqrt{\frac{G}{m}}\,\sin\frac{ka}{2}\,[\,1 + 3\varepsilon^2 B|R|^2\,]. \tag{9.34}$$

For $B = 0$, that is, for very low amplitudes, this *nonlinear dispersion relation* reduces to a linear dispersion relation. Setting $X = \varepsilon x$ and $T = \varepsilon t$, and proceeding as in Chap.4, 5, and 8, one can associate to (9.34) a nonlinear Schrödinger equation of the form

$$i\frac{\partial R}{\partial T} + P\frac{\partial^2 R}{\partial X^2} + Q|R|^2 R = 0, \tag{9.35}$$

where the dispersion coefficient P and nonlinear coefficient Q,

$$P = \frac{1}{2}\frac{\partial^2\omega}{\partial k^2} = -\frac{a^2}{4}\sqrt{\frac{G}{m}}\,\sin\frac{ka}{2}, \qquad Q = -\frac{\partial\omega}{\partial|R|^2} = -6B\sqrt{\frac{G}{m}}\,\sin\frac{ka}{2}, \tag{9.36}$$

are evaluated at carrier wavenumber $k = k_c$, which corresponds to carrier frequency $\omega = \omega_c$.

In the present case we have $Q/P > 0$ and the nonlinear Schrödinger equation (9.35) has an envelope-soliton solution (see (4C.15) with c=0) of the form

$$R = R_0 \,\mathrm{sech}\,[\sqrt{\frac{Q}{2P}}\,R_0\,\varepsilon(x - v_{gt})]\,\exp\left\{-i\,\varepsilon^2\frac{|Q|}{2}\,R_0^2 t\,\right\}, \tag{9.37}$$

where R_0 is the amplitude. Using (9.33), the longitudinal displacement is expressed as a wave packet or lattice envelope soliton of the form

$$r_n = 2R_0\,\mathrm{sech}\,[\,\varepsilon R_0\sqrt{\frac{Q}{2P}}\,(x - v_{gt})]\,\cos\,[k_c na - (\omega_c + \varepsilon^2\frac{|Q|}{2}\,R_0^2)\,t\,]. \tag{9.38}$$

Solution (9.38) represents a soliton with an envelope moving at velocity v_g and a phase moving at velocity Ω/k, as shown in Fig.9.6.

We note that for $k_c = k_p = \pi/a$, we have $v_g = 0$ and $|Q| = 6B\sqrt{G/m}$. To order ε (c=0) the soliton does not move; it oscillates at a frequency

$$\Omega_p = (\omega_m + \varepsilon^2\frac{6B}{2}\sqrt{\frac{G}{m}}\,R_0^2),$$

234

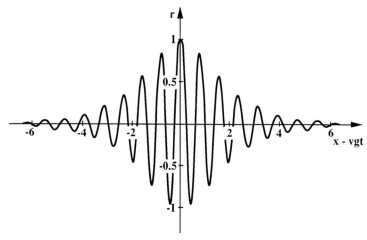

Fig. 9.6. Sketch of a lattice envelope soliton for $R_0=1/2$, $Q/P=1/2$, and $k_ca=0.8$.

which is higher than the cut-off frequency $\omega_m = 2\sqrt{G/m}$ of the lattice. Nonlinear modes with similar properties are often called *intrinsic localized modes* (Sievers and Takeno 1988) in nonlinear lattice dynamics or *gap solitons* (Mills 1991; Martin de Sterke and Sipe 1988) in the field of nonlinear optics.

Ever since the discovery of solitons, there has been considerable interest (Takeno 1992; Flach 1994) in localized modes for spatially extended systems. In this context, intrinsic collapse to self-localized states is increasingly studied because of its wide significance in a great variety of physical systems (see Sects 9.10 and 9.11). Contrary to the impurity induced modes, which can exist in linear lattices with impurities and disorder, these intrinsic localized modes can exist in nonlinear lattices without defects or disorder. By extension, these modes are called breathers (McKay and Aubry 1994), in any system where they could exist, because they present some properties similar to the breather soliton (see Sect. 6.9) of the Sine–Gordon system.

9.5 The one-dimensional lattice with transverse nonlinear modes

In the previous sections we considered longitudinal motions of particles in one-dimensional lattices. Now, we examine transverse nonlinear waves in a one-dimensional lattice (Gorbatcheva and Ostrowsky 1983; Cadet 1987). Specifically, the model consists of a chain of identical particles of mass m which can move in transverse planes normal to a direction Ox, as represented in Fig.9.7 the lattice spacing is a. To a first approximation we neglect the longitudinal motions. The forces of interaction between nearest-neighbor particles are assumed to be linear with interaction constant G; they are balanced by a constant longitudinal tension T.

Thus, in a plane which contains the x axis and the chain, the equation of motion governing the motion of the nth particle is

$$m\,\mathbf{u}\,\frac{d^2 U_n}{dt^2} = \mathbf{F}_{n+1} + \mathbf{F}_{n-1}, \tag{9.39a}$$

where $\mathbf{F}_{n\pm1}$ are the forces acting on the nth particle. Working these forces explicitly we obtain

$$m\mathbf{u}\,\frac{d^2 U_n}{dt^2} = \mathbf{u}\,[T + G\,(d_n - a)]\,\sin\theta_{n+1} - \mathbf{u}\,[T + G\,(d_{n-1} - a)]\,\sin\theta_{n-1}. \tag{9.39b}$$

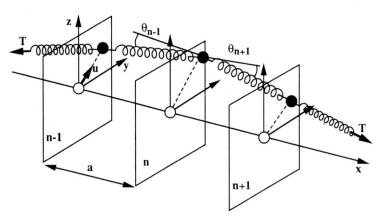

Fig. 9.7. A one-dimensional lattice with transverse displacements in the plane zOy.

Here, U_n is the transverse displacement in the direction of unit vector $\mathbf{u}$, d is the distance between two neigboring particles, and θ_{n-1} and θ_{n+1} are angles as defined on Fig.9.7. In terms of the components y_n and z_n of the displacement vector $\mathbf{u}U_n$ the above equation becomes

$$m\,\frac{d^2 y_n}{dt^2} = [T + G\,(d_n - a)]\,\frac{y_{n+1} - y_n}{d_n} - [T + G\,(d_{n-1} - a)]\,\frac{y_n - y_{n-1}}{d_{n-1}}, \tag{9.40a}$$

$$m\,\frac{d^2 z_n}{dt^2} = [T + G\,(d_n - a)]\,\frac{z_{n+1} - z_n}{d_n} - [T + G\,(d_{n-1} - a)]\,\frac{z_n - z_{n-1}}{d_{n-1}}, \tag{9.40b}$$

where

$$d_n = \sqrt{a^2 + (y_{n+1} - y_n)^2 + (z_{n+1} - z_n)^2}\,.$$

We note that, *although the interactions between particles are linear, a nonlinearity arises, owing to the geometry of the motion.* Next, we assume that the amplitude of the motion is small and that we can make the approximation

236

$$d_n = a + \frac{1}{2}\left[\frac{(y_{n+1}-y_n)^2}{a} + \frac{(z_{n+1}-z_n)^2}{a}\right] a. \tag{9.41}$$

The obvious symmetry of (9.40) allows us to simplify the writing by introducing the complex quantity

$$H_n = y_n + i\, z_n, \tag{9.42}$$

which describes completely the transverse displacement. Then, inserting (9.42) and (9.41) into (9.40), we get

$$\frac{d^2 H_n}{dt^2} = \frac{C_T^2}{a^2}(H_{n+1} + H_{n-1} - 2H_n)$$

$$+ \frac{C_L^2 - C_T^2}{2a^4}\left[\,|H_{n+1} - H_n|^2 (H_{n+1} - H_n) - |H_{n-1} - H_n|^2 (H_{n-1} - H_n)\right] \tag{9.43}$$

where

$$C_L^2 = Ga^2/m, \qquad C_T^2 = Ta/m, \tag{9.44}$$

represent the velocities of longitudinal and transverse linear (sound) waves, respectively. In practice, one often has $C_L > C_T$, as will be assumed in the following. The set of lattice equations (9.43) cannot be solved, and, as in Sect.9.1, we use the continuum approximation to get

$$H_{tt} = C_T^2 H_{xx} + \frac{C_T^2 a^2}{12} H_{xxxx} + \frac{C_L^2 - C_T^2}{2}\left[\,|H_x|^2 H_x\,\right]_x, \tag{9.45}$$

where $x = na$. Differentiating (9.45) with respect to x and setting $V = H_x$, yields

$$V_{tt} = C_T^2 V_{xx} + \frac{C_T^2 a^2}{12} V_{xxxx} + \frac{C_L^2 - C_T^2}{2}\left[\,|V|^2 V\,\right]_{xx}. \tag{9.46}$$

We look for solutions with a permanent profile propagating with velocity v; that is, as previously, we assume that $V(x-vt) = V(s)$. Then we find after some calculations (see Appendix 9A) the two components

$$y(x,t) = \pm \frac{a}{\sqrt{3}} \frac{C_T}{\sqrt{C_L^2 - C_T^2}} \tan^{-1}\left[\sinh\frac{x-vt}{L}\right] \cos\phi_0, \tag{9.47a}$$

and

$$z(x,t) = \pm \frac{a}{\sqrt{3}} \frac{C_T}{\sqrt{C_L^2 - C_T^2}} \tan^{-1}\left[\sinh\frac{x-vt}{L}\right] \sin\phi_0. \tag{9.47b}$$

of the transverse displacement, which corresponds to a kink-shaped solitary wave with a transverse polarization in a direction given by $\phi_0 = C^{te}$.

9.6 Motion of dislocations in a one-dimensional crystal

In condensed matter the dislocations correspond to defects or breakings of the regular crystal structures along some direction. The dislocations are not immobile: they can move inside the crystal. The first model describing the propagation of such objects was developed by Frenkel and Kontorova (1939). In their one-dimensional approach they considered an atomic chain of mass m which can move and a second one which is assumed to be fixed in place, as shown in Fig.9.8.

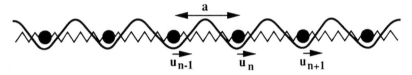

Fig. 9.8. An atomic chain with linear couplings (springs) interacting with a periodic nonlinear substrate potential.

This rigid "substrate chain" creates *a substrate potential* or *on site potential* V which is periodic and has the form

$$V(u_n) = V_0 \left(1 - \cos \frac{2\pi u_n}{a}\right). \tag{9.48}$$

Here, V_0 represents the amplitude of the potential, u_n is the displacement of the nth atom from its equilibrium position, and a is the lattice spacing. The system is nondissipative and the Hamiltonian is written as

$$H = \sum_n \left[\frac{1}{2} m \left(\frac{du_n}{dt}\right)^2 + \frac{G}{2}(u_{n+1} - u_n)^2 + V(u_n)\right], \tag{9.49}$$

where G is the elastic force constant between neighboring atoms, and the number of particles is assumed to be infinite. The equation of motion of the nth atom is

$$m\frac{d^2 u_n}{dt^2} = G(u_{n+1} + u_{n-1} - 2u_n) - \frac{2\pi V_0}{a} \sin \frac{2\pi u_n}{a}. \tag{9.50}$$

This equation is identical to (6.3) and as in Chap.6 we can use the continuum approximation $[u_n(t) \to u(x, t)]$ to reduce (9.50) to a Sine–Gordon equation,

238

$$\phi_{tt} - c_0{}^2\phi_{xx} + \omega_0{}^2\sin\phi = 0, \tag{9.51}$$

where

$$\phi = 2\pi u/a, \quad c_0 = a\sqrt{G/m}, \quad \omega_0 = (2\pi/a)\sqrt{V_0/m}. \tag{9.52}$$

Here, c_0 represents the linear wave or sound velocity in the crystal and ω_0 is a characteristic frequency. The kink solution of (9.51) is given by (see (6.19b))

$$\phi = 4\arctan\left[\exp\left(\frac{x - vt}{d\sqrt{1 - \dfrac{v^2}{c_0{}^2}}}\right)\right]. \tag{9.53}$$

This represents a kink dislocation propagating in the crystal with velocity v. Its profile or width is determined by the quantity $d\sqrt{1 - v^2/c_0{}^2}$, where

$$d = c_0/\omega_0 = (a^2/2\pi)\sqrt{G/V_0}, \tag{9.54}$$

and d measures the characteristic length (compare to (6.4b)) over which the atomic displacement changes. Proceeding exactly as in Sect.6.3 one can calculate the energy of a moving kink dislocation and its rest mass. In the above model the dissipation was not taken into account. However, if the substrate chain is allowed to move because of thermal fluctuations, during its motion the dislocation will experience losses (Seeger and Schiller 1966).

9.7 The one-dimensional lattice model for structural phase transitions

Lattice dynamics is useful for approaching the dynamics of crystals exhibiting structural phase transitions. Such transitions result from the shifts of the equilibrium positions of the molecules. They apparently take place owing to the instability of some lattice displacement pattern, which drives the system from some stable high-temperature phase $T > T_c$ to a different low-temperature $T < T_c$ lattice configuration, where T_c is a critical temperature. The dynamics of such transitions is frequently characterized by a vibrational or phonon mode whose frequency sharply decreases as the temperature approaches the critical temperature T_c from above. As a result, the restoring force corresponding to that displacement pattern softens, and one calls this particular mode a *soft mode*.

Altough in nature many systems present such behavior, the study of such transitions is particularly well documented for ferroelectric compounds. In these materials the displacement of a group of atoms in each unit cell results in a net dipole moment in the cell, which represents the so-called *order parameter*. Owing to preparation, impurity and defects content, the crystal may be in a state of coexisting *domains* separated by *walls*, or entirely in a state of uniform polarization (one domain).

At the transition temperature T_c, the lattice displacements become large and *the dynamics is highly nonlinear. Consequently, theoretical approaches based on an anharmonic phonon perturbation, which uses some set of self-consistent high-temperature lattice phonons as a basis, are inadequate. One needs a fully nonlinear and nonperturbative approach.* Significant approaches in this direction were made by Krumhansl and Schrieffer (1975) and Aubry (1976), who considered a one-dimensional lattice model. Such a model has proved to be very useful in dealing with the problem of phase transitions, as it provides a nonperburbative approach for strongly nonlinear behaviors, in spite of the fact that it does not show a phase transition at finite temperature (strictly speaking, the transition occurs only at zero temperature).

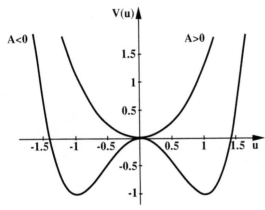

Fig.9.9. Representation of the potential (9.56) for A=2, B=4 (simple well) and for A=- 4 , B = 4 (double well)

Let us consider this one-dimensional model, which has similarities with that of Sect.9.6.The model Hamiltonian is given by

$$H = \sum_{n}^{N} [\frac{1}{2} m(\frac{du_n}{dt})^2 + \frac{G}{2}(u_{n+1}-u_n)^2 + V(u_n)].$$ (9.55)

However, the substrate potential is nonperiodic, it has the form

$$V(u_n) = \frac{A}{2} u_n^2 + \frac{B}{4} u_n^4.$$ (9.56)

The Hamitonian (9.55) represents an enormous oversimplification of any real three-dimensional systems, particularly of symmetry restrictions and long-range forces which play an important role in ferroelectric compounds. Nevertheless, it presents

240

interesting features. Taking A<0 and B>0, the potential (9.56) develops a double-well character with two degenerate minima at $u_o = \pm [A/B]^{1/2}$ and a barrier height $E_o = -|A|^2/4B$, as represented in Fig.9.9. For A>0 and B>0, the potential has a simple minimum at u =0. This potential can be thought of as a microscopic analog of the free energy. It is a function of the order parameter (polarization P for ferroelectrics), which represents the basis of the Landau–Ginzburg theory of phase transitions.

Leaving the thermodynamics aside, we confine ourselves to the interesting case: A = -|A| and B >0. The Hamiltonian (9.55) is rewritten as

$$H = m \sum_n [\frac{1}{2}(\frac{du_n}{dt})^2 + \frac{1}{2}\frac{c_o^2}{a^2}(u_{n+1}-u_n)^2 - \frac{\omega_o^2}{2}(u_n^2 - \frac{1}{2}\frac{u_n^4}{u_o^2})], \qquad (9.57)$$

where

$$c_o = a\sqrt{G/m}, \qquad \omega_o = \sqrt{|A|/m},$$

are the linear wave or phonon velocity and a characteristic frequency, respectively. The corresponding set of equations of motion is

$$\frac{d^2u_n}{dt^2} = \frac{c_o^2}{a^2}(u_{n+1} + u_{n-1} - 2u_n) + \omega_o^2(u_n - \frac{u_n^3}{u_o^2}). \qquad (9.58)$$

From (9.57) we can expect two limiting physical regimes: the *order–disorder* and *displacive* regimes, which depend essentially on the relative magnitudes of the coupling energy between nearest-neigbor particles and the potential-barrier height.

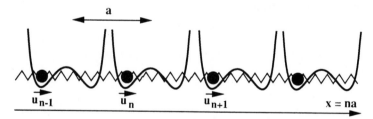

Fig. 9.10. Sketch of an atomic chain interacting with a double well potential

9.7.1 The order–disorder transition

First, if the coupling energy between nearest neighbors displaced to opposite wells $(u=u_o)$ is not sufficient to lift the particle over the barrier $E_o = m\omega_o^2 u_o^2/4$, one has

$$\frac{m\frac{c_o^2}{a^2}4u_o^2}{E_o} = \frac{16}{a^2}\frac{c_o^2}{\omega_o^2} \ll 1. \qquad (9.59)$$

In this case the thermal energy kT (k = Boltzman constant) is small compared to the barrier height E_0, $kT \ll E_0$, the jumps to neighboring wells occur with a small probability; they can occur only for large thermal fluctuations. One has a collection of weakly coupled nonlinear oscillators, randomly displaced to $u \approx \pm u_0$. The system is disordered: a transition to this regime occurs when condition (9.59) is satisfied. At each site, the particles vibrate in one of the minima of the potential well, and one can set

$$u_n = u_0 \sigma_n + \delta_n, \tag{9.60}$$

where $\sigma_n = \pm 1$ represents the *pseudospin* of the nth particle, and δ_n is its displacement near the position u_0, such as $\delta_n \ll u_n$. Substituting (9.60) in (9.57), to a first approximation we obtain

$$H \approx NGu_0{}^2 + \sum_n^N 4\frac{E_0}{u_0{}^2}\,\delta_n{}^2 - J\sum_n^N \sigma_n\sigma_{n+1}, \tag{9.61}$$

where $J = Gu_0{}^2$. In Hamiltonian (9.61) the second term represents the potential energy of independent harmonic oscillators, and the third term represents the energy of an *Ising model* with coupling constant J. Note that all the nonlinear degrees of freedom are included in this last term. In the extreme limit the coupling between the displacements can be neglected and equations of motion (9.58) become independent, they describe the motion of N uncoupled nonlinear oscillators.

9.7.2 The displacive transition

If the coupling between nearest-neighboring sites is large, the displacement u_n does not change rapidly and the standard continuum appoximation can be used to reduce (9.57) and (9.58) to

$$H = \frac{m}{a} \int_{-\infty}^{+\infty} [\frac{1}{2}(\frac{\partial u}{\partial t})^2 + \frac{1}{2}c_0{}^2(\frac{\partial u}{\partial x})^2 - \frac{\omega_0{}^2}{2}(u^2 - \frac{1}{2}\frac{u^4}{u_0{}^2})]\ dx, \tag{9.62}$$

$$u_{tt} - c_0{}^2 u_{xx} = \omega_0{}^2(u - u^3/u_0{}^2). \tag{9.63}$$

We first examine the small-amplitude limit when the particle oscillates near one of the two equilibrium positions. We put

$$u(x, t) = \pm u_0 + \varepsilon u_1, \tag{9.64}$$

242

where u_l is the weak dimensionless displacement or phonon variable and $\varepsilon \ll 1$. Substituting (9.64) in (9.63), to first order in ε we find a linear dispersive equation:

$$u_{ltt} - c_o^2 u_{lxx} = -2\omega_o^2 u_l. \tag{9.65}$$

Equation (9.65) admits solutions with amplitude u_m, frequency ω and wavevector k, of the form

$$u = u_m \sin(\omega t - kx), \tag{9.66}$$

with the linear phonon dispersion relation

$$\omega^2 = 2\omega_o^2 + c_o^2 k^2. \tag{9.67}$$

Next we consider the large-amplitude solutions with permanent profiles propagating at velocity v, the kink and antikink solutions, respectively. In the original physical variables these kink (+sign) and antikink (-sign) have the form (Appendix 9B)

$$u = \pm u_o \tanh\left[\frac{\gamma(x - vt)}{\sqrt{2}d}\right], \tag{9.68}$$

where

$$v = cc_o, \qquad \gamma = (1 - v^2/c_o^2)^{-1/2}, \qquad d = c_o/\omega_o. \tag{9.69}$$

This kink-soliton solution (9.68) describes a transition from one minimum of the double-well potential - u_o at $x \to -\infty$ to the other one + u_o at $x \to -\infty$, as shown in Fig.9.11. In other words, the structural or physical character of the system is changed by the passage of the kink-soliton which represents two regions ($u=u_o$ and $u=-u_o$) and the *domain wall* between them. This one-dimensional kink or domain wall through which the transition takes place moves with velocity v and has a thickness proportional to $d\sqrt{2}/\gamma$. The kink energy E_k is concentrated in this region. Proceeding as in Chap.6 we can calculate this energy by substituting (9.68) in (9.62) and integrating. We find that

$$E_k = \frac{2}{3}\frac{m}{a}\gamma \omega_o c_o u_o^2 = m_o c_o^2 \gamma, \tag{9.70}$$

where the kink mass m_o is given by

$$m_o = \frac{2}{3}\frac{m}{a}\frac{u_o^2}{d}. \tag{9.71}$$

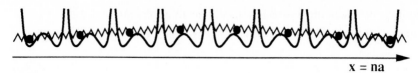

$$x = na$$

Fig 9.11. Sketch of a transition described by a kink or domain wall of the form (9.68).

In contrast to the periodic substrate potential considered in Sect.9.6 the present potential presents only two minima and one cannot get a sequence of successive kinks but rather an alternate sequence of kinks and antikinks. Moreover, owing to this topology of the potential it is worth noting that a kink and an antikink cannot pass (as for the Sine–Gordon system) through each other when propagating: they are not solitons in the strict sense. Nevertheless, it can be shown that these localized waves are stable and transparent to collisions with weak extended excitations such as phonons. Krumhansl and Schrieffer (1975) have studied the thermodynamic properties of this system. They have shown that at low temperature the thermodynamic behavior agrees with that found from a phenomenological model which considers a thermal mixture of the two kinds of excitation: phonons (extended) and domain walls, that is, localized kink solitons. At low temperature, these domain walls can be regarded as weakly interacting elementary excitations (see Bishop and Schneider 1978) which are distributed at random in low concentration along the one-dimensional system.

9.8 Kink-soliton solutions for generalized on-site potentials

A number of studies have been made for systems obeying the Sine-Gordon equation (9.51), for which the potential (9.48) is sinusoidal and also for systems modeled by (9.63), where the potential (9.56), with A<0, often called the "ϕ-four potential" (with $\phi \equiv u$), has a double well. For these systems a given particle can migrate from minimum to minimum rather than being confined to its well. Equations (9.51) and (9.63) are of the form

$$\frac{\partial^2 \phi}{\partial t^2} - c_0^2 \frac{\partial^2 \phi}{\partial x^2} = - \omega_0^2 \frac{dV(\phi)}{d\phi} . \tag{9.72}$$

Using the transformation (6.9) we reduce (9.72) to the normalized form

$$\frac{\partial^2 \phi}{\partial T^2} - \frac{\partial^2 \phi}{\partial X^2} = - \frac{dV(\phi)}{d\phi} . \tag{9.73}$$

This wave equation corresponds to a conservative system (with no energy dissipation) whose total energy (see Sect.6.3) or Hamiltonian is

244

$$H = E = \int_{-\infty}^{+\infty} \mathscr{H}(\phi) \, dX, \tag{9.74}$$

where the energy density (Hamiltonian density) is given by

$$\mathscr{H}(\phi) = A\omega_o c_o [\frac{1}{2}(\frac{\partial \phi}{\partial T})^2 + \frac{1}{2}(\frac{\partial \phi}{\partial X})^2 + V(\phi)] \tag{9.75}$$

Here, A is a constant which depends of the physical parameters (see 6.22 and 9B.2) and the quantity $A\omega_o c_o$ sets the energy scale.

In addition to the Sine-Gordon (sinusoidal) and ϕ-four potentials considered in Sects. 9.7 and 9.6, there are some other useful potentials that are listed hereafter.
(i) *The Double Sine-Gordon* (Condat et al. 1983) potential (DSG) :

$$V(\phi) = (\cos\phi + r \cos \frac{\phi}{2}), \tag{9.76a}$$

a periodic potential with multiple wells, where r is a parameter ($0 \le r < 1$).
(ii) *The deformable periodic* potential (Remoissenet and Peyrard 1981, Peyrard and Remoissenet 1982) :

$$V(\phi,r) = (1-r)^2 \frac{1-\cos\phi}{1+r^2+2r\cos\phi}, \tag{9.76b}$$

whose shape can be varied continuously as a function of the parameter r. As this parameter varies in the range -1<r<1, the amplitude of the potential remains constant while its shape changes from sharp wells separated by flat tops to wells with flat bottoms separated by very sharp peaks. For r=0, this potential reduces to the SG potential, for small values of r it reduces to the DSG potential considered above.
(iii) *The double-quadratic* potential

$$V(\phi) = \frac{1}{2}(|\phi|^2 - 1), \tag{9.76c}$$

is a double well potential .
(iv) *The assymmetric double-well* potential (Konwent et al. 1996) :

$$V(\phi) = \frac{1}{2}\{A [\cosh(r\phi) - 1)]^2 + B \sinh(r\phi)\}, \tag{9.76d}$$

which becomes symmetric for B=0.
(v)*The parametric double-well potential* (Dikande and Kofane 1991) :

$$V(\phi) = \frac{a}{8} \left[\frac{1}{\mu^2} \sinh^2(\mu\phi) - 1 \right]^2 \qquad (9.76e)$$

which reduces to the ϕ-four potential for small values of ϕ : $V(\phi, \mu \rightarrow 0) = (a/8)$ $(\phi^2-1)^2$.

For these potentials, localized waves of permanent profile, that is, kink-solitons, can be calculated (see Appendix 6A and 9B). The boundary conditions are :

$$\phi\ (s=\infty) = \phi_1 \text{ and } \phi\ (s= -\infty) = \phi_2, \ d\phi/ds\ (s=\pm\infty) = 0, \ V[\phi\ (s=\pm\infty)\ ]= 0.$$

From (9.73) we get

$$\frac{d\phi}{ds} = \sqrt{\frac{2}{1-u^2}} \sqrt{V(\phi)}, \qquad (9.77)$$

and

$$s = \pm \sqrt{\frac{1-u^2}{2}} \int_{\phi_1}^{\phi_2} \frac{d\phi}{\sqrt{V(\phi)}} ,$$

Inserting the expression for $d\phi/ds$ in (9.75) we obtain

$$\mathcal{H}\ (\phi) = A\omega_o c_o \frac{2V(\phi)}{1-u^2} . \qquad (9.78)$$

From this equation and $d\phi/ds$ (= $d\phi/dX$) given by (9.77) we transform (9.74) into

$$E = \frac{A\sqrt{2}}{\sqrt{1-u^2}} \int_{\phi_1}^{\phi_2} \sqrt{V(\phi)}\ d\phi. \qquad (9.79)$$

Then taking account of the transformation (6.9) we get

$$E = \frac{A\sqrt{2}}{d} \frac{c_o^2}{\sqrt{1-\frac{v^2}{c_o^2}}} \int_{\phi_1}^{\phi_2} \sqrt{V(\phi)}d\phi, \qquad (9.80)$$

where $d=c_0/\omega_0$, is defined in (6.4b). The total energy E can be expressed in the relativitic form (see (6.24))

$$E = \frac{m_0 c_0^2}{\sqrt{1 - \dfrac{v^2}{c_0^2}}} ,$$

(9.81)

where the mass of the kink-soliton given by

$$m_0 = \frac{A\sqrt{2}}{d} \int_{\phi_1}^{\phi_2} \sqrt{V(\phi)}\,d\phi.$$

(9.82)

depends on the shape of the on-site potential.

9.9 A lattice model with an exact kink-soliton solution

In many physical problems the width of the kink-soliton or domain wall is only of a few lattice spacings, that is, the discretness parameter d is of the order of the lattice spacing a (see Sect. 6.10), so that lattice effects become important. The continuum limit breaks down. In order to gain understanding of kink-soliton motions in discrete systems it is desirable to investigate lattice models with exact solutions. In this regard, it was pointed out (Schmidt 1979) that if the ϕ-four potential (9.56) is suitably modified the single soliton (9.68) becomes an exact solution to the discrete model (9.58).

In the following, generalizing Schmidt's approach we show (Remoissenet 1998) that this double-well model, with additional external force and dissipation terms, can also admit an exact solution. We consider the general discrete equation of motion

$$m \frac{d^2 u_n}{dt^2} + \Gamma \frac{du_n}{dt} = - \frac{dU(u_n)}{du_n} - F + K\,(u_{n+1} + u_{n-1} - 2u_n),$$

(9.83)

where $\Gamma du_n/dt$ is a dissipative term (see Sect 6.8) and F a constant term which may be an external force. The energy gain due to F is compensated by the loss mechanism. We assume that a soliton solution of (9.83) has a form similar to (9.68) :

$$u_n(t)= u_0 \tanh\,(wt-kna),$$

(9.84)

where u_0 is the amplitude, a the lattice spacing and w and k are two constants ; one notes that $w/k=v$ the velocity of the kink-soliton. The procedure is to construct a total potential, including the external potential, which has the form

$$V(u_n) = U(u_n) + Fu_n = Fu_n - \frac{1}{2} Au_n{}^2 + \frac{1}{3} B_ou_n{}^3 + \frac{1}{4} Bu_n{}^4$$

$$+ \frac{1}{5} C_ou_n{}^5 + \frac{1}{6} Cu_n{}^6 + \frac{1}{7} D_ou_n{}^7 + \frac{1}{8} Du_n{}^8 + ..., \quad (9.85)$$

such that (9.84) *becomes an exact solution even in the discrete case.* Dividing (9.85) by $Ku_o{}^2$ and introducing the dimensionless variable $\phi_n = u_n/u_o$, after some calculations (see Appendix 9C) one finds

$$\frac{m}{K} \frac{d^2\phi_n}{dt^2} + \frac{\Gamma}{K} \frac{d\phi_n}{dt} = - \frac{dV(\phi_n)}{d\phi_n} + (\phi_{n+1} + \phi_{n-1} - 2\phi_n), \quad (9.86)$$

where the potential is given by

$$V(\phi) = - \frac{w\Gamma}{K} \phi + (\frac{mw^2}{K} - 1) \phi^2 + \frac{w\Gamma}{3K} \phi^3 - \frac{mw^2}{2K} \phi^4 + (1 - \frac{1}{\alpha}) \, Ln \, (1 - \alpha\phi^2),$$

$$(9.87)$$

with $\alpha = \tanh^2(ka)$. Hence, the discrete system of equations (9.86) has the exact solution

$$\phi_n(t) = \tanh \, [k(vt - na)], \quad (9.88)$$

which propagates at velocity $v = w/k = aw/\tanh^{-1}(\sqrt{\alpha})$.

(i) When $\Gamma = 0$ (F=0: see (9C.5)) the coefficients of ϕ and ϕ^3 become zero, the symmetry breaking due to F disappears and the potential becomes symmetric. The system is conservative and the total energy is constant, to each (w,k) combination corresponds a different well shape and a velocity.

(ii) When $m = 0$, Eq (9.86) becomes purely diffusive (see Sect.11.4). The coefficient of ϕ^2 reduces to -1 and the coefficient of ϕ^4 becomes zero: the potential is assymmetric with two nondegnerate minima. The value of w is imposed (see (9C.5)) by the relation : $F = -wu_o\Gamma$. Consequently, the potential well shape and the velocity both depend uniquely on the choice of k, that is, α.

(iii) When $\Gamma \neq 0$ and $m \neq 0$ both inertial and dissipative effects are present. The same restrictions apply to the potential well shape and the velocity. This general case which remains to be explored will not be considered in the following.

Let us leave case (ii) for chapter (11) and discuss case (i). In this case, in the continuum limit (ka<<1, $\alpha = \tanh ka \cong ka$) the potential (9.87) can be expanded with respect to ka and reduces to a symmetric double-well potential of the form (9.56):

248

$$V_{cl}(\phi) = -\, A' \frac{\phi^2}{2} + B' \frac{\phi^4}{4} ,$$

where $A' = B' = 2(\alpha - \frac{mw^2}{K})$. The potential barrier height is $V(\phi{=}0) - V(\phi{=}1) = A'/4$.

In the discrete regime the potential has the form depicted in Fig. 9.12 for $\alpha = 0.5$. As α further increases, the kink-soliton solution (9.88) departs from the continuum form and the minima of the potential at $\phi = \pm 1$ become sharper, becoming abrupt in the limit $\alpha \to 1$. In this limit, the very narrow kink will consist of a single particle moving from one minimum to another in a given well. When α exceeds 1 the potential looses its double-well nature.

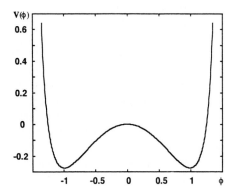

Fig. 9. 12 Representation of double-well potential $V(\phi)$ given by (9C.7) for : $mw^2/K = 0.04$ and $\alpha = \tanh(ka) = 0.5$.

In fact, this potential model introduced by Schmidt is interesting for exploring the behavior of discrete systems. Nevertheless, its limitation is that to each (w,k) combination corresponds a different well shape.

Remark

It has been shown (Jensen et al. 1983) that in the static limit, potential (9.87) admits periodic solutions or unpinned lattices of regularly spaced solitons.

9.10 Energy localization in nonlinear lattices

9.10.1 Self-trapped states: polaron and conformon

In previous chapters we have seen that the soliton concept plays a significant role in understanding the dynamical behavior of localized states in different kinds of systems in nature. Solitons provide rigourous examples of self-localized states in completely integrable systems (see Chap.10). In fact, the concept of nonlinear localization or self-trapping (see Sects. 4.4 and 8.8) which plays an important role in real systems (Eilbeck et al. 1985, Scott 1985, 1992) seems to emerge with the quasi-particle state or *polaron* (Pekar 1954) which had been first suggested by Landau (1933) in the context of condensed matter physics. In the polaron case an electron moves through a crystal lattice as a localized wave function rather than an extended Bloch function. Since the electron is localized, it polarizes the crystal in its vicinity thereby lowering its energy, which keeps it localized as sketched in Fig. 9.13. Other examples of self-trapping phenomena can be found when a hole, a defect or a domain wall propagating through a lattice interacts with the deformation or reaction field (phonons) which in turn, modifies the dynamics of the particle. In many cases, one can think the particle being "dressed" by its self-created polarization field (potential well created self-consistently: see Sect 4.4), that is, one has a quasi-particle. When the particle-field interaction (electron-phonon coupling) is strong, both the particle wave function and lattice deformation are localized, leading to a localized entity known as a *small polaron*. (Holtstein 1959). Like the soliton, the polaron is an excitation (see Alexandrov and Mott 1998) that maintains integrity owing to the dynamical balance between the dispersion (exchange inter-site interaction) and the nonlinearity (particle-phonons coupling).

Fig. 9.13. Rough sketch of a polaron : a particle with a given wave function deforms the lattice inducing a potential well which in turn traps the particle

Another example of self-trapped state is the *conformon* (Volkenstein 1972; Green and Ji 1972; Kemeny and Goklany 1973). Since it was defined in different ways by these different authors, here we just mention the definition of Volkenstein who introduced the conformon in order to model the electronic-conformational interaction in biopolymers. It consists in the following. The displacement of an electron or of the electronic density in a macromolecule produces the deformation of the lattice, i. e. the conformational change; thus it looks like quite similar to a polaron. Nevertheless, the conformon differs from the polaron because of the lack of periodicity and homogeneity in the macromolecule. Therefore, it cannot be considered as a real quasi-particle able to move for long distances.

250

9.10.2 Intrinsic localized modes or discrete breathers

The self-trapping phenomenon in discrete systems such as optical waveguide arrays has been briefly discussed in Sect. 8.8. Here, we consider *intrinsic localized modes* or *discrete breathers* in lattices. The discovery of such modes and their active study is an important step towards understanding the role played by nonlinear spatial nonlinear localization in realistic, usually nonintegrable and discrete, physical models, many years after the blossom of the theory of solitons in continuous models described by integrable nonlinear equations (see previous chapters and Chap 10).

Intrinsic collapse to self-localized states is increasingly studied because of its wide significance in a great variety of physical systems. In lattices, intrinsic localized modes with internal (envelope) oscillations, mentioned in Sect. 9.4, are often called *discrete breathers*. (for recent reviews see: Aubry 1997, Flach and Willis 1998). In fact, an envelope mode or nonlinear wavepacket can be considered as the small amplitude limit of a breather-soliton as shown in Sect.6.9 for the continuum Sine-Gordon system and also for other systems (Remoissenet 1986) in the semi-discrete appproximation, where the oscillations (discrete carrier wave) vary rapidly inside the slow (continuous) envelope. In the framework of this approximation, it is now well understood that discrete breathers can be linked to envelope solitons (Yoshimura and Watanabe 1991; Kivshar et al 1998) that are described in Sect. 9.4. Contrary to continuous breathers which are known to exist only in some particular systems, discrete breathers have been shown to be structurally stable as soon as nonlinear oscillators are coupled sufficiently weakly and locally.

Historically, the possibility of existence of localized long-lived molecular vibrational states in simple molecular crystals was pointed out by Ovchinnikov (1970). Then, the existence of discrete breathers was explored by Kosevich and Kovalev (1974) for a lattice with on-site potential (see Sect. 6.11.1) and linear inter-particle coupling. Later, Dolgov (1986) proposed a model for self-localization of vibrations in one-dimensional lattice with nonlinear inter-particle coupling without on-site potential. In 1988 Sievers and Takeno investigated the same kind of lattice. They proposed that large amplitude vibrations in perfectly periodic 1D lattices can localize because of nonlinearity and discreteness, and clearly suggested that intrinsic localized modes or discrete breathers should be quite general and robust solutions, that exist in many models. Contrary to the case of purely harmonic lattices where spatially localized modes can occur only when defects or disorder are present (see Fig.9.14a), so that the translational invariance of the underlying lattice is broken, discrete breathers may be created anywhere in a perfect homogenous nonlinear lattice (see Fig. 9.14b). In this case one has an *odd parity mode* (Sievers and Takeno 1988) that is centered at a lattice site: the corresponding longitudinal displacement pattern is sketched in Fig 9.15a. One may also have an even parity mode (Page 1990; Kiselev 1990) with inversion symmetry with respect to the midpoint between lattice sites, as represented in Fig. 9.15b.

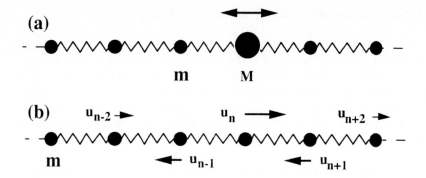

(a)

m   M

(b) u_{n-2} → u_n → u_{n+2} →

m   ← u_{n-1} ← u_{n+1}

Fig.9.14a,b. (a) Schematic reprentation of a mass defect induced localized mode, which oscillates mainly in the vicitnity of the mass M, in a chain where particles of mass m are linearly coupled.
(b) Schematic representation of an *intrinsic nonlinear localized mode* or *discrete breather* where five atoms are oscillating with an amplitude which decreases spatially from the center (particle with longitudinal displacement : u_n) of the mode. The neighboring particles vibrate out-of-phase with respect to each other

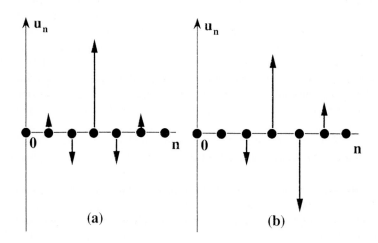

(a)   (b)

Fig. 9.15a,b. (a) Schematic representation of the longitudinal displacement pattern correspon-ding to the odd parity mode of Fig. 9.14b. **(b)** Displacement pattern of an even parity mode

The basic properties (Kiselev et al. 1995) of the discrete breather or intrinsic localized mode can be summarized as follows.
(i) *It is a time periodic solution localized in space : in one, two or three dimensions*
(ii) *It may be centered on any lattice site giving rise to a configurational entropy .*
(iii) *It extend over a few lattice sites .*
(iv) *It has an amplitude dependent frequency (compare to Sect. 9.4) .*
(v) *It may be mobile.*

The first rigourous proof for the existence of intrinsic localized modes or discrete breathers in a wide class of lattice models was given by Mac Kay and Aubry (1994) using the approach of *antiintegrable or anticontinuous limit* (Aubry and Abramovici 1990). Specifically, this proof was obtained by first considering a limit where the lattice model reduces to a discrete array of uncoupled nonlinear oscillators. Thus, breathers corresponding to one oscillator vibrating freely trivially exist and can be continued up to nonzero values of the coupling. The conditions of existence and stability of these breathers have been studied (Flach 1994, 1995; Aubry 1997) , they may also exist in sytems with spatial disorder (Sepulchre and Mac Kay 1998). A particular important condition of breather existence is that all multiples of the fundamental frequency lie outside of the linear excitations spectrum. Small discrete breathers can emerge from the modulational instabitlity of a continuous wave in the lattice (Peyrard 1998). Their interactions lead then to the growth of large amplitude breathers. The creation and possible annihilation process of discrete breathers has been explored (Rasmussen et al. 1998), such an approach should represent a step towards understanding the role played by these localized modes in non-equilibrium properties of real systems.

Contrary to solitons, discrete breathers do not require integrability for their existence and stability. They are excitations that can exist in two or three dimensions and should exist in real systems. In this regard, nonlinear features associated with intrinsic localized spin wave modes in one-dimensional periodic ferromagnets and antiferromagnets have been explored (for a review see: Lai and Sievers 1998).

Breather solutions have been observed numerically; until now no analytical solution is known. Moreover, these nonlinear excitations have been observed in analog systems (see Sect. 9.11) but they have not yet been observed experimentally in real systems.

9.11 Observation of discrete breathers

In this section we present observations of localized modes that can be simply performed with the mechanical chains described in Sects. 6.10 and 6.11.5, in the regime where the coupling between neighboring units is weak. We also present a chain with magnetic coupling.

9.11.1 Discrete pendulums chain

We consider the chain of pendulums with weak coupling (d<<a) described in Sect. 6.10.2 which corresponds to a discrete Sine-Gordon chain. One can create a discrete breather by an initial perturbation ($\theta \approx \pi$ or less) of one pendulum. As represented in Fig. 9.16, several large amplitude oscillations of the central pendulum and small oscillations of its two neighbors can be observed. However, owing to dissipation effects, the lifetime of the discrete breather is finite.

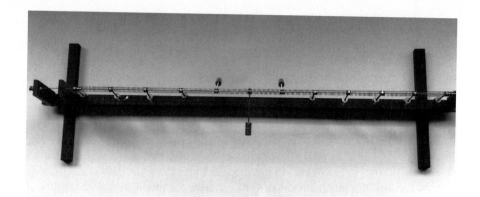

Fig. 9.16. A stationary *discrete breather* or *nonlinear localized mode*, produced by exciting a pendulum of the chain, which executes large amplitude oscillations ($-180°<\theta_{0,max}<180°$); the two neigbboring pendulums execute small amplitude oscillations ($-15°<\theta_{\pm1,max}<15°$) (Photo R. Chaux).

With this chain one can also easily observe (not represented here) modulational instability in a system which is discrete and finite. Each pendulum is launched from an initial position ($\theta_{in} \approx 70°$) and the lattice is let to evolve. The spatial inhomogeneities play the role of a spatial noise which perturbs the initial condition (wave with zero wavenumber). As a result some pendulums execute very large amplitude oscillations, much larger than the amplitude of the original perturbation, as predicted by the investigations (see Kivshar and Peyrard 1992). Some of these modes that emerge from the modulational instability phenomenon, are discrete breathers.

Remark

Modulational instabilities in discrete lattices have been studied analytically and numerically by Kivshar and Peyrard (1992). Reducing a discrete Klein-Gordon equation to a discrete Nonlinear Schrödinger equation they have shown that discreteness can drastically change the conditions for modulational instability discussed in Sect 4.6. Specifically, at small wave numbers a nonlinear carrier wave is unstable with respect to all possible modulations of its amplitude if the wave amplitude exceeds a certain treshold value. A direct investigation of the modulational instability phenomenon on the Klein-Gordon lattice was considered by Burlakov and coworkers (1996).

9.11.2 Mechanical chain with torsion and gravity

In Sect. 6.11 we have considered a mechanical chain, with torsion and gravity, whose dynamics can be described by (6.62) in the continuum approximation which is valid when the coupling is strong (stiff springs). Now, if we use the same chain with soft springs the dynamics of the chain is described by (6.F1) in appendix 6F,

254

that we rewrite here for convenience:

$$I\frac{d^2\theta_n}{dt^2} = -K\theta_n + mgd\sin\theta_n + M_{n-1,n} - M_{n,n+1} \tag{9.89}$$

where $M_{n-1,n}$, $M_{n,n+1}$ are the torque exerted by pendulum n-1 on pendulum n and pendulum n on pendulum n+1, respectively. In the full discrete regime these torques are given by the complicated relation (6.F6 in Appendix 6F). No approximation can be made and (6.F1) cannot be solved analytically. Nevertheless, with such a chain, whose springs can be rapidly replaced by other ones, we can observe and illustrate the properties of discrete breathers. For example, we can observe (see Fig 9.17) a discrete breather where the nth pendulum oscillates, over the potential barrier, with large amplitude ($\left|\theta_{n,\,max}\right| \le \pi/2$), and the two adjacent pendulums oscillate in one of the potential well.

With this chain, one can also vizualise discrete breathers in the presence of spatial disorder, as predicted by Sepulchre and Mac Kay (1998). Specifically, along the chain one can first create an arbitrary sequence of highly discrete kinks or antikinks, whose width is a few lattice spacings, that are completely pinned on the lattice and cannot propagate. Then, by exciting one pendulum, in a kink, one can observe a nonlinear localized mode oscillating on a disordered structure, as represented in Fig. 9.18. It is important to note that both phenomena of spatial disorder an spatial localization are *intrinsic* in the sense that no defect or impurity is present in the lattice. Such phenomena should lead to new effects when investigating the dynamics of condensed matter systems.

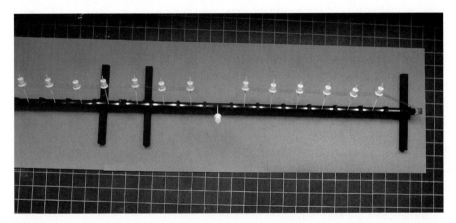

Fig. 9.17. A stationary discrete breather on the mechanical chain with torsion and gravity is produced by perturbing a pendulum of the chain. This pendulum executes large amplitude oscillations ($-70°\le\theta_0\le70°$) over the potential barrier and the two neigbboring pendulums execute small amplitude oscillations ($-15°\le\theta_{\pm1}\le15°$) around one equilibrium point (Photo R. Chaux)

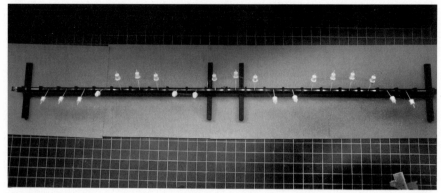

Fig. 9.18. A disordered structure containing an arbitrary number of discrete kinks and antikinks. Starting from the left, one can observe the 4rth pendulum and also the 12th one which oscillate inside a kink (Photo R. Chaux)

9.11.3 A chain of magnetic pendulums

Interest in the possibility that moving breathers play a role in the transfer of energy from atoms or ions moving at relatively high speed to atoms in a solid, with which they collide, arose from the discovery of tracks in doped muscovite mica crystals (Russel and Collins 1995). It was found that the tracks were associated with particular crystalline directions in the lattice. To investigate the dynamical behavior in these preferred directions an analog model of coupled magnetic pendulums, with similar nonlinearity of force as in the muscovite system, was constructed (Russel et al. 1997, see also Zolotaryuk 1998).

Following Russell, a magnetic lattice, represented in Fig. 9.19, can be constructed. It consists of 50 dipole magnets, freely suspended by rigid struts from pivots spaced at equal intervals to form a linear chain. In this model, if one neglects the dissipation effects, the total energy, or Hamiltonian, is the sum of the kinetic energy, the potential energy owing to gravitation and the inter-site potential energy owing to (nonlinear) dipole-dipole interaction of adjacent magnets. The corresponding equations of motion contain an on-site gravitational force term and a nonlinear interaction force term (for details see Russel et al. 1997; Zolotaryuk 1998), they cannot be solved analytically. However, numerical simulations show that discrete breathers can be generated and can propagate along the chain.

Such predictions are confirmed by experiments on the magnetic chain. If the initial disturbance of the first pendulum has a large amplitude, it rapidly evolves into a discrete breather which can travel along the chain , as represented in Fig. 9.19. More experiments can be performed with this analog model demonstrating: the creation of oppositely directed pairs of breathers and the creation of stationary breathers, the survival of oppositely propagating breathers after their mutual interactions, their reflection from boundaries and discontinuities.

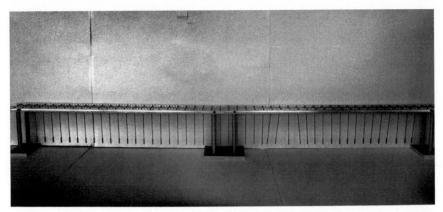

Fig. 9.19. A discrete breather, with spatial extension D= 3a, traveling along the magnetic chain The magnetic chain consists of two units of 25 dipole magnets, each with length L=75cm. In the present case, the lattice spacing is a= 3x1cm nevertheless it can be easily modified (Photo R. Chaux)

Appendix 9A. Solutions for transverse displacements

We look for solutions with permanent profiles propagating with velocity v, that is, as previously, we assume that $V(x -vt) =V(s)$. After two integrations with respect to s, (9.46) becomes

$$(v^2- C_T^2)V = \frac{C_T^2 a^2}{12} V_{ss} + \frac{C_L^2 - C_T^2}{2} [|V|^2 V],$$ (9A.1)

where the two integration constants have been chosen to be zero. The complex quantity V can be written in the form

$$V = A\, e^{i\phi}.$$ (9A.2)

Replacing (9A.2) in (9A.1) and separating the real and imaginary parts, we obtain

$$p (A_{ss} - A\phi_s^2) - g A + q |A|^2 A = 0,$$ (9A.3a)

$$2A_s\phi_s + A\phi_{ss} = 0,$$ (9A.3b)

where

$$p = C_T^2 a^2/12 , \quad q = (C_L^2 - C_T^2)/2 , \quad g = (v^2- C_T^2).$$

Equation (9A.3b) can be simply integrated to give

$$A^2\, \phi_s = K,$$ (9A.4)

where K is an integration constant. Inserting (9A.4) into (9A.3) leads to

257

$$A_{ss} = -\alpha A^3 + \beta A + \frac{K^2}{A^3},$$

(9A.5a)

where the coefficients

$$\alpha = 6 \frac{C_L{}^2 - C_T{}^2}{C_T{}^2 a^2}, \qquad \beta = 12 \frac{v^2 - C_T{}^2}{C_T{}^2 a^2}$$

(9A.5b)

are positive. Equation (9A.5a) can be integrated to give

$$(A_s)^2 = -\frac{1}{2}\alpha A^4 + \beta A^2 - \frac{K^2}{A^2} + K',$$

(9A.6)

where K' is an integration constant. We remark that (9A.6) is identical to (4C.9). The first interesting case is for K = K' = 0. In this case, we will have localized pulse solutions. Proceeding exactly as in Appendix 4C we obtain

$$A^2 = (2\beta/\alpha)\ \text{sech}^2\ (\sqrt{\beta}\ s).$$

Using (9A.5b), from this equation we find that

$$A = \pm 2 \sqrt{\frac{v^2 - C_T{}^2}{C_L{}^2 - C_T{}^2}}\ \text{sech}\ (\frac{x - vt}{L}),$$

(9A.7a)

where

$$\frac{1}{L} = \frac{2\sqrt{3}}{a} \sqrt{\frac{v^2 - C_T{}^2}{C_T{}^2}}.$$

(9A.7b)

For K = 0, (9A.4) yields $\phi = \phi_0 = $ Cte. This means that V maintains a constant direction in space; in other words we have a transversely polarized wave. Remembering that $V = H_x$ and using (9.42), from (9A.7a) we find the two components

$$y\,(x,t) = \pm \frac{a}{\sqrt{3}} \frac{C_T}{\sqrt{C_L{}^2 - C_T{}^2}} \tan^{-1}\ [\ \sinh \frac{x-vt}{L}\ ]\ \cos\phi_0,$$

(9A.8a)

and

$$z(x,t) = \pm \frac{a}{\sqrt{3}} \frac{C_T}{\sqrt{C_L{}^2 - C_T{}^2}} \tan^{-1}\ [\ \sinh \frac{x-vt}{L}\ ]\ \sin\phi_0,$$

(9A.8b)

of the transverse displacement, which corresponds to a kink-shaped solitary wave with a transverse polarization in a direction given by ϕ_0. For K≠0 and K'≠0, (9A.6) can be solved to give periodic solutions in terms of elliptic functions. This case will not be discussed here.

258

Appendix 9B. Kink soliton or domain-wall solutions

As in Sect.6.2 we introduce the dimensionless variables

$$X = (\omega_0/c_0)\, x, \qquad T = \omega_0 t, \qquad u = u_0 F, \tag{9B.1}$$

and transform (9.62) and (9.63) into

$$H = \frac{m}{a}\, \omega_0 c_0 u_0{}^2 \int_{-\infty}^{+\infty} [\frac{1}{2}(\frac{\partial F}{\partial T})^2 + \frac{1}{2}(\frac{\partial F}{\partial X})^2 + (\frac{F^4}{4} - \frac{F^2}{2})]\, dX, \tag{9B.2}$$

$$F_{TT} - F_{XX} = F - F^3. \tag{9B.3}$$

Then, proceeding as in many other places in this book, we look for waves with permanent profiles, traveling at velocity c, of the form

$$u(s) = u\,(X - cT).$$

Thus, we reduce (9B.3) to an ordinary differential equation:

$$(1 - c^2)\,\frac{d^2 F}{ds^2} = F^3 - F.$$

Integrating this equation yields

$$(1 - c^2)\,(\frac{dF}{ds})^2 = \frac{F^4}{2} - F^2 + C, \tag{9B.4}$$

where C is an arbitrary integration constant. If we look for localized solutions such as $F \rightarrow \pm 1$ (that is $u \rightarrow \pm u_0$) and $dF/ds \rightarrow 0$ when $s \rightarrow \pm \infty$, we get $C = 1/2$. Substituting in (9B.4) and integrating yields the simple solution

$$F = \pm\tanh\,[\frac{s}{\sqrt{2(1-c^2)}}]. \tag{9B.5}$$

Here, (as in Sect.6.2) the ($\pm$) signs correspond to the kink and antikink solutions, respectively. In the original physical variables, (9B.5) is rewritten as

$$u = \pm\, u_0 \tanh\,[\frac{\gamma(x - vt)}{\sqrt{2}d}], \tag{9B.6}$$

where

$$v = c c_0, \qquad \gamma = (1 - v^2/c_0{}^2)^{-1/2}, \qquad d = c_0/\omega_0. \tag{9B.7}$$

259

Appendix 9C. Construction of a double-well potential

Differentiating (9.85) with respect to u_n leads to

$$\frac{d}{du_n}[U(u_n) + Fu_n] = F - Au_n + B_o u_n^2 + Bu_n^3 + C_o u_n^4 + Cu_n^5 \qquad (9C.1)$$
$$+ D_o u_n^6 + Du_n^7 + ...,$$

From (9.84) and the first two terms of equation of motion (9.83) one obtains

$$m\frac{d^2 u_n}{dt^2} + \Gamma\frac{du_n}{dt} = 2mw^2 u_o T_n (T_n^2 - 1) + \Gamma u_o w (1 - T_n^2). \qquad (9C.2)$$

The last term of (9.83) yields

$$K(u_{n+1} + u_{n-1} - 2u_n) = Ku_o\{\tanh [wt-k(n+1)a] - \tanh [wt-k(n-1)a]$$

$$- 2\tanh [wt-kna]\}$$

$$= 2Ku_o T_n (1-\alpha)(1 + \alpha T_n^2 + \alpha^2 T_n^4 + \alpha^4 T_n^6 + ...) - 2Ku_o T_n,$$

where we have set $T_n = \tanh(wt-kna)$ and $\alpha = \tanh^2(ka)$. With these substitutions, equation of motion (9.83), upon dividing by $u_o K$ and dropping the subscript n, becomes

$$-2\frac{m}{K}w^2 T (1-T^2) + \frac{\Gamma}{K} w (1-T^2) - 2T(1-\alpha)(1 + \alpha T^2 + \alpha^2 T^4 + \alpha^4 T^6 + ...) + 2T$$

$$+ \frac{1}{K}[\frac{F}{u_o} - AT + B_o u_o T^2 + Bu_o^2 T^3 + C_o u_o^3 T^4$$

$$+ Cu_o^4 T^5 + D_o u_o^5 T^6 + Du_o^6 T^7 + ...] = 0. \qquad (9C.3)$$

Writing this equation as a power series in T, we get

$$\frac{1}{K} (\Gamma w + \frac{F}{u_o}) + (2\alpha - \frac{2m}{K}w^2 - \frac{A}{K})T + \frac{1}{K}(B_o u_o - \Gamma w)T^2$$

$$+ [\frac{1}{K} (2mw^2 + Bu_o^2) - 2(1-\alpha)\alpha]T^3 + \frac{C_o}{K}u_o^3 T^4 + [\frac{C}{K}u_o^4 - 2(1-\alpha)\alpha^2] T^5$$

$$+ \frac{D_o}{K} u_o^5 T^6 + [\frac{D}{K} u_o^6 - 2(1-\alpha)\alpha^4]T^7 + ... = 0. \qquad (9C.4)$$

Eq (9C.4) is satisfied for all t and n only if each coefficient of the series vanishes, yielding

260

$$F = -wu_0\Gamma, \qquad A = 2K\alpha - 2mw^2, \qquad B_0 = \frac{\Gamma w}{u_0}, \qquad B = \frac{2K}{u_0^2}(1-\alpha)\alpha - 2mw^2,$$

$$C_0 = 0, \qquad C = \frac{2K}{u_0^4}(1-\alpha)\alpha^2 \qquad D_0 = 0, \qquad D = \frac{2K}{u_0^6}(1-\alpha)\alpha^4.$$

$$(9C.5)$$

Let us divide the potential expression (9.85) by Ku_0^2 and introduce the dimensionless variable $\phi_n = u_n/u_0$. We obtain

$$V(\phi) = \frac{F}{Ku_0}\phi - \frac{1}{K}\left(\frac{1}{2}A\phi^2 + \frac{1}{3}B_0u_0\phi^3 + \frac{1}{4}Bu_0^2\phi^4 + \right.$$

$$\left. \frac{1}{5}C_0u_0^3\phi^5 + \frac{1}{6}Cu_0^4\phi^6 + \frac{1}{7}D_0u_0^5\phi^7 + \frac{1}{8}Du_0^6\phi^8 + ... \right), \qquad (9C.6a)$$

where the index n has been dropped. Then, replacing the coefficients by expressions (9C.5) gives

$$V(\phi) = \frac{-w\Gamma}{K}\phi + \left(\frac{mw^2}{K} - \alpha\right)\phi^2 + \frac{w\Gamma}{3K}\phi^3 - \frac{mw^2}{2K}\phi^4$$

$$+ \frac{1}{2}(\alpha - \alpha^2)\phi^4 + \frac{1}{3}(\alpha^2 - \alpha^4)\phi^6 + \frac{1}{4}(\alpha^4 - \alpha^6)\phi^8 + ..., \qquad (9C.6b)$$

This expression can be rearranged to give

$$V(\phi) = -\frac{w\Gamma}{K}\phi + \left(\frac{mw^2}{K} - 1\right)\phi^2 + \frac{w\Gamma}{3K}\phi^3 - \frac{mw^2}{2K}\phi^4$$

$$- \left(1 - \frac{1}{\alpha}\right)\left(\alpha\phi^2 + \frac{1}{2}\alpha^2\phi^4 + \frac{1}{3}\alpha^4\phi^6 + \frac{1}{4}\alpha^6\phi^8 + ...\right). \qquad (9C.6c)$$

Finally, the potential takes the form

$$V(\phi) = -a_0\phi + b\phi^2 + c\phi^3 - d\phi^4 + \left(1 - \frac{1}{\alpha}\right)\text{Ln}\,(1 - \alpha\phi^2), \qquad (9C.7)$$

with

$$a_0 = \frac{w\Gamma}{K}, \qquad b = \left(\frac{mw^2}{K} - 1\right), \qquad c = \frac{w\Gamma}{3K}, \qquad d = \frac{mw^2}{2K}. \qquad (9C.8)$$

10 A Look at Some Remarkable Mathematical Techniques

The nonlinear equations that we have encountered in the previous chapters can be solved by using mathematical techniques such as the powerful inverse scattering transform (IST) (Gardner et al. 1967) and the remarkable Hirota method (Hirota 1971). Specifically, in addition to the one-soliton solutions, explicit multisoliton solutions representing the interaction of any number of solitons can be constructed. Moreover, in several cases a precise prediction, closely related to experiments, can be made by the IST of the nonlinear response of the physical system, that is, of the number of solitons that can emerge from a finite initial disturbance (Zakharov, 1980. Ablowitz and Segur 1981; Calogero and Degasperis 1981; Newell 1985).

In this chapter, rather than drown the reader in extensive calculations, we give the main steps which are necessary to understand how the inverse spectral problem can be treated (Ablowitz and Clarkson 1992). This method was first used for the KdV equation and now solves a great variety of nonlinear-evolution equations. In the following we will emphasize the ingredients which are useful for the physicist and then point the reader to the relevant papers in the literature. Then we will illustrate the Hirota method, which is useful for the physicist, with a few examples.

10.1 Lax equations and the inverse scattering transform method

The basic idea of the inverse scattering transform (IST) is the same that is used in any transform method: one defines a transformation of the original problem in a space of functions where the time dependence is particularly simple (Fig.10.1). Then, after having determined the transformed data at a later time, the inverse transform is applied to find the solution. Each step in this method is linear and the whole procedure can be thought of as a generalization to certain nonlinear problems of the Fourier-transform technique for solving linear problems. In fact, at time $t = 0$ the initial data is mapped into scattering (Fourier) space by postulating an associated linear eigenvalue problem (by taking the Fourier transform). The direct scattering problem $\mathcal{S}$ (Fourier transform $\mathcal{F}$) thus gives the initial data at time $t = 0$. The evolution of the scattering data is now determined by a set of linear equations. The solution of the original evolution equation is reconstructed by mapping back into physical space via the inverse problem $\mathcal{S}^{-1}$ (inverse Fourier transform $\mathcal{F}^{-1}$).

10.1.1 The Fourier-transform method for linear equations

Let us first briefly review the Fourier-transform method for linear evolution equations. The nonlinear KdV equation expressed in dimensionless variables has the form (5C.19); it is rewritten as

$$\phi_t - 6\phi\phi_x + \phi_{xxx} = 0, \tag{10.1}$$

where, for convenience, we have made the substitutions $\tau \to t$ and $\xi \to x$, and have chosen the nonlinear coefficient to be -6.

Linearization of (10.1) yields the linear evolution equation

$$\phi_t + \phi_{xxx} = 0.$$

This equation can be solved by the Fourier transform method. In this approach one first maps the initial data $\phi(x, 0)$ onto its Fourier transform

$$\phi(k) = \mathcal{F}[\phi(x,0)] = \int_{-\infty}^{\infty} \phi(x,0)\ e^{-ikx}\ dx.$$

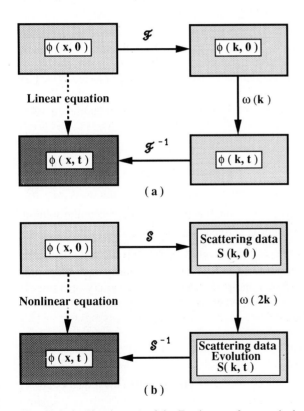

(a)

(b)

Fig. 10.1a,b. Sketches (**a**) of the Fourier transform method, and (**b**) of the inverse scattering method

As time t changes, $\phi(x,t)$ evolves according to the above linear partial-differential equation. One can directly look for solutions of the form

$$\phi(x,t) = \mathcal{F}^{-1}[\phi(k)] = \frac{1}{2\pi} \int_{-\infty}^{\infty} \phi(k) \, e^{i(kx - \omega(k)t)} \, dk.$$

Then, substituting this equation into the linear evolution equation, we obtain the linear dispersion relation

$$\omega(k) = -k^3.$$

Summarizing, with the Fourier transform method we have the following steps

$$\phi(x, 0) \xrightarrow{\mathcal{F}} \phi(k) \xrightarrow{\omega(k)} \phi(k) \, e^{i(kx - \omega(k)t)} \xrightarrow{\mathcal{F}^{-1}} \phi(x, t).$$

10.1.2 The Lax pair for nonlinear evolution equations

We next consider nonlinear evolution equations. Historically, Gardner, Greene, Kruskal and Miura (1967) succeeded in showing how it is possible to solve the initial-value problem for the KdV equation through a succession of linear calculations. This discovery marked the beginning of a series of papers which made more and more systematic the investigations on solitons. In the following, to present these results, we use the elegant and general procedure introduced by Lax (1968).

Equation (10.1) is a special example of a general nonlinear evolution equation of the form

$$\phi_t = G(\phi), \tag{10.2}$$

where G is a nonlinear operator on some suitable space of functions, say $\mathcal{R}$. Suppose that we can associate to any ϕ of the function space $\mathcal{R}$, where the solution of (10.1) is sought, a linear operator L defined in a suitable Hilbert space $\mathcal{L}$ such as $\phi \in R \to L \in \mathcal{L}$ (see Fig.10.2). Moreover, we assume that when $\phi = \phi(t)$ changes with time according to (10.2), L = L(t) remains unitarily equivalent to L (0). In other words, we can define a one-parameter family of unitary operators U (t), with $UU^{-1} = I$ and $UU^+ = U^+U = I$, where U^+ is the conjuguate of U and I is the identity operator, such that

$$U(t)^{-1}L(t)U(t)] = L(0). \tag{10.3}$$

Equivalently, this fact can be expressed by setting the time derivative of (10.3) equal to zero:

$$-U^{-1}U_t U^{-1}LU + U^{-1}L_t U + U^{-1}LU_t = 0. \tag{10.4}$$

264

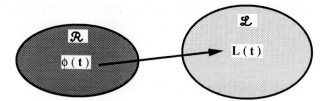

Fig. 10.2. To any ϕ of $\mathcal{R}$ one associates an operator L in $\mathcal{L}$

Multiplying (10.4) by i and setting

$$B = i \, U_t U^{-1},$$ (10.5)

we obtain

$$i\frac{\partial L}{\partial t} = BL - LB = [\, B, L\,].$$ (10.6)

Since $U^+ = U^{-1}$, one has

$$L - L^+ = i \, U_t U^{-1} + iU_t^{-1} = i \, (UU^{-1})_t = 0,$$

and the operator L is self-adjoint or Hermitian; this implies that the eigenvalues E of L, which appear in the equation

$$L(t)\psi(t) = E\psi(t),$$ (10.7)

are independent of time. This looks surprising because since L changes with time according to (10.6), one would expect that E would also change with time. However, E is actually constant, because it is assumed that L remains unitarily equivalent to a constant operator L (0). Now, if we differentiate (10.7) with respect to time we get

$$L_t\psi + L\psi_t = E\psi_t + \psi \, E_t.$$ (10.8)

Then, multiplying both sides of (10.8) by i, after using (10.6) and (10.7) we get

$$(L - E) (i\psi_t - B\psi) = i \, \psi \, E_t.$$

This equation indicates that the quantity $(i\psi_t - B\psi)$ belongs to an eigenspace of the operator L. If the eigenfunction ψ satisfies (10.7) initially and is allowed to develop with time t according to

$$i\frac{\partial\psi}{\partial t} = B\psi,$$ (10.9)

265

then ψ continues to satisfy (10.7) with unchanged E, that is, dE/dt =0. The discovery of Gardner et al. (1967) was to show that it is possible to associate a linear operator L with the KdV equation. Let us consider this spectral problem.

10.2 The KdV equation and the spectral problem

We consider the KdV equation (10.1) and set $\phi(x, 0) = \phi_0(x)$. Following Gardner et al. (1967), we guess an operator L of the form

$$L(t) = -\frac{\partial^2}{\partial x^2} + \phi(x, t) I, \tag{10.10}$$

where $\phi(x, t)$ is a solution of the KdV equation and I is the identity operator. Thus, the spectral problem of the linear operator L is represented by (10.7), which in this case takes the form

$$- \psi_{xx} + \phi(x,t) \psi = E\psi. \tag{10.11}$$

Here, the eigenfunctions $\psi [x, E(t), t]$ and the eigenvalues E(t) of L depend on a parameter which is the time t. Nevertheless, when t is fixed, (10.11) is the well-known time-independent linear Schrödinger equation of quantum mechanics for a particle in the potential $\phi(x,t)$, as represented in Fig.10.3.

We note that if ϕ evolves according to the KdV equation and, if we choose

$$B= -4i \frac{\partial^3}{\partial x^3} + 3 (\phi \frac{\partial \phi}{\partial x} + \frac{\partial}{\partial x} \phi), \tag{10.12}$$

the operator equation (10.6) is satisfied. Under this condition L remains unitarily equivalent to its own initial value as time goes on. We assume that the solutions $\phi(x,t)$ of the KdV equation are continuous, bounded, and tend to zero for x -> ±∞.

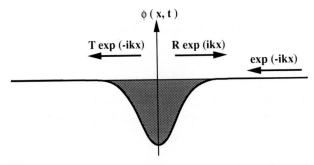

Fig. 10.3. Scattering solutions (waves) corresponding to the continuous spectrum of the linear operator L

266

Consequently, the Schrödinger equation (10.11) will have a finite number of bound states for discrete and negative eigenvalues E_n and a continuous spectrum for positive eigenvalues, respectively:

$$E_n = - K_n^2, \qquad (n = 1, 2, ...,N), \qquad E = k^2 \qquad (k \text{ real}).$$

Now, when time t is fixed one can define, as in quantum mechanics, the bound-state solutions of (10.11) by the boundary conditions

$$\psi_n (x, K_n(t), t) \approx \exp [-K_n(t) x], \qquad\qquad \text{for } x \rightarrow + \infty; \qquad (10.13a)$$

$$\psi_n (x, K_n(t), t) \approx C_n (K_n(t), t) \exp [-K_n(t) x], \qquad \text{for } x \rightarrow - \infty. \qquad (10.13b)$$

On the other hand, one can associate generalized eigenfunctions with the continuous spectrum; these scattering solutions are defined by the boundary conditions and have the following behavior:

$$\psi (x, k, t) \sim e^{- ikx} + R (k, t) e^{ikx} , \qquad \text{for } x \rightarrow + \infty; \qquad (10.13c)$$

$$\psi (x, k, t) \sim T(k, t) e^{-ikx} , \qquad \text{for } x \rightarrow - \infty. \qquad (10.13d)$$

The above equations represent (see Fig.10.3) a wave coming from $+\infty$ which is partly reflected and partly transmitted, that is, $R(k, t)$ and $T(k, t)$ are the reflection and transmission coefficients. Note that if one considers a wave coming from $-\infty$, one has to replace k by -k in the above equations. In summary, the parameters E_n, $C_n(t)$, $R(k, t)$, and $T(k, t)$ *represent the so-called scattering data*.

10.3 Time evolution of the scattering data

We rewrite (10.3) in the form

$$L(t) = U(t) L(0) U^+(t) \tag{10.14}$$

and now use this equation to calculate the time evolution of the scattering data.

10.3.1 Discrete eigenvalues

We first consider an arbitrary eigenstate at time t =0 with eigenfunction $\psi_{no} = \psi_n(x, K_n(o), 0)$ and corresponding eigenvalue $- K_n^2(0))$. From (10.7) we get

$$L(0) \psi_{no} = - K_n^2(0) \psi_{no}. \tag{10.15}$$

Then letting U(t) operate on (10.15) gives

$$U(t) \, L(0) \psi_{no} = - K_n^2(0) \, U(t) \psi_{no}.$$

This equation can be transformed into

$$U(t) \, L(0) \, U^+(t) U(t) \psi_{no} = L(t) \, U(t) \psi_{no} = - K_n^2(0) \, U(t) \, \psi_{no}.$$

Then, using the relation $U(t) \, \psi_{no} = \psi_n(t)$, we obtain

$$L(t) \psi_n(t) = - K_n^2(0) \, \psi_n(t), \tag{10.16a}$$

where $\psi_n(t)$ corresponds to a bound state at time $t \neq 0$ with the same energy

$$E_n = -K_n^2 \, (t) = - K_n^2(0) \tag{10.16b}$$

as at time $t = 0$. One can also establish the reciprocal result: if $\psi_n(t)$ is a bound state with eigenvalue $(-K_n^2)$ at time t, $U^+(t) \, \psi_n(t)$ is a bound state with the same eigenvalue at time $t = 0$. Consequently, *the number N of bound states and corresponding eigenvalues does not change with time when $\phi(x,t)$ evolves according to the KdV equation.* Now putting (10.5) in the form

$$iU_t = BU, \tag{10.17}$$

and applying (10.17) to ψ_{no} yields

$$i \frac{\partial}{\partial t} \psi_n = B \psi_n. \tag{10.18}$$

For $x \rightarrow -\infty$, this equation can be transformed by using (10.13b) and by noting that the operator B defined by (10.12) asymptotically tends to $B \approx - 4i \partial^3 / \partial x^3$, since ϕ and ϕ_x vanish. Under these assumptions, (10.18) becomes

$$\frac{\partial C_n}{\partial t} = -4 K_n^3 C_n.$$

This equation is simply solved to give

$$C_n \, (K_n, t) = C_n \, (K_n, 0) \exp \, (- 4 K_n^3 \, t), \tag{10.19}$$

where $C_n \, (K_n, 0)$ is determined by the initial data $\phi(x,0)$ of the KdV equation.

10.3.2 Continuous spectrum

We now have to determine the time evolution of the reflection coefficient $R(k, t)$, which is defined by the boundary condition (10.13c). Let us consider

$$\psi (x, k, t) = U(t)\, \psi(x, k, 0),$$

which is an eigenfunction of $L(t)$ with corresponding eigenfunction $E = k^2$ such that

$$L(t)\, \psi (x, k, t) = k^2\, \psi (x, k, t). \tag{10.20}$$

The wavefunction ψ evolves with time according to (10.9); moreover, since it satisfies (10.20), as $x\to\infty$ it behaves asymptotically as

$$\psi (x, k, t) \approx a(k,t)\, e^{ikx} + b (k,t)\, e^{-ikx}. \tag{10.21}$$

If we compare (10.21) to the boundary condition (10.13c) we obtain

$$R(k, t) = a(k, t)/b(k, t). \tag{10.22}$$

Putting (10.22) into (10.9) and replacing, as previously, the operator B by its asymptotic limit for large x, we get

$$\frac{\partial}{\partial t} a(k,t)\, e^{ikx} + \frac{\partial}{\partial t} b (k,t)\, e^{-ikx} = - 4(ik)^3 \frac{\partial^3}{\partial x^3} [\, a(k,t)\, e^{ikx} - b (k,t)\, e^{-ikx}\,].$$

According to the linear independence of the exponential functions we have

$$\frac{\partial}{\partial t} a(k,t) = 4ik^3 \frac{\partial^3}{\partial x^3} a(k,t), \qquad \frac{\partial}{\partial t} b (k,t) = - 4ik^3 \frac{\partial^3}{\partial x^3} b (k,t).$$

These equations are integrated to give

$$a(k,t) = a (k,0)\, \exp (4ik^3 t), \qquad b(k,t) = b (k,0)\, \exp (- 4ik^3 t). \tag{10.23}$$

Consequently, from (10.22) and (10.23) we obtain

$$R(k, t) = R(k,o)\, \exp (8ik^3 t). \tag{10.24}$$

Recapitulating the above results, when $\phi(x, t)$ evolves with time according to the KdV equation, (10.16b), (10.19), and (10.24) allow us to obtain the scattering data at $t \neq 0$ based on their values at $t =0$. The eigenvalues remain constant when time passes, that is, the bound states are well specified for a given initial potential $\phi(x, 0)$.

10.4 The inverse-scattering problem

The final step in the inverse scattering transform method is to find the desired function, or potential $\phi(x, t)$, in terms of the scattering data. Such an inverse problem for the Schrödinger equation is solved by using the method developed by Gelfand and Levitan (1951), and Marchenko (1955). In the following, rather than bombard the reader with extensive calculations we give the main steps which are necessary to understand how the inverse spectral problem can be solved. The above authors showed that the desired function $\phi(x, t)$ is defined by

$$\phi(x, t) = -2 \frac{d}{dx} g(x, y=x, t), \tag{10.25}$$

where the function $g(x, y, t)$ is the solution of the linear integral equation

$$g(x, y, t) + F(x + y, t) + \int_x^\infty dz\, F(y + z, t)\, g(x, z, t) = 0. \tag{10.26}$$

In this equation, known as the Gelfand–Levitan–Marchenko equation, the function $F(x+y)$ is related to the scattering data $[K_n, C_n(t), R(k, t)]$ by

$$F(x+y, t) = \sum_{n=1}^N C_n^2(K_n, t)\exp[-K_n(x+y)]$$

$$+ \frac{1}{2\pi} \int_{-\infty}^{+\infty} dk R(k, t)\exp[ik(x+y)], \tag{10.27}$$

where the last term of (10.27) is the inverse Fourier transform of $R(k, t)$. Inserting (10.19) and (10.24) into (10.27), we obtain

$$F(x+y, t) = \sum_{n=1}^N C_n^2(0)\exp[-K_n(x+y)+8K_n^3 t]$$

$$+ \frac{1}{2\pi} \int_{-\infty}^{+\infty} dk R(k, t)\exp[8ik^3 t + ik(x+y)]. \tag{10.28}$$

Thus, in the general case this linear integral equation, which corresponds to the KdV equation, is not easy to evaluate owing to the presence of the last term (see Witham 1974, for a very clear discussion of the Gel'fand-Levitan equation).

10.4.1 Discrete spectrum only: soliton solution

If the initial potential $\phi(x, 0)$ has only a discrete spectrum, that is, if the potential is reflectionless, then (10.28) contains only the first term. Moreover, if we consider the case where $n = N = 1$ with $R(k, 0) = 0$, one has $E = -K^2$, and with the use of (10.28), (10.26) may be reduced to

$$g(x,y,t) + c \, \exp[-K(x+y)+8K^3t]$$
$$+ c \, \exp(8K^3t) \int_X^\infty dz \, \exp[-K(y+z)] \, g(x,z,t) = 0, \qquad (10.29a)$$

where we have set $C^2(0) = c$. We note that (10.29a) can be put in the form

$$g(x, y, t) = - c \, \exp(8K^3t) \, \{ \exp(-Kx)$$
$$+ \int_X^\infty dz \, \exp(-Kz) \, g(x, z, t) \} \, \exp(-Ky), \qquad (10.29b)$$

or

$$g = A(x, t) \, e^{-Ky}. \qquad (10.30)$$

We then substitute (10.30) in (10.29b) to get

$$A(x, t) = - c \, \exp(8K^3t) \, [\exp(-Kx) + A(x, t) \int_X^\infty dz \, \exp(-Kz) \,].$$

Solving this equation in terms of $A(x, t)$ and using (10.30) yields

$$g(x, y, t) = - \frac{c \, \exp(8K^3t - Kx)}{1 + \dfrac{c}{2K} \exp(8K^3t - 2Kx)} \, \exp(-Ky). \qquad (10.31)$$

With the condition $y = x$ we put (10.31) in the form

$$g(x, x, t) = - \frac{c \, \exp[K(x - x_0) - 4K^3t)}{\cosh [K(x - x_0) - 4K^3t]}, \qquad (10.32)$$

where

$$x_0 = 1/2K \, \text{Ln} \, (c/2K). \qquad (10.33)$$

We then substitute (10.32) in (10.25) to finally obtain

$$\phi(x, t) = - 2K^2 \, \text{sech}^2 \, [\, K(x - x_0 - 4K^2t)]. \qquad (10.34)$$

The maximum of the function (10.34), which represents a *one soliton* solution, corresponds to the coordinate $x = x_0 + 4K^2t$ and propagates with velocity $V = 4K^2$

and $x = x_0$ for time $t = 0$. Expressing (10.34) in terms of this velocity, we get

$$\phi(x, t) = -\frac{V}{2} \operatorname{sech}^2 [\frac{\sqrt{V}}{2} (x - x_0 - Vt)]. \tag{10.35}$$

By restoring the original physical variables the reader will find that this solution coincides with the soliton solution obtained previously (see Sects. 3.3 and 5.5) for the KdV equation.

Next, we consider the case where the initial potential has a discrete spectrum associated with N bound states such that $E_n = -K_n^2$. In this case (10.26) and (10.28) yield

$$g(x,y, t) = -\sum_{n=1}^{N} C_n^2 \exp(8K_n^3 t) \, [\exp(-K_n x)$$

$$+ \int_{x}^{\infty} dz \exp(-K_n z) \, g(x, z, t)] \exp(-K_n y). \tag{10.36}$$

For example, for N = 2 one will have a solution of the form

$$g(x, y, t) = A_1(x, t) \exp(-K_1 x) + A_2(x, t) \exp(-K_2 x). \tag{10.37}$$

After substitution of (10.37) in (10.36) one can calculate A_1 and A_2, and using (10.25) one gets the solution where two solitons are interacting. This two-soliton solution is not represented here; however, it is calculated in Sect. 10.6 by using the Hirota method. As time asymptotically goes to infinity the two solitons become separated (see Fig. 10.6) from each other and have waveforms similar to (10.34), respectively:

$$\phi(x, t) = -2\{K_1^2 \operatorname{sech}^2 [K_1(x - x_{o1} - 4K_1^2 t)] + K_2^2 \operatorname{sech}^2[K_2(x - x_{o2} - 4K_2^2 t)]\}. \tag{10.38}$$

Thus for two bound states of the initial potential $\phi(x,0)$ with eigenvalues $E_1 = -K_1^2$ and $E_2 = -K_2^2$ one has two solitons. More generally, an N-soliton solution is obtained for N bound states, with the following asymptotic form:

$$\phi(x, t) = -2 \sum_{n=1}^{N} K_n^2 \operatorname{sech}^2[K_1(x - x_{on} - 4K_n^2 t)]. \tag{10.39}$$

The permanence of these solitons is ensured by the fact that the discrete eigenvalues $(-K_n^2)$ are time independent. Each soliton has a positive velocity, and the bigger solitons travel faster. For detailed calculations related to these solutions and to know more about the IST, the reader is advised to consult the literature cited in the introduction of this chapter and the books by Whitham (1974), Lamb (1980), Dodd et al. (1982), Novikov et al (1984), and Drazin and Johnson (1989).

272

10.5 Response of the KdV model to an initial disturbance

The IST is important in calculating the N soliton solutions. Moreover, from a physical point of view it is very interesting because it allows one to calculate the response of the system, that is, to predict exactly the number of solitons and also the small-amplitude oscillations that can emerge from a finite initial disturbance. Zabusky and Kruskal (1965) solved numerically the initial-value problem for the KdV equation. They found that the initial disturbance $\phi(x, 0) = \cos \pi x$ breaks into solitons. Here we give a brief outline of some important results obtained for the KdV equation when one chooses particular initial potentials whose eigenvalue problem may be found in standard books on quantum mechanics (for example, Landau and Lifshitz 1965).

10.5.1 The delta function potential

One chooses a delta potential

$$\phi(x, 0) = - U_o \, \delta(x), \tag{10.40}$$

where U_o is a constant and $\delta(x)$ is the Dirac distribution. If $V_o > 0$, there is one bound state with corresponding eigenvalue $K = U_o/2$ and a continuous spectrum with reflection coefficient $R(k,0)$. In this case in the function (10.27) the discrete eigenvalue will produce a soliton solution; the continuous spectrum, which in a crude sense represents a part of the solution that behaves almost as if the problem were linear, will produce oscillations as represented in Fig.10.4.

If $U_o < 0$, there is no discrete eigenvalue and no soliton. However, one has $R(k,0) \neq 0$, and oscillations corresponding to the continuous spectrum are present.

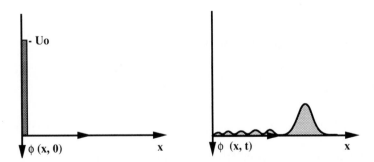

Fig.10.4. Time evolution of an initial Dirac delta disturbance. The soliton corresponds to the discrete eigenvalue K, and the oscillations are associated with the continuous spectrum R(k, t)

10.5.2 The rectangular potential well

Another interesting example is the rectangular well potential. For instance, one may choose an initial disturbance of the form

$$\phi(x, 0) = - U_o, \qquad \text{for } 0 < x < L; \qquad (10.41a)$$

$$\phi(x, 0) = 0, \qquad \text{for } x < 0, x > L. \qquad (10.41b)$$

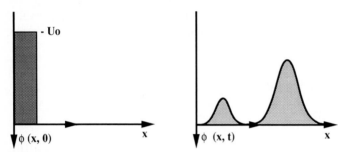

Fig.10.5. Time evolution of an initial rectangular disturbance which breaks into solitons depending on the surface of the pulse

We know from quantum mechanics that the discrete eigenvalues corresponding to this problem are given by $-K_n^2 = E_n = \pi^2 n^2 / L^2$. Moreover, one can show (Landau and Lifshitz 1965) that the number of eigenvalues is controlled by the parameter $1/\gamma = L\sqrt{U_0}$. Consequently, as $1/\gamma$ increases, corresponding to stronger initial disturbances, the number of solitons increases as shown in Fig.10.5. By comparing this figure to Figs.3.10 and 5.11 the reader should note that the theory remarkably predicts the experimental behavior.

10.5.3 The sech-squared potential well

Another important example which can be found in the literature is the potential with a sech-squared inital profile:

$$\phi(x, 0) = - U_0 \text{ sech}^2 x. \qquad (10.42)$$

This potential belongs to the class of reflectionless potentials if $U_0 = N(N+1)$, where N is an integer (N = 1, 2,...), so that one has N discrete eigenvalues (see Landau and Lifschitz 1965) and no continuous spectrum. Thus, an initial state will disintegrate with time into N independent solitons of the form (10.39). If U_0 does not satisfy the above condition, the continuous spectrum exists and oscillations will be present.

10.6 The inverse-scattering transform for the NLS equation

Next, we outline the inverse scattering transform for the nonlinear Schrödinger equation, which was first developed by Zakharov and Shabat (1972). They discovered that the NLS equation (see Chaps. 4, 5, and 8),

$$i \frac{\partial u}{\partial t} + \frac{\partial^2 u}{\partial x^2} + \kappa |u|^2 u = 0, \tag{10.43}$$

can be written in the form (10.6) if the Lax operators are given by

$$L = i \begin{pmatrix} 1+p & 0 \\ 0 & 1-p \end{pmatrix} \frac{\partial}{\partial x} + \begin{pmatrix} 0 & u^* \\ u & 0 \end{pmatrix} \tag{10.44a}$$

and

$$B = -p \begin{pmatrix} 1 & 0 \\ 0 & 1 \end{pmatrix} \frac{\partial^2}{\partial x^2} + i \begin{pmatrix} \frac{|u|^2}{(1+p)} & iu_x^* \\ -iu_x^* & -\frac{|u|^2}{(1+p)} \end{pmatrix}, \tag{10.44b}$$

where $1-p^2 = 2/\kappa$. By comparing (10.44a) to (10.10) we remark that L is not a Schrödinger operator as for the KdV equation, and now (10.7) is written

$$L \begin{pmatrix} \psi_1 \\ \psi_2 \end{pmatrix} = E \begin{pmatrix} \psi_1 \\ \psi_2 \end{pmatrix} \tag{10.44c}$$

By making the change of variables

$$\psi_1 = \sqrt{1-p} \, \exp(-i \frac{E}{1-p^2} x) v_2, \quad \psi_2 = \sqrt{1+p} \, \exp(-i \frac{E}{1-p^2} x) v_1,$$

Zhakarov and Shabat showed that equations (10.44) can be transformed into the following system:

$$\frac{\partial v_1}{\partial x} + i \zeta v_1 = q v_2, \qquad \frac{\partial v_2}{\partial x} - i \zeta v_2 = q^* v_1, \tag{10.45}$$

with eigenvectors v_1 and v_2, and eigenvalues $\zeta = \xi + i\eta$. Here $q(x,0) = iu(x, 0)/\sqrt{1-p^2}$ and $\zeta = Ep/(1-p^2)$. Thus, system (10.45) corresponds to the NLS equation (10.43) with an initial state, given by an arbitrary function $u(x, 0)$ decreasing sufficiently rapidly at infinity. Although system (10.45) is not self-adjoint, the problem of scattering for this system parallels in many respects the problem of scattering of the linear Schrödinger equation associated with the KdV equation, discussed in the previous sections. When there is only one bound state, the one-soliton solution is found to be

$$u(x, t) = \sqrt{2\kappa} \, \eta \, \text{sech} \, [2\eta(x - x_0) + 8\eta\xi t] \exp [-4i(\xi^2-\eta^2)t -2i\xi x +i\phi]. \tag{10.46}$$

This envelope-soliton solution is the same as the one found in Chap.4. It is characterized by four constant parameters: η, ξ, x_0 and ϕ, and represents the envelope of a single wavepacket which propagates with velocity (-4ξ). Unlike the KdV soliton, the parameter η which characterizes the amplitude and the parameter ξ which determines the velocity of the soliton are independent and can be chosen arbitrarily. N bound states generate a solution corresponding to N interacting solitons with 4N arbitrary parameters: η_n, ξ_n, x_{0n}, and ϕ_n. When time goes to infinity if the solitons have different velocities, the N-soliton solution breaks up asymptotically into a sum of N individual solitons of the form (10.46).

The IST allows one to determine, as for the KdV equation, the soliton solution which arises from an initial disturbance. If, for example, the initial pulse shape has the form

$$u(x, 0) = N \text{ sech } x, \tag{10.47}$$

where N is an integer, the number of eigenvalues is given by (10.45). In this case, it was shown by Satsuma and Yajima (1974) that for N = 1 the pulse remains forever with the same shape : one has a one soliton state. For higher N one has N interacting solitons. The solutions have period $T = \pi/2$; that is, they pulsate with this period (see Sect.10.8). In fact, the solitons travel with the same velocity and the resulting waveform oscillates owing to phase interference between the solitons. This remarkable effect predicted by IST, was observed experimentally in optical fibers as discussed in Sect.8.4.

If the shape of the initial disturbance does not coincide with the soliton profile, for example, if it is a rectangular pulse, the IST can show (Hyman et al. 1981; Davydov 1985) that there exists a certain threshold in order to create envelope solitons.

Remark

Stimulated Raman scattering (SRS) (see Yariv 1989; Agrawal 1989) is one of the most studied three-wave interactions processes (compare to four wave interactions in remark (ii) of Sect.4.6)) not only because it retains all the ingredients of any other stimulated process, but also because it has revealed many striking and sometimes unexplained phenomena. In this context the stimulated Raman scattering equations have been solved by using the IST (Claude and Leon 1995; Claude et al. 1995). The results were used to interpret the experiments (Drühl et al. 1983), in particular the creation of a spike of anomalous pump radiation. Although the observed spikes of radiation were generally referred to as *Raman solitons*, Leon and coworkers have shown that the related nonlinear Fourier spectrum does not contain discrete eigenvalues, hence they conclude that the the Raman spikes are not solitons. Further investigations (Leon and Mikhailov 1998) have shown that the SRS equations written with careful consideration of the group velocity dispersion and solved as a boundary value problem, do induce the generation of solitons by pair, and that after the passage of the pulses, the solitons are static in the medium.

Although the solitons are not seen in the ouput intensity, where only the Raman spikes are observed, the signature of the soliton birth lies in the phase of the pump output, which remains to be experimentally measured.

10.7 The Hirota method for the KdV equation

A method introduced by Hirota (1971) allows us to obtain the N-soliton solutions of several equations (Hirota 1976). We first illustrate this method by showing how one can obtain the two-soliton solution of the KdV equation. Thus, we consider the KdV equation in the form (10.1). We then set

$$\phi = -2\, u_x \tag{10.48}$$

and integrate by imposing the condition that u and its derivatives tend to zero when x->±∞. We transform (10.1) into

$$u_t + 6(u_x)^2 + u_{xxx} = 0. \tag{10.49}$$

Then, the nonlinear transformation

$$u = \frac{\partial}{\partial x} \log F = \frac{F_x}{F}, \tag{10.50}$$

known as the Cole–Hopf transformation (Cole 1951; Hopf 1950), reduces (10.49) to

$$FF_{xxxx} - 4F_xF_{xxx} + 3(F_{xx})^2 + FF_{xt} - F_xF_t = 0. \tag{10.51}$$

By using the transformations (10.50) and (10.48), we first note that the solution

$$F = 1 + \exp\theta, \qquad \theta = 2K(x - x_0) - 8K^3t, \tag{10.52}$$

of (10.51) corresponds to the one-soliton solution (10.34) of (10.1):

$$\phi(x,t) = -2K^2 \operatorname{sech}^2 [K(x - x_0) - 4K^3t].$$

Although (10.52) is nonlinear and the solutions cannot be superposed, this result suggests looking for a more general solution of the form

$$F = 1 + \varepsilon F_1 + \varepsilon^2 F_2 + \varepsilon^3 F_3 + ..., \tag{10.53}$$

where ε is an expansion parameter. Inserting (10.53) into (10.51) yields

$$O(\varepsilon) : (F_{1xxx} + F_{1t})_x = 0, \tag{10.54a}$$

277

$$O(\varepsilon^2) : (F_{2t} + F_{2xxx})_x = -3 \, [(F_{1xx})^2 - F_{1x}F_{1xxx}], \qquad\qquad (10.54b)$$

and so on. At order ε one recovers solution (10.52), but, since (10.54a) is linear, one can also take a solution with two exponential terms, such as

$$F_1 = \exp\theta_1 + \exp\theta_2, \qquad \theta_i = 2K_i \, (x - x_{oi}) - 8K_i{}^3 t, \qquad i = 1,2. \quad (10.55)$$

Under these conditions, we substitute (10.55) in (10.54b) and obtain

$$(F_{2t} + F_{2xxx})_x = 48 \, K_1 K_2 \, (K_1 - K_2)^2 \exp(\theta_1 + \theta_2).$$

This linear equation can be solved to give

$$F_2 = \frac{(K_1 - K_2)^2}{(K_1 + K_2)^2} \exp(\theta_1 + \theta_2). \qquad\qquad (10.56)$$

Then, substituting the expressions for F_1 and F_2 in the expansion in the next term $O(\varepsilon^3)$ of the expansion shows remarkably that the right-hand side of this term is zero. Moreover, all the remaining equations in the expansion have zero right-hand side. Consequently, the series self-truncates, and the solution

$$F = 1 + \exp\theta_1 + \exp\theta_2 + \frac{(K_1 - K_2)^2}{(K_1 + K_2)^2} \exp(\theta_1 + \theta_2) \qquad\qquad (10.57)$$

is an exact solution which represents two interacting solitons. In (10.57) we have set $\varepsilon = 1$. By using (10.50) and (10.48) one has

$$\phi(x, t) = -2 \frac{\partial^2}{\partial x^2} \log F$$

$$= -8 \, \frac{K_1{}^2 f_1 + K_2{}^2 f_2 + 2(K_1 - K_2)^2 f_1 f_2 + (K_1 - K_2)^2 / (K_1 + K_2)^2 (K_1{}^2 f_1 f_2{}^2 + K_2{}^2 f_2 f_1{}^2)}{(1 + f_1 + f_2 + (K_1 - K_2)^2 / (K_1 + K_2)^2 f_1 f_2)^2},$$

$$\qquad\qquad (10.58a)$$

where

$$f_1 = \exp(\theta_1), \qquad f_2 = \exp(\theta_2).$$

As time goes to infinity, one can check analytically that the two solitons with different velocities far apart and (10.58) reduce asymptotically to the solution (10.38) where the solitons no longer interact.

As an example let us consider the two-soliton solution such $K_1 = 1$, $K_2 = 2$, and $x_{o1} = 0$, $x_{o2} = -10$ at time $t = 0$. In this case, one has

$$F = 1 + \exp(2x - 8t) + \exp[4(x+10) - 64t] + \frac{1}{9}\exp[(2x - 8t) + 4(x+10) - 64t].$$

(10.58b)

Then, by using a software package (like Mathematica or Maple) for doing mathematical calculations, one can easily plot (10.58a) or (10.58b) as a function of x for different times t. This time evolution of the two-soliton solution is represented in Fig.10.6. At initial time t =0 the solitons are well separated: they interact weakly. Then as time evolves the taller soliton, which moves with velocity $V_2 = 32$, catches the smaller one which has velocity $V_1 = 4$, to form a single wave (at time t = 0.85). Then the taller soliton reappears to the right and moves away from the smaller one as time increases, until they become far apart. Note that we have chosen to plot $-\phi(x, t)$ rather than $\phi(x, t)$. This allows a direct comparison to be made with the application of the KdV model for electrical transmission lines (Chap.3) and shallow water waves (Chap.5).

The Hirota method can be extended to find the N-soliton solution. In fact, Hirota introduced special bilinear operators D_t and D_x and products of them defined as

$$D_t{}^m(a.b) = (\frac{\partial}{\partial t} - \frac{\partial}{\partial t'})^m\, a(x, t)\, b(x',t'), \quad \text{at } t' = t;$$

(10.59a)

$$D_x{}^n (a.b) = (\frac{\partial}{\partial x} - \frac{\partial}{\partial x'})^n\, a(x, t)\, b(x',t'), \quad \text{at } x' = x,$$

(10.59b)

where m and n are positive integers. For example, in the simple case m = n = 1, one writes the product

$$D_t{}^m D_x{}^n (\frac{\partial}{\partial t} - \frac{\partial}{\partial t'})(\frac{\partial}{\partial x} - \frac{\partial}{\partial x'})\, a(x, t)\, b(x',t') = a_{xt}b - a_t b_{x'} - a_x b_{t'} + ab_{x't'}.$$

For x' = x, t' = t, it becomes

$$D_t D_x (a.b) = a_{xt}b + ab_{xt} - a_t b_x - a_x b_t.$$

By making a = b = F, one can use these operators to put (10.51) in the form

$$D_x (D_t + D_x{}^3) (F.F) = 0,$$

(10.60)

which Hirota used to find the N-soliton solution.

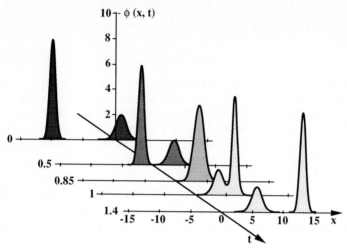

Fig. 10.6. Evolution of the two-soliton solutions (10.58b) for $K_1=1$, $K_2 = 2$ and $x_{01}= 0$ and $x_{02} = -10$ at initial time $t=0$

10.8 The Hirota method for the NLS equation

Next, we calculate a two-soliton solution of the NLS equation that is of physical interest (see Chap.8). For this purpose we rewrite (10.43) as

$$i \frac{\partial q}{\partial t} + \frac{1}{2} \frac{\partial^2 q}{\partial x^2} + |q|^2 q = 0, \tag{10.61}$$

by making the transformations

$$q = \sqrt{Q}\, u, \qquad x \to x\sqrt{2}. \tag{10.62}$$

Then, setting $q = G/F$, where F is a real function, and using operators (10.59) we transform (10.61) into

$$i \frac{D_t G.F}{F^2} + \frac{1}{2} \left[\frac{D_x^2 G.F}{F^2} - \frac{G}{F} \frac{D_x^2 F.F}{F^2} \right] + \frac{G}{F} \frac{GG^*}{F^2} = 0, \tag{10.63}$$

where G^* is the complex conjugate of G. Therefore, F and G will be solutions of (10.63), and hence u is a solution of (10.61), if they are chosen to satisfy the following coupled bilinear equations

$$(i\, D_t + \frac{1}{2} D_x^2)\, G.F = 0, \qquad D_x^2 F.F = 0. \tag{10.64}$$

As before (see Sect.10.7) to solve the above equations we expand F and G as

280

power series in a parameter ε:

$$F = 1 + \varepsilon^2 f_2 + \varepsilon^4 f_4 + \dots \tag{10.65a}$$

$$G = \varepsilon g_1 + \varepsilon^3 g_3 + \dots \tag{10.65b}$$

Substituting (10.65) in (10.64) and collecting terms with the same power of ε, we get

$$O(\varepsilon) : (2i\partial_t + \partial_x^2) g_1 = 0,$$

$$O(\varepsilon^2) : \partial_x^2 f_2 = 2g_1 g_1{}^*,$$

$$O(\varepsilon^3) : (2i\partial_t + \partial_x^2) g_3 + (2iD_t + D_x^2) g_1.f_2 = 0,$$

$$O(\varepsilon^4) : 2\partial_x^2 f_4 + D_x^2 f_2.f_2 = 2 (g_1{}^* g_3 + g_1 g_3{}^*),$$

$$O(\varepsilon^5) : (2i\partial_t + \partial_x^2) g_5 + (2iD_t + D_x^2) (g_3.f_2 + g_1.f_4) = 0,$$

and so on. One solution of the equation at $O(\varepsilon)$ is

$$g_1 = 4 (e^{it/2+x} + 3 e^{9it/2 +3x}). \tag{10.66a}$$

Substituting (10.66) in equation at $O(\varepsilon^2)$ yields

$$f_2 = 4 (e^{2x} + e^{6x}) + 3e^{4x} (e^{4it} + e^{-4it}). \tag{10.66b}$$

Then, substituting (10.66a) and (10.66b) in equation at $O(\varepsilon^3)$ gives

$$g_3 = 4 e^{it/2+7x} + 12 e^{9it/2+5x}. \tag{10.66c}$$

Finally, the equation at $O(\varepsilon^4)$ requires that

$$f_4 = e^{8x}. \tag{10.66d}$$

One can verify that all the remaining equations (for order $> O(\varepsilon^4)$) yield f_n and g_n equal to zero. Consequently, the series self-truncates and we have obtained an exact two-soliton solution. By using (10.65) and (10.66) and rearranging, this two-soliton solution is written as

$$q (x, t) = \frac{G}{F} = 4 e^{it/2} \frac{\cosh 3x + 3 e^{4it} \operatorname{ch} x}{\cosh 4x + 4 \cosh 2x + 3 \cos 4t}. \tag{10.67}$$

Now, if for g_1 we choose the first term of (10.66a) such that

$$g_1 = 4\ e^{it/2+x},$$

we then find at $O(\varepsilon^2)$ that

$$f_2 = 4\ e^{2x}.$$

In this case, for $O(\varepsilon^n)$ with $n > 2$, all the f_n and g_n are equal to zero. The series self truncates and one gets an exact solution of the form

$$q_1 = e^{-\,it/2}\ \text{sech}\ (x - x_0), \qquad x_0 = \log\ (1/2), \tag{10.68}$$

which represents one soliton with zero velocity (compare to (10.46)). Similarly, if we choose

$$g_1 = 12\ e^{9it/2 + 3x},$$

we get another one-soliton solution

$$q_2 = 3\ e^{9it/2}\ \text{sech}\ (x - x_0), \qquad x_0 = \log\ (1/2). \tag{10.69}$$

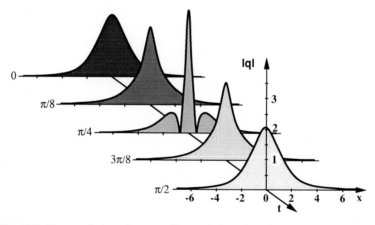

Fig. 10.7. Time evolution of a two-soliton solution, given by (10.67): it pulsates with period $T = \pi/2$

We have

$$q(x, 0) = 2\ \text{sech}\ x, \qquad q_1(x, 0) = \text{sech}\ (x - x_0). \tag{10.70}$$

This shows that, at time $t = 0$, the two-soliton solution has an amplitude which is two times the amplitude of one soliton. However, in contrast to the one-soliton

282

solution, as time evolves, the profile of solution (10.67) is not constant: as mentioned in Sect.10.6 it pulsates with period $T = \pi/2$. This remarkable behavior, which was observed in optical experiments (see Sect.8.5), is represented in Fig.10.7, where we have plotted the time evolution of $|q|$. Moreover, an initial condition (10.70) with an amplitude which is two times that of one isolated soliton will evolve into a state given by (10.67) where two solitons are bound.

11 Diffusive Solitons

In previous chapters we have considered reversible or quasi-reversible systems where dispersion dominates and dissipative effects are small enough to be neglected or in some cases to be considered weak perturbations. The model equations supported soliton solutions owing to the dynamical balance between the linear dispersion and nonlinearity, and an initial disturbance could break up into solitons. In the present chapter we investigate irreversible systems, that have become known under the name *excitable media*. A system *is excitable if an initial perturbation or stimulus of sufficient amplitude can initiate a traveling wave which will propagate through the medium* (Keener 1980, Grinrod 1991, Murray 1993). Dissipation and nonlinearity dominate the physical behavior of such systems leading to nonlinear diffusive processes. Typically, dissipation can balance the effects of nonlinearity, leading to *a localized wave traveling a unique constant velocity* that we will call a *diffusive solitary wave or diffusive soliton*. Such a wave which preserves its shape when propagating, but often annhilates upon a collision with an identical wave, is not a soliton in the strict sense of Chap.10. Nevertheless, as pointed out in the introduction (Chap.1) we keep the word soliton to describe such a robust localized nonlinear wave which can play an important role in the physical behavior of many systems in nature. We restrict ourselves to the simplest patterns that may form in a real system, that is, localized waves. We do not consider the variety of more complex spatio-temporal patterns (Meron 1992, Cross and Hohenberg 1993) in non-equilibrium systems whose study has opened the way to an intensive interdisciplinary study.

We introduce the diffusive soliton by considering a nonlinear electrical transmission line, where kink-shaped diffusive solitons can be predicted, and present recent experiments on these localized waves. Second, we introduce reaction-diffusion processes by considering a chemical model for first order non-equilibrium phase transitions. In this model the *power balanced solitary wave or diffusive soliton* represents a moving domain wall or interface layer between two phases. We then present an electrical lattice which has been used as an analog model for the transmission of information along nerve fibers, and can be approximated by a reaction diffusion equation. In the next section we present a mechanical analog which allows to observe diffusive solitons. The last section is devoted to a discussion of reaction-diffusion processes in lattices. The models we present have been deliberately kept simple for pedagogical purposes.

11. 1 Combined effects of dissipation and nonlinearity

11.1.1 A diffusive electrical transmission line

In this section, following Takagi (1983) and Watanabe et al. (1997a, 1997b) we consider a a nonlinear and strongly dissipative LCR transmission line depicted in Fig. 11.1.

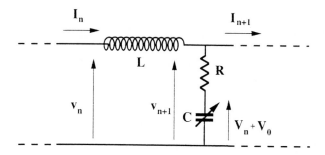

Fig. 11.1. Equivalent circuit of an electrical circuit with linear inductance L, linear resistance R in series with a nonlinear capacitance $C(V_n)$, in each unit section

The circuit equations are given by

$$v_n - v_{n+1} = L \frac{dI_n}{dt}, \qquad (11.1a)$$

$$v_n = V_0 + V_n + R (I_{n-1} - I_n), \qquad (11.1b)$$

$$I_{n-1} - I_n = \frac{dQ_n}{dt}, \qquad (11.1c)$$

where the voltage v_n and V_n and the current I_n are defined in Fig. 11.1 for the nth section of the electrical lattice. V_0 represents the bias voltage of the capacitance. We assume that the charge voltage relation is approximated by a relation similar to (3.12) that we rewrite hereafter

$$Q_n = C_0 (V_n - bV_n^2). \qquad (11.2)$$

From relations (11.1) we obtain a set of equations

$$L \frac{d^2Q_n}{dt^2} = v_{n+1} + v_{n-1} - 2v_n, \qquad (11.3)$$

which is transformed into

$$L\frac{d^2Q_n}{dt^2} = (V_{n+1} + V_{n-1} - 2V_n) + RC_o \frac{d}{dt} [(V_{n+1} + V_{n-1} - 2V_n)$$

$$-b(V^2_{n+1} + V^2_{n-1} - 2V^2_n)]. \tag{11.4}$$

Using the continuum limit expansion (2.35b) and keeping terms up to fourth order derivatives, we approximate (11.4) by the following partial differential equation

$$\frac{\partial^2 V}{\partial t^2} - b\frac{\partial^2 V^2}{\partial t^2} = c_o^2 \left[\frac{\partial^2 V}{\partial x^2} + \frac{1}{12}\frac{\partial^4 V}{\partial x^4} \right] + \frac{R}{L}\frac{\partial}{\partial t}\left[\frac{\partial^2 V}{\partial x^2} - b\frac{\partial^2 V^2}{\partial x^2} \right], \tag{11.5}$$

where $c_o = 1/\sqrt{LC_o}$. Here, for convenience we have set x=n (δ=1: see Sects 2.4.2 and 3.2). We note that if R=0, (11.5) reduces to (3.13) and can also be approximated by a KdV equation (see Sect 3.3). On the other hand, if R>>L the dissipative effects are much larger than the dispersive effects and (11.5) can be approximated by a diffusive equation, as we shall see hereafter.

Proceeding as in Sect 3.3, we introduce a small parameter ε<<1 and perform the following change of variables

$$s = \varepsilon(x - c_o t), \qquad T = \varepsilon^2 t, \tag{11.6}$$

and express the voltage in a perturbation series

$$V = \varepsilon V_1 + \varepsilon^2 V_2 + ... \tag{11.7}$$

In terms of the slow space and time variables s and T, to order ε^4, we obtain (see Appendix 11A)

$$\frac{\partial V_1}{\partial T} - D\frac{\partial^2 V_1}{\partial s^2} + \lambda V_1 \frac{\partial V_1}{\partial s} = 0, \tag{11.8}$$

where

$$\lambda = bc_o \text{ and } D = R/2L.$$

This simple nonlinear diffusive evolution equation combining both a nonlinear mode coupling term characterized by λ and a linear diffusive term controlled by damping or diffusion coefficient D is the so-called Burgers equation (Burgers 1948, Witham 1974, Fogedby 1998).The Burgers equation first arose in a fluid mechanics context in order to describe one dimensional turbulence.

286

11.1.2 Linear diffusive waves

In the low amplitude limit, nonlinear effects can be ignored and (11.8) reduces to a linear diffusion equation

$$\frac{\partial V_1}{\partial T} - D \frac{\partial^2 V_1}{\partial s^2} = 0. \tag{11.9}$$

The general solution of (11.9) is well known and can be handled by a variety of methods. Eq (11.9) admits a particular solution of the form

$$V_1 = V_{o1} e^{i(\omega T - ks)} + V^*_{o1} e^{-i(\omega t - ks)}, \tag{11.10a}$$

if

$$\omega = \pm i \, Dk^2. \tag{11.10b}$$

Thus, the linear diffusive modes characterized by (11.10b) have the linear decaying form

$$V_1 = 2V_{o1} \exp(-Dk^2 t) \cos(\omega t - ks). \tag{11.11a}$$

As a result, a linear or Fourier superposition of these modes, as for example an initial step function, will smoothen and decay when diffusing. In terms of the original laboratory coordinates (11.6), one gets

$$V = 2\varepsilon V_{o1} \exp[-(R/2L) \varepsilon^2 k^2 t] \cos[\omega \varepsilon^2 t - k\varepsilon(x - c_o t)], \tag{11.11b}$$

for the linear diffusive modes on the electrical transmission line.

Remark

Interestingly, the Burgers equation (11.8) can be reduced to a standard linear diffusion equation by a nonlinear transformation $V_1 = -(2D/\lambda)(\phi_s/\phi)$ introduced independently by Hopf (1950) and Cole (1951).
Specifically, one first introduces

$$V_1 = \frac{\partial u}{\partial s},$$

and integrates (11.8) in order to get

$$\frac{\partial u}{\partial T} + \frac{\lambda}{2} \left(\frac{\partial u}{\partial s}\right)^2 = D \frac{\partial^2 u}{\partial s^2}.$$

Then, through the relation

$$u = -\frac{2D}{\lambda} \ln \phi,$$

one gets a linear diffusion equation

$$\frac{\partial \phi}{\partial T} = D \frac{\partial^2 \phi}{\partial s^2} .$$

From the solution of this linear diffusion equation one can then obtain an exact solution of the Burgers equation (11.8) through the Cole-Hopf transformation (see Witham 1974). For example, one can determine the evolution of an initial disturbance (Witham 1974).

11.1.3 Kink-shaped diffusive solitons

Let us look for a localized solution of (11.8) with a permanent profile (compare to Sect 6.2.2) of the form

$$V(s, T) = V(s - vT), \tag{11.12}$$

propagating at velocity v. Here, we have set $V=V_1$ to simplify the writing. Using $\partial/\partial T = -v\partial/\partial x = d/dS$ with $S = s-vT$, from (11.8) one obtains

$$-v\frac{dV}{dS} - D\frac{d^2V}{dS^2} + \lambda V \frac{dV}{dS} = 0.$$

Hence

$$\frac{\lambda}{2} V^2 - vV + C = D \frac{dV}{dS} , \tag{11.13}$$

where C is an arbitrary constant of integration. With the appropriate boundary conditions $V \rightarrow V_+$, V_-, $dV/dS \rightarrow 0$, as $S \rightarrow \pm \infty$, for a localized solution, (11.13) yields the following conditions

$$V_+ + V_- = \frac{2v}{\lambda} , \qquad\qquad C = \frac{\lambda}{2} V_+ V_- .$$

Equation (11.13) is written as

$$-\frac{dV}{dS} = \frac{\lambda}{2D}(V-V_+)(V_--V).$$

(11.14)

The solution of (11.14) is given by

$$\frac{\lambda}{2D}S = \frac{2}{V_--V_+}Ln\frac{V_--V}{V-V_+}.$$

(11.15)

It may be put in the form of a diffusive soliton

$$V = \frac{V_++V_-}{2} - \frac{V_+-V_-}{2}\tanh\left[\frac{\lambda}{4D}(V_+-V_-)S\right].$$

(11.16)

The width of the soliton is related to its amplitude and is of order $\dfrac{4D}{\lambda(V_++V_-)}$.

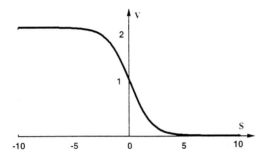

Fig. 11.2 Sketch of a diffusive kink-shaped soliton solution (11.16): for t=0, $V_+=2$, $V_-=0$ and $\lambda/4D = 0.25$

Replacing S by (s-vT) and using (11.6) we obtain the voltage $V = \varepsilon V_1$ in terms of the original laboratory coordinates

$$V = \varepsilon\frac{V_++V_-}{2} - \varepsilon\frac{V_+-V_-}{2}\tanh\left[\varepsilon\frac{V_+-V_-}{2}\frac{b}{R}\sqrt{\frac{L}{C_o}}[x-(c_o+\varepsilon v)]t\right].$$

(11.17)

The small parameter ε, which can be dropped, reminds us that space and time vary slowly and the amplitude of the diffusive soliton is weak. This diffusive soliton is often called a shock wave because for vanishing diffusion coefficient $D=R/2L \to 0$ its width becomes very narrow. Note that, when $R \to 0$, eq (11.8) which was derived under the assumption R>>L is not a good model anymore. Instead, (11.5) can be reduced to a KdV equation (see Sect 11.1.1).

289

Remark

(i) Note that *contrary to linear waves* (see Sect 11.1.2) *which decay and smoothen when diffusing, a nonlinear localized wave or soliton can travel at constant velocity.*
(ii) The Burgers equation is intrinsically dissipative and an initial disturbance will eventually decay owing to dissipation, unless energy is fed into the system. In this regard the soliton (11.16) or (11.17) is a dissipative localized wave that owes its stability to the energy flux corresponding to the current entering at the input of the transmission line.

11.1.4 Experiments on electrical diffusive solitons

Recently, experiments were performed (Watanabe et al. 1997b) on diffusive solitons or shock waves in a nonlinear LCR network as represented in Fig 11.1. The network contains 55 unit sections, constructed with the following components: $L=15\mu H$, $R=100\ \Omega$. The charge voltage relation is given (see relations (4.10) and (4.6)) by $\ln[1+V/F_o] \approx V/F_o - V^2/2F_o^2 = C_o(V - bV^2)$ where : $F_o=2.07$ V (for $V_o=0$ V) and $C_o=0.889$nF.
The response to an initial step voltage was studied. For R=0, an oscillatory structure is observed, owing to the competition between dispersion and nonlinearity, it is the KdV regime (see also Watanabe et al. 1978).

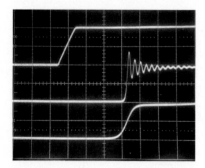

Fig. 11.3. Response of the diffusive transmission line to an initial step voltage. The upper trace shows the initial disturbance. For $R{\approx}0$, an oscillatory shock is observed (middle trace). In the dissipative or diffusive transmission line with R=100 Ω, a smooth shock front or diffusive soliton is formed (lower trace). (Reproduced by kind permission of S. Watanabe and the Journal of the Physical Society of Japan)

The oscillations in the wavefront evolve asymptotically to KdV solitons. If one progressively increases the resistance R, the number of oscillations diminishes and disappears, as it is observed in Fig 11.3 when $R=100\ \Omega$. In this case, the experiment shows that the soliton (smooth shock wave) with constant amplitude travels stably in the transmission line with dissipation. The velocity and width of the observed soliton are consistent with the theoretical predictions (for more details see Watanabe et al. 1997b).

Remarks

(i) The Burgers equation (11.8) was derived in the lowest order approximation. Nevertheless, in the next order a higher order equation which brings corrections to (11.8) has been obtained (Watanabe et al. 1997a , 1997b), allowing to improve the fitting of the experimental results.
(ii) Recently, Malfliet (1998) has proposed a different approach from Watanabe et al. to solve (11.5) and get higher order corrections to (11.17).

11.2 Reaction diffusion processes

11.2.1 Reaction diffusion equations

Reaction diffusion equations may be used to model a vast variety of phenomena. They arise in many areas of physics, chemistry, biology and ecology . Let us derive a general diffusion equation. We first derive the conservation equation in a way similar to the derivation of the equation of mass conservation in fluid mechanics (see Appendix 5A.1). Let c(**r**,t) the concentration of some species : concentration of chemicals, number of cells or animals, at point **r** and time t. Let J (**r**,t) the flux or current of these species and the scalar product **J.n** the net rate at which the species crosses a unit surface in a plane perpendicular to **n** (positive in the **n** direction). We can then write a general conservation equation which states that the rate of change of the amount of material in an arbitrary volume V is equal to the rate of flow across the surface (boundary) S enclosing the volume, plus any that is created or destroyed within the volume.

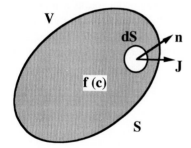

Fig.11.4 . Sketch of an arbitrary volume V of material, current flux **J** and source f of species

That is

$$\frac{\partial}{\partial t} \int_V c(\mathbf{r},t)\, dV = - \int_S \mathbf{J.n}\, dS + \int_V f\, dV, \qquad (11.18)$$

where f which represents the source may be a function of c, **r** and t. The surface integral can be transformed into a volume integral by using Green's theorem

$$\int_V \left[\frac{\partial c}{\partial t} + \text{div}(\mathbf{J}) - f \, dV \right] = 0. \tag{11.19}$$

Since the volume V is arbitrary, this equation must hold for any volume, and the integrand must vanish, yielding an equation that expresses the continuity of material or *continuity equation*:

$$\frac{\partial c}{\partial t} + \text{div } \mathbf{J} = f. \tag{11.20}$$

This equation holds for a general flux transport J, whether by diffusion or some othe process. However, for a given model we must specify **J** and f. For example, according to Fick's law, for classical diffusion, the flux J is proportional to the gradient in the concentration. Thus

$$\mathbf{J} = - D \, \mathbf{grad} \, c. \tag{11.21}$$

Here, D, which may be a function of c, r and t, is the diffusion coefficient or diffusivity. D is positive, the minus sign indicates that diffusion transports matter from high to low concentrations. Using Fick's law (11.21) becomes

$$\frac{\partial c}{\partial t} = \text{div } (D \, \mathbf{grad} \, c) + f. \tag{11.22}$$

Eq (11.22) is a *reaction diffusion equation*. It was proposed as a model for the chemical basis of morphogenesis by Turing (1952) in one of the most important papers in theoretical biology. This paper opened up an enormous range of study spanning the fields of developmental biology, chemistry, applied mathematics, and engineering.
If D is constant, (11.22) reduces to

$$\frac{\partial c}{\partial t} = D \, \Delta c + f. \tag{11.23}$$

We restrict ourselves now to one spatial dimension and assume that the source term is a function of c, only. Thus, we can reduce (11.23) to

$$\frac{\partial c}{\partial t} = D \frac{\partial^2 c}{\partial t^2} + f \, (c). \tag{11.24}$$

Equation (11.23) or its one-dimensional version (11.24) are *reaction diffusion equations*: in addition to diffusion they contain a function f(c) which in many models describes the reaction or kinetic properties of the system, as we shall illustrate in the next section.

11.2.2 A chemical model with reaction diffusion

In 1972 Schlögel introduced two interesting and simple chemical reaction models which exhibit the behavior of the first and second order non-equilibrium phase transitions (Haken 1978), respectively. Schlögel's second model which exhibits first order behavior is briefly presented in this section. It includes the existence of two spatially separate phases and is connected with the autocatalytic chemical reactions between four chemical species or molecules A, B, C and X

$$A + 2X \rightarrow 3X, \tag{11.25a}$$

$$B + X \rightarrow C. \tag{11.25b}$$

The reactions may be possible in both directions and the concentrations a, b, c of the species or molecules A, B, C and the reaction rates k_1, k_1', k_2, k_2' are kept fixed externally and only the concentration n of the intermediate product X can vary in time and space owing to chemical reactions and diffusion. Under these conditions, in general, no chemical equilibrium is possible. For example in process (11.25a) the number of molecules X produced per unit time is proportional to the concentration a of molecules A and to that of the molecules X: n^2. The proportionality factor is just the reaction rate: k_1. Thus, the reaction rates of the two reactions are

$$r_1 = k_1 \, a \, n^2 - k_1' \, n^3, \qquad\qquad r_2 = k_2 \, b \, n - k_2'c.$$

They do not vanish. Now, by adequate choice of units and parameters

$$k_1'=1, \qquad k_1 a = 3, \qquad k_2 b = \beta, \qquad k_2'c = \gamma,$$

we can write the time rate of change of the concentration n in the form

$$\frac{dn}{dt} = r_1 - r_2 = f(n), \tag{11.26a}$$

where

$$f(n) = -n^3 + 3n^2 - \beta n + \gamma, \tag{11.26b}$$

represents the reaction function. The concentration of species X can change not only by the above chemical reactions (11.25a) and (11.25b) but by diffusion of the molecules X as well. As a result it is space and time dependent: $n(x,t)$. The concentrations of the other components may remain constant in time and space. In this case we get the evolution equation, similar to (11.24), describing the diffusion and reaction of species X

$$\frac{\partial n}{\partial t} = D \frac{\partial^2 n}{\partial x^2} + f(n),$$ (11.27a)

where $f(n)$ is given by (11.26b). Note that $f(n)$ can be replaced by a cubic polynomial (see Fig 11.5) in n with three positive zeros $n_1 \leq n_3 \leq n_2$ (compare to Appendix 4A and Fig 4A1) which can be calculated (see Schlögel 1972). We rewrite (11.27a) as

$$\frac{\partial n}{\partial t} = D \frac{\partial^2 n}{\partial x^2} + \alpha (n - n_1)(n - n_2)(n_3 - n).$$ (11.27b)

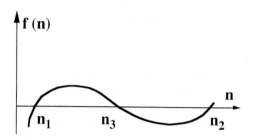

Fig. 11.5a. Sketch of cubic polynomial $f(n)$ with the three possible roots

Note that in the homogeneous reactor without diffusion n_1 and n_2 represent two stable states, and n_3 an unstable state. If diffusion is included, the two phases n_1, n_2, can coexist in different domains in space. Altough the system corresponding to (11.27a,b) is not conservative it is convenient to introduce a potential $V(n)$ such as

$$f(n) = -\frac{\partial V}{\partial n}.$$ (11.28a)

Thus, if $f(n)$ has the form (11.26b) the corresponding double-well or *bistable* potential is

$$V(n) = \frac{n^4}{4} - n^3 + \beta \frac{n^2}{2} - \gamma n + C,$$ (11.28b)

with two non-degenerate minima. Here, C is an integration constant which can be set equal to zero. As in section (11.1.3) we can look for a localized solution of (11.27b) with a permanent profile propagating at velocity v, such as

$$V(s) = V(x - vT), \qquad n \to n_1, n_2, \qquad dn/ds \to 0, \text{ as } s \to \pm \infty.$$ (11.29)

The calculations (see Murray 1993 and Appendix 11B) yield

294

$$n = \frac{n_2 + n_1}{2} - \frac{n_2 - n_1}{2} \tanh\left[\frac{K}{2}(n_2 - n_1)(x - vt)\right], \tag{11.30}$$

where

$$K = \sqrt{\frac{\alpha}{2D}}, \qquad\qquad v = \sqrt{\frac{\alpha D}{2}}(n_1 - 2n_3 + n_2). \tag{11.31}$$

Solution (11.30) has a kink shape (like the Burgers soliton represented in Fig 11.2), that is, it connects the two stable non-degenerate minima (corresponding to n_1 and n_2) of "potential" $V(n)$. It describes a domain wall or interface layer, between two different phases of the species X, moving at velocity v. This velocity is unique because it corresponds to the case where the potential difference between the two stable states is exactly compensated by dissipation. Thus, solution (11.30) is a *power balanced solitary wave or diffusive soliton* (see Fig 11.5b).

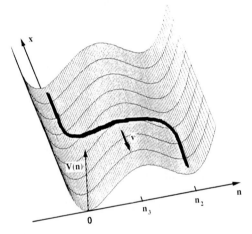

Fig. 11.5b. Schematic representation of a *power balanced solitary wave or diffusive soliton* traveling at unique velocity v on the "potential" surface $V(n,x)$ (Compare to the representation in Fig. 6.19 for a conservative system). Here we have set $n_1 = 0$ and $C = 0$

If $\alpha = 1$, $\beta = \gamma + 2$, with $0 < \gamma < 1$, comparison of coefficients in $f(n)$ given by (11.26b) and its cubic form yields (Schlögl 1972):

$$n_1 = 1 - \sqrt{1 - \gamma}, \qquad n_2 = 1 + \sqrt{1 + \gamma}, \qquad n_3 = 1. \tag{11.32}$$

Combining relations (11.31) and (11.32) one finds $v = 0$ and (11.30) reduces to a static soliton solution

$$n = 1 - \sqrt{1 - \gamma} \tanh\left[\sqrt{\frac{1 - \gamma}{2D}}\, x\right]. \tag{11.33}$$

first found by Schlögl (1972). In this steady state, which corresponds to $\partial n/\partial t = 0$,

295

the variation of concentration n owing to the chemical reactions is instantaneously balanced in each elementary volume of the system by its variation owing to diffusion.

Remarks

(i) The general reaction-diffusion system described by eq (11.27b) has attracted considerable mathematical and physical interest since it occurs in a wide field of natural sciences such as biology and ecology (see the excellent book of Murray 1993). Equation (11.27b) is often referred to as the reduced Nagumo equation (Nagumo et al. 1962) (see Sect. 11.2.3).
 (ii) More general traveling wave fronts solution of (11.27b) have been found (see Schlögel and Berry 1980, Magyari 1982, Dehrman 1982)
(iii) In many applications of reaction-diffusion systems, interest is centered upon so-called *excitable systems* (Keener 1980) and their wave-like solutions.

11. 2. 3 An electrical lattice with reaction diffusion

An important class of systems that display nonlinear diffusion is found in nerve fibers. The axon or outgoing fiber of a nerve cell is covered by a voltage sensitive membrane with a nonlinear characteristic somewhat similar to that of a tunnel diode. In 1962, Nagumo and coworkers demonstrated that the dissipative and nonlinear electrical lattice, represented in Fig 11.6, could serve as useful a model to study the transmission of information along nerve fibers. This model, developed by Figtzhugh (1961), is a simplification of the Hodgkin-Huxley equation (1952) related to electric signalling or firing by individual nerve cells or neurons. Also Crane (1962) discussed in detail the concept of *neuristor* or the electronic analog (Nagumo et al. 1962, 1965, Scott 1965, Parmentier 1967, Marquié et al. 1998) of a nerve axon.

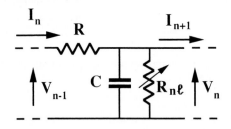

Fig. 11.6 Unit section of a dissipative electrical lattice with a linear resistance R in series and a nonlinear resistance $R_{n\ell}$ in parallel with a linear capacitance

Each unit section of this electrical lattice (carefully analyzed in the book of Scott, 1970) contains a linear resistance R in series and a nonlinear resistance $R_{n\ell}$ in parallel with a linear capacitance C. From Kirchoff's laws we obtain

$$V_{n-1} - V_n = RI_n, \tag{11.34a}$$

$$I_n - I_{n+1} = C \frac{dV_n}{dt} + j_n(V_n), \tag{11.34b}$$

where $j_n(V_n)$ is the current in $R_{n\ell}$. The current-voltage characteristics is, like for a Gunn diode, of the form

$$j_n(V_n) = \frac{V_n}{R_0} \left(1 - \frac{V_n}{V_1} \right) \left(1 - \frac{V_n}{V_2} \right). \tag{11.35}$$

Here, R_0 is a constant resistance and V_1 and V_2 are constant voltages : $0 < V_1 < V_2$. Combining the above equations yields

$$\frac{dV_n}{dt} = \frac{1}{RC}(V_{n+1} + V_{n-1} - 2V_n) - \frac{V_n}{R_0C}\left(1 - \frac{V_n}{V_1}\right)\left(1 - \frac{V_n}{V_2}\right), \tag{11.36}$$

which can be transformed into

$$\frac{dV_n}{d\tau} = D(V_{n+1} + V_{n-1} - 2V_n) - \alpha V_n (V_n - V_1)(V_n - V_2), \qquad n = 1, 2, \dots, \tag{11.37}$$

by setting $\tau = t/R_0C$, $D = R_0/R$ and $\alpha = 1/V_1V_2$. Using the continuum limit expansion (2.35b) (with $\delta = 1$: see Sect. 2.4.3) and keeping terms up to second order derivatives, we approximate (11.37) by the reaction diffusion equation

$$\frac{\partial V}{\partial \tau} = D \frac{\partial^2 V}{\partial x^2} - \alpha V(V - V_1)(V - V_2). \tag{11.38}$$

Equation (11.38) which is similar to (11.28) can be solved to give a diffusive soliton solution

$$V = \frac{V_2}{2}\left\{ 1 \pm \tanh\left[\frac{V_2}{2}\sqrt{\frac{\alpha}{2D}}(x - vt)\right] \right\}, \tag{11.39}$$

where the propagation velocity has the value

$$v = \pm\sqrt{\frac{\alpha D}{2} \frac{(V_2 - 2V_1)}{R_0C}}. \tag{11.40}$$

Note that in (11.38) the diffusion coefficient D plays the role of a discreteness parameter, as defined in Sect. 6.1.1. For d>>1 the continuum approximation holds, for d<<1 it is not valid anymore.

11.2.4 Experiments in an electrical lattice

In this section we describe recent experiments that have been performed by Binczak et al. (1998) on a nonlinear electrical lattice of the form sketched in Fig 11.6. The lattice that contains 48 unit sections, is constructed with the following components. One has C= 33nF, R_0 = 17kΩ and the nonlinear resistance $R_{n\ell}$ is realized through the electrical scheme described in Fig. 11.7, the voltages are V_1= 0.68V and V_2 = 1.23V ($2V_1 > V_2$).

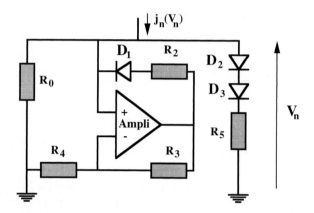

Fig.11.7 Electrical circuit (see Binczak at al. 1998) simulating the nonlinear resistance $R_{n\ell}$

In the case where R=680Ω, one has D=25 and the system can be considered continuous, that is, (11.37) can be approximated by Eq (11.38). In fact, the response to an initial step voltage (the five first cells are set to V_n=0V and all the other cells are set to $V_n = V_2$) shows that such an initial condition evolves into a wave front or diffusive soliton, depicted in the oscillogram of Fig. 11.8, which travels at constant velocity v_{exp}= 2125 cell/s (the velocity calculated from (11.40) is v_{theo}= 2240 cell/s).

Decreasing the value of R allows to decrease D: lattice effects become dominant and the continuum approximation is not valid anymore as can be checked with the experimental lattice. Moreover, below some critical value D_c the wavefront cannot propagate anymore: this is the *propagation failure phenomenon* (see Sect. 11.4.1) For example for D = 0.11 the experiments (Binczak et al. 1998) show that the propagation fails.

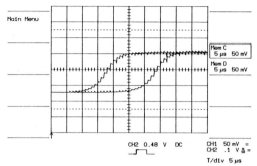

Fig. 11.8 Oscillograms showing a traveling wavefront (velocity v = 2125 cell/s) or diffusive soliton at two different times : t =6ms and t = 14.6ms (see Binczak et al.)

11.3 A mechanical analog with diffusive solitons

11.3.1 Chain with flexion and gravity

In order to observe diffusive solitons we have constructed a mechanical analog which consists of an experimental chain where each basic unit (see Appendix 11C) is an *Euler strut* (Pippard 1979, Jackson 1992). An Euler strut is a classic example of a system which may be bistable, depending on the forces of flexion and gravity in opposition. In the following, we consider the bistable configuration with *two non-degenerate equilibrium positions* (see Appendix 11C). Each unit has an overdamped motion and is coupled to its nearest neighbors by springs like for the mechanical chain (see Appendix 6F) presented in Sect. 6.11. When the difference between angular displacements of neighboring pendulums are small enough, the equation of motion of the nth unit in the chain is given by

$$\beta \frac{d\theta_n}{dt} = -K\theta_n + mgd\sin(\theta_n+\phi) + C_{o,\ell}(\theta_{n+1}+\theta_{n-1}-2\theta_n)$$
$$- C_{o,n\ell}(\theta_n-\theta_{n+1})^3 - C_{o,n\ell}(\theta_n-\theta_{n-1})^3. \quad (11.41)$$

Equation (11.32) is a *discrete reaction-diffusion equation* where each cell is a bistable unit coupled diffusively, linearly and *nonlinearly*, to its nearest neighbors. These coupling terms are the same as in (6.62). ϕ is a symmetry-breaking term, defined in Appendix 11C, which corresponds to the tilting angle of the mechanical support of the whole chain. Proceeding as in Sect. 6.11.2, in the strong coupling limit, using the continuum approximation and setting X=x/a we can approximate (11.41) by

$$\frac{\partial\theta}{\partial t} = D_\ell \frac{\partial^2\theta}{\partial X^2} + D_{n\ell} (\frac{\partial\theta}{\partial X})^2 \frac{\partial^2\theta}{\partial X^2} + f(\theta), \quad (11.42a)$$

299

where

$$G= mgd/\beta, \qquad\qquad D_\ell =C_{o,\ell}/\beta, \qquad\qquad D_{n\ell} = 3C_{o,n\ell}/\beta, \qquad (11.42b)$$

and

$$f(\theta) = -\frac{\partial V}{\partial \theta} = -\kappa\theta + G\sin (\theta+\phi), \qquad\qquad (11.42c)$$

with

$$V(\theta) = \kappa\frac{\theta^2}{2} + G \cos (\theta +\phi), \qquad\qquad \kappa=K/\beta. \qquad (11.42d)$$

Here, $V(\theta)$ is the asymmetric or tilted double-well potential (compare with 11.28a) corresponding to $f(\theta)$. This potential has *two non-degenerate minima* (compare to (6.67)) and can be fitted approximatively by a potential of the form (11.28b). Thus, (11.42a) is a reaction diffusion equation similar to (11.27a,b) but, with an additional nonlinear diffusion term. In the static case ($\partial\theta/\partial t =0$): for $D_\ell =0$ and $\phi =0$ one can obtain compacton solutions similar to those obtained in Sect. 6.11.4; for $D_\ell \neq 0$ and $\phi = 0$, we cannot solve (11.33a) but we expect a kink-shaped static soliton solution which connects the two degenerate potential minima. In the general case we expect a soliton solution of (11.33a) diffusing at a unique velocity v.

11.3.2 Experimental chain

In this subsection we describe the mechanical chain we have constructed and comment briefly the qualitative observations that can be made with it. The chain is constructed with 40 Euler struts attached vertically to an horizontal support. The lattice spacing is a=4cm. A basic unit consists of a flexible metal strip (width =10mm, thickness=0.1mm and d=97mm. Two rectangular pieces (total mass, m=18.5g) of metal are fixed along the strip. The mass m must be as weak as possible in order to minimize the inertial effects compared to the dissipative ones which originate from the flexion process and the friction experienced by the strips (if their width is large enough) when they move in the air. Each unit is connected to its neighbors by springs. When the support is horizontal, the on-site potential (11.42d) is a double well potential with two degenerate equilibrium states. When the support of the chain is tilted, the symmetry is broken and the double-well potential has two non-degnerate equilibrium states. In this case, kink-shaped diffusive solitons can be easily created as represented in Fig. 11.9.

We stress that the observations we can make with this mechanical chain are relatively crude. Nevertheless, we think that, in spite of the approximations we have made, this analog can be very useful to illustrate qualitatively nonlinear reaction diffusion processes in one dimensional systems.

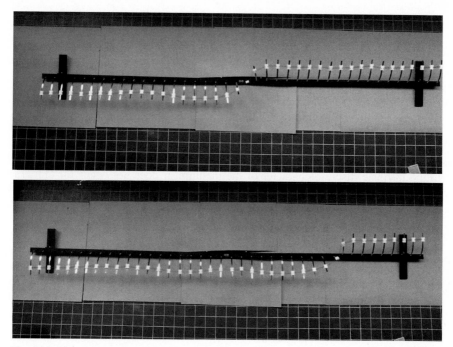

Fig. 11.9. Photographs showing a kink-shaped wave-front or diffusive soliton at two different positions when it travels on the mechanical chain with flexion and gravity (Photos R. Chaux)

11.4 Reaction diffusion processes in lattices

11.4.1 Propagation failure

In the previous sections we have considered reaction diffusion processes in continuum systems or in the continuum approximation (strong coupling limit) of discrete systems. However, continuous reaction diffusion of the general form (11.27a) provide an inadequate desciption of the behavior of weakly coupled systems where the interplay between spatial discreteness and nonlinearity can lead, like for conservative systems considered in previous chapters, to novel effects or lattice effects not present in the continuum models. For example, a series of recent studies in the area of celllular neurobiology have shown that a continuous reaction-diffusion description of the celllular population does not lead to a satisfactory explanation of *propagation failure* (Pauwelussen 1982, Keener 1987, Rinzel 1990) which was mentioned in Sect. 11.2.4.

In fact, in many important cases, diffusive processes in a nonlinear lattice may be modeled by a discrete equation identical to (11.37) that we rewrite into the form

$$\frac{du_n}{dt} = D\ (u_{n+1} + u_{n-1} - 2u_n) + f(u_n). \tag{11.43}$$

Here, u_n represents the, physical, chemical, biological or ecological, state at the nth lattice site and $f(u_n) = \alpha u_n\ (1 - u_n)\ (u_n - a)$ is a cubic polynomial which corresponds to a bistable potential with two non-degenerate stable states at $u = 0$, $u = 1$ separated by an unstable state at $u = a$ ($0 < a < 1$). When the units are weakly coupled, that is, for D small, there exists a critical value D_c of the coupling strength below which propagation is impossible (Erneux and Nicolis 1993, Fath 1998). Instead, a series of stable steady state solutions emerge. This propagation failure phenomenon is due to the intrinsic discrete character of the medium, it does not occur for (11.27a) which represents the continuum approximation of (11.43).

11.4.2 Discrete reaction diffusion model with exact solution

In order to gain insight into this propagation failure phenomenon and to reduce the critical value D_c it is desirable to investigate nonlinear discrete reaction diffusion equations with exact solutions. In this regard, in Sect .9.9 we have shown that a general equation of the form (9.86) with both inertial and dissipative terms can admit exact solutions. Setting m = 0, that is, in the overdamped limit one can reduce Eq (9.86) to a discrete reaction diffusion equation

$$\frac{d\phi_n}{dt} = D\ (\phi_{n+1} + \phi_{n-1} - 2\phi_n) + f(\phi_n), \tag{11.44}$$

where

$$f(\phi) = w + 2\ D\ \phi - w\phi^2 + 2\ D\ (\ 1 - \frac{1}{\alpha}\ )\ \frac{\alpha\phi}{1 - \alpha\phi^2}. \tag{11.45}$$

Here, we have set $D = K/\Gamma$. The double-well on-site potential with two nondegenerate minnima corresponding to $f(\phi)$ is

$$V(\phi) = -\ w\phi - \phi^2 + w\frac{\phi^3}{3} + D\ (\ 1 - \frac{1}{\alpha}\ )\ Ln\ (1 - \alpha\phi^2). \tag{11.46}$$

As shown in Sect. 9.9, (11.44) admits an exact discrete kink- soliton solution

$$\phi_n = \tanh\ (wt - kna), \tag{11.47}$$

which diffuses at unique velocity $v = w/k$. Such an exact traveling wave solution was also found recently by Bressloff and Rowlands (1997). They used a construction different from the one we presented in Sect. 9.9; their numerical results show that the soliton speed decreases monotonically with D until propagation failure occurs a at critical value D_c of the diffusion coefficient which is very low compared to a model with a pure polynomial potential.

Appendix 11A. Derivation of the Burgers equation

Expressing dV in terms of the variables (n,t) and the new variables (s, T) defined by (11.6), we have

$$dV = \frac{\partial V}{\partial n} dn + \frac{\partial V}{\partial t} dt = \frac{\partial V}{\partial s} \varepsilon(dn - c_0 dt) + \frac{\partial V}{\partial T} \varepsilon^2 dt, \qquad (11A.1a)$$

and identifying the coefficients of dt and dn we get

$$\frac{\partial}{\partial n} = \varepsilon \frac{\partial}{\partial s}, \qquad \frac{\partial}{\partial t} = -\varepsilon c_0 \frac{\partial}{\partial s} + \varepsilon^2 \frac{\partial}{\partial T}. \qquad (11A.1b)$$

Substituting these differential operators and the voltage expansion (11.7) in (11.5) yields

$$\left[\varepsilon^2 c_0^2 \frac{\partial^2}{\partial s^2} - 2\varepsilon^3 c_0 \frac{\partial^2}{\partial s \partial T} + \varepsilon^4 \frac{\partial^2}{\partial T^2} \right] (\varepsilon V_1 - b\varepsilon^2 V_1^2) \qquad (11A.2)$$

$$= c_0^2 \varepsilon^3 \frac{\partial^2 V_1}{\partial s^2} + \frac{c_0^2}{12} \varepsilon^5 \frac{\partial^4 V_1}{\partial s^4} + \frac{R}{L} \left[-\varepsilon c_0 \frac{\partial}{\partial s} + \varepsilon^2 \frac{\partial}{\partial T} \right] (\varepsilon V_1 - b\varepsilon^2 V_1^2).$$

Keeping terms to lowest order O (ε^4), we find that

$$2c_0 \frac{\partial^2 V_1}{\partial s \partial T} + c_0 b \frac{\partial^2 V_1^2}{\partial s^2} - \frac{R}{L} \frac{\partial^3 V_1}{\partial s^3} = 0, \qquad (11A.3)$$

Integrating with respect to s and rearranging, we find the Burgers'equation

$$\frac{\partial V_1}{\partial T} - D \frac{\partial^2 V_1}{\partial s^2} + \lambda V_1 \frac{\partial V_1}{\partial s} = 0. \qquad (11A.4)$$

Appendix 11B. Solution of the reaction diffusion equation

Substituting boundary conditions (11.29) in (11.27b) gives

$$v \frac{dn}{ds} + D \frac{d^2 n}{ds^2} - \alpha (n - n_1) (n - n_2) (n - n_3) = P(n) = 0. \qquad (11B.1)$$

With the experience gained from the solution of (11.14), as an ansatz we try (see Murray 1993) making n satisfy

$$\frac{dn}{ds} = K (n - n_1) (n - n_2), \qquad (11B.2)$$

303

where a is a constant to be determined. The solution of (11B.2) tends exponentially to n_1 and n_2 as $s \to \pm \infty$. Thus, substituting (11B.2) into (11B.1) yields

$$P(n) = vK(n - n_1)(n - n_2) + DK^2(n - n_1)(n - n_2)(n - n_2 + n - n_1)$$

$$- \alpha(n - n_1)(n - n_2)(n - n_3),$$

which is rearranged into

$$P(n) = (n - n_1)(n - n_2)\{(2DK^2 - \alpha)n - [DK^2(n_1 + n_2) - \alpha n_3 - vK]\}.$$

$$(11B.3)$$

$P(n)$ will be zero if

$$2DK^2 - \alpha = 0, \qquad\qquad DK^2(n_1 + n_2) - \alpha n_3 - vK = 0. \qquad (11B.4 \text{ ab})$$

These two relations determine K and the wave velocity v as

$$K = \pm\sqrt{\frac{\alpha}{2D}}, \qquad\qquad v = \pm\sqrt{\frac{\alpha D}{2}}(n_1 - 2n_3 + n_2). \qquad (11B.5)$$

The solutions of (11B.2) satisfy the full equation (11.27b) if K and v are given by relations (11B.5). The actual solution n is then obtained by solving (11B.2) which, in fact, is similar to equation (11.14) (see Sect 11.1.3). We obtain

$$n = \frac{n_2 + n_1}{2} - \frac{n_2 - n_1}{2} \tanh\left[\frac{K}{2}(n_2 - n_1)s\right], \qquad (11B.6)$$

where $s = x - vt$, and K (>0) and v (>0) are given by (11B.4 ab).

Appendix 11C. Equation of motion of an Euler strut

An Euler strut consists (Jackson 1992) of a light elastic (metal) strip with its bottom edge attached to a support (see Fig 11C.1(a)) and two masses on opposite sides of the strip (to maintain symmetry) at a distance d, which acts as a control parameter like for the pendulum considered in Appendix 6E. We assume that the torque generated by the strip about the bottom axis, when it is displaced by an angle θ, is simply proportional to this angle: the restoring torque due to flexion is $-K\theta$. It acts in opposition with the torque $-Mgd \sin\theta$ produced by gravity. Second, to a first approximation we assume that the distance d (and consequently the moment of inertia $I = Md^2$) does not depend on θ. The strip also experiences a damping force $-\beta d\theta/dt$, where β is the damping coefficient: dissipation is inherent to the flexion

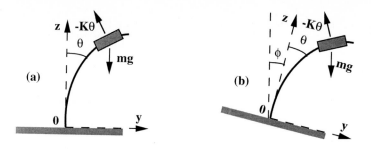

Fig.11C.1a,b. Schematic representation of an Euler strut. **(a)** When the axis Oz perpendicular to the support is vertical . **(b)** When the support is tilted Oz makes an angle φ with the vertical axis.

Under these assumptions we obtain (see Jackson 1992) an equation of motion similar to (5E.1) :

$$I\frac{d^2\theta}{dt^2} = -K\theta + mgd\sin\theta - \beta\frac{d\theta}{dt} .$$ (11C.1)

If damping effects are large compared to inertial effects, that is, if $\beta d\theta/dt \gg Id^2\theta/dt^2$, and if the support of the strut is tilted by an angle φ (see Fig. 11C.1 (b)) which breaks the symmetry, (11C.1) is approximated by an overdamped equation of motion

$$\beta\frac{d\theta}{dt} = -K\theta + mgd\sin(\theta + \phi).$$ (11C.2)

References

Abdullaev, F., Darmanyan, S., and Khabibullaev, P. (1993) *Optical Solitons.* (Springer)

Ablowitz, M. J., and Clarkson, P. (1992) *Solitons, Nonlinear Evolutions and Inverse Scatttering* (Cambridge University Press)

Ablowitz, M. J., and Segur, H. (1981) *Solitons and the Inverse Scattering Transform,* Siam Studies in Applied Mathematics, Vol 4 (Philadelphia)

Aceves, A. B., De Angelis, C., Trillo, S., and Wabnitz, S. (1994) Storing and steering of self-trapped discrete solitons in nonlinear waveguide arrays, Opt. Lett, **19**, 332–334

Agrawal, G. S. (1989) *Nonlinear Fiber Optics* (Academic Press)

Airy, G. B. (1845) *Tides and Waves,* Encyc. Metrop. (Fellowes)

Akhiezer, A.I., and Borovik, A.E. (1967), Sov. Phys. JETP, **25**, 332–338

Akmanov, S. A., Shukorukov, A. P., and Khoklov, R. V. (1968) Self-focusing and diffraction of light in a nonlinear medium, Sov. Phys. Usp. **93**, 609–636

Akmediev, N. N., and Ankiewicz, A. (1997) *Solitons, Nonlinear Pulses and Beams* (Chapman and Hall)

Alexandrov, A. S., and Mott, N. F. (1998) *Polarons and Bipolarons* (World Scientific)

Anderson, P. W., and Rowell, J. M. (1963) Probable observation of the Josephson superconductive effect, Phys. Rev. Lett. **10**, 230–232

Aubry, S. (1997) Breathers in nonlinear lattice: existence, linear stability and quantization, Physica D **103**, 201–250

Aubry, S., and Abramovici, G. (1990) Chaotic trajectories in the standard map, the concept of antiintegrability, Physica D **43**, 199–249

Aubry, S. (1976) A unified approach to the interpretation of displacive or order-disorder systems, J. Chem. Phys. **64**, 3392–3402

Baldock, G. R., and Bridgeman, T. (1981) *Mathematical Theory of Wave Motion* (Ellis Horwood , Wiley)

Barone, A., and Paterno, G. (1982) *Physics and Application of the Josephson Effect* (Wiley)

Barone, A., Esposito, F., Magee, C. J., and Scott, A. C. (1971) Theory and applications of the Sine–Gordon equation, Riv. Nuovo. Cimento. **1**, 227–267

Batchelor, G. K. (1967) *An Introduction to Fluid Mechanics* (Cambridge University Press)

Bazin, H. (1865) Recherches expérimentales relatives aux remoux et à la propagation des ondes, in *Recherches Hydrauliques,* ed. by Darcy H., and Bazin H., (Imprimerie Impériale)

Benjamin, T. B., and Feir, J. E. (1967) The disintegration of wavetrains on deep water. Part 1 Theory, J. Fluid Mech. **27**, 417–430

Benney, D. (1962) Nonlinear gravity waves interactions, J. Fluid Mech. **14**, 577–584

Benney, D. J., and Newell, A. C. (1968) The propagation of nonlinear waves envelope, J. Math. Phys. **46**, 133–139

Bespalov, V. I., and Talanov, V. I. (1966) Filamentary structures of light beams in nonlinear liquids, Zh. Eksp. Teor. Fiz. Pis'ma. **3**, 471–476 [(1966) JETP Lett, **3**, 307–309]

Bettini, A., Minelli, T. A., and Pascoli, D. (1983) Solitons in undergraduate laboratory, Am. J. Phys. **51**, 977–984

Bhatnagar, P. L. (1979) *Nonlinear Waves in One-Dimensional Dispersive Systems* (Clarendon)

Bilbault, J. M., Marquié, P., and Michaux, B. (1995) Modulational instability of two counterpropagating waves in the experimental transmission line, Phys. Rev. E. **51**, 817–820

Binczak, S., Comte, J. C., Michaux, B., Marquié, P., and Bilbault, J. M. (1998) Experimental nonlinear electrical reaction-diffusion lattice, Electron. Lett. **34**, 1061–1062

Bishop, A. R., and Schneider, T. (eds) (1978) *Solitons and Condensed Matter Physics*, Proc. Symp. Nonlinear Structure and Dynamics in Condensed Matter, Oxford, June 1978 (Springer)

Boussinesq, J. (1871) Théorie de l'intumescence liquide appelée onde solitaire ou de translation, se propageant dans un canal rectangulaire, C. R. Acad. Sci. Paris. **72**, 755–759

Boyd, J. P. (1989) Weakly nonlocal solitary waves, in *Mesoscale/Synoptic Coherent Structures in Geophysical Turbulence*, ed by Nihoul, J. C. J., and Janart, B. M. (Elsevier)

Boyd, J. P. (1990a) A numerical calculation of a weakly nonlocal solitary wave: the ϕ^4 breather, Nonlinearity **3**, 177–195

Boyd, J. P. (1990b) New directions in solitons and nonlinear periodic waves, in *Advances in Applied Mechanics,* ed by Hutchinson, J. W., and T. Y. Wu (Academic)

Braun, O. M., and Kivshar, Y. S. (1998) Nonlinear dynamics of the Frenkel–Kontorova model, Phys. Rep. **306**, 1–108

Bressloff, P. C., and Rowlands, G. (1997) Exact travelling wave solutions of an "integrable" discrete reaction-diffusion equation, Physica D **106**, 255–269

Bullough, R. K., and Caudrey, P. (Eds) (1980) *Solitons* (Springer)

Burgers, J. M. (1929) *The Nonlinear Diffusion Equation* (Riedel)

Burgers, J. M. (1948) A mathermatical model illustrating the theory of turbulence, Adv. Appl. Mech. **1**, 171-199

Burlakov, V. M., Darmanyan, S. A., and Pyrkov, V. N. (1996) Modulational instability and recurrence phenomena in anharmonic lattices, Phys. Rev E **54**, 3257–3265

Buryak, V. (1995) Stationary soliton bound states existing in resonance with linear waves, Phys. Rev E **52**, 1156–1163

308

Cadet, S. (1987) Transverse envelope solitons in an atomic chain, Phys. Lett. A **1321**, 177–181

Cai, D., Bishop, A. R., and Gronbech-Jensen, N. (1994) Localized states in discrete nonlinear Schrödinger equations, Phys. Rev. Lett. **72**, 591–595

Calogero, F., and Degasperis, A. (1981) *Solitons and Inverse Scattering Transforms* (North-Holland)

Carman, E., Case, M., Kamegawa, M., Yu, R., Giboney, K., and Rodwell, M.J.W.(1992) V-band and W-broadband, monolithic frequency multpliers, IEEE Microwave Guided-Wave Lett. **2**, 253–254

Cavalcanti, S. B., Agrawal, G. P., and Yu, M. (1995) Noise amplification in dispersive nonlinear media, Phys. Rev. A **51**, 4087–4092

Charru, F. (1997) A simple mechanical system mimicking phase transitions in a one-dimensional medium, Eur. J. Phys. **18**, 417–424

Chen, J. T., Finnegan, T. F., and Langenberg, D. N. (1971) Physica **55**, 413–420

Chiao, R. Y., Garmire, E., and Townes, C. H. (1964) Self trapping of optical beams, Phys. Rev. Lett, **13**, 479–482

Christodoulides, D. N., and Joseph, R. I. (1988) Discrete self-focusing in nonlinear arrays of coupled waveguides, Opt. Lett. **13**, 794–796

Claude, C., and Leon, J. (1995) Theory of pump depletion and spike formation in stimulated Raman scattering, Phys. Rev. Lett. **74**, 3479–3482

Claude, C., Ginovart, F., and Leon, J. (1995) Nonlinear theory of transient stimulated Raman scattering an its application to long-pulse experiments, Phys. Rev. A **52**, 767–782

Cole, J. D. (1951) On a quasilinear parabolic equation occuring in aerodynamics. Q. Appl. Math. **9**, 225–236.

Collins, M. A, (1981) A quasi-continuum approximation for solitons in an atomic chain, Chem. Phys. Lett. **77**, 342–347

Collins, M. A. (1983) Solitons in chemical physics, Adv. Chem. **53**, 225–340

Condat. C.A., Guyer. R. A. and Miller, M.D. (1983) Double Sine-Gordon chain, Phys. Rev. B **27**, 474–489

Costabile, G., Parmentier. R. D., and Savo, B. (1978) Exact solutions of the Sine-Gordon equation describing oscillations in a long (but finite) Josephson Junction, Appl. Phys. Lett. **32**, 587–589

Coste, C., Falcon, E., and Fauve, S. (1997) Solitary wave in a chain of beads under Hertz contact, Phys. Rev. E **56**, 6105–6117

Craik, A. D. (1985) *Waves Interactions and Fluid Flows* (Cambridge University Press)

Crane, H.D. (1962) Neuristor. A novel device and system concept, Proc. IRE **50**, 2048–2060

Crapper, G. D. (1984) *Introduction to Water Waves* (Ellis Horwood)

Crawford, D. R., Lake, B. M., Saffman, P. G., and Yuen, H. C. (1981) Stability of deep nonlinear water waves in two and three dimensions, J. Fuid. Mech. **105**, 177–191

Crawford, F. S. Jr (1965) *Waves*, Berkeley Physics Course (McGraw Hill)

Cross, M. C., and Hohenberg, P. C. (1993) Pattern formation outside of equilibrium, Rev. Mod. Phys. **65**, 851–1112

Currie, J. F., Trullinger, S. E., Bishop, A. R., and Krumshansl, J. A. (1977) Numerical simulations of Sine–Gordon dynamics in the presence of perturbations, Phys. Rev. B **15**, 5567–5580

Davidson, A., Ducholm, B., and Pedersen, N. F. (1986) Experiments on soliton motion in annular Josephson Junctions, J. Appl. Phys. **60**, 1447–1454

Davidson, C. W. (1978) *Transmission Lines for Communications* (Macmillan)

Davydov, A. S. (1985) *Solitons in Molecular Systems* (Reidel)

De Sterke, C. M., and Sipe, J. E. (1988) Envelope function approach for the electrodynamics of nonlinear periodic structures, Phys. Rev. A **38**, 5149–5165

Debye, P. (1914) In *Vorträge über die kinetische der Materie und der Elektrizität*, ed. by Planck, M. (Teubner)

Dehrmann, T. (1982) Travelling kinks in Schlögl's second model for non-equilibrium, J. Phys. Math. Gen. **15**, L649–L652

Dey, B. (1998) Compactons solutions for a class of two parameter generalized odd-order Korteweg-de Vries equations, Phys. Rev. E **57**, 4733–4738

Dey, B., and Khare, A. (1998) Stability of compacton solutions, Phys. Rev E **58**, R2741–R2744

Dikande A. M., and Kofane, T.C. (1991) Exact kink solutions in a new nonlinear hperbolic double-well potential, J. Phys: Condens. Matt. **3**, 5203–5206

Dodd, R. K., Eilbeck, J. C., Gibbon, J. D., and Morris, H. C. (1982) *Solitons and Nonlinear Wave Equations* (Academic Press)

Doderer, T. (1997) Microscopic imaging of Josephson junction dynamics, Int. J. Mod. Phys. **11**, 1979–2042

Dolgov, A. S. (1986) The localization of vibrations in a nonlinear crystalline structure, Fiz. Tverd. Tela, **28**, 1641–1644 [Sov. Phys. Solid State **28**, 907–909]

Doran, N. J., and Blow, K. J. (1983) Solitons in optical communications, IEEE. J. Quant. Electron. QE-**19**, 1883–1888

Dragoman, M., Kremer, R., and Jäger, D. (1993) Pulse generation and compression on a travelling-wave MMIC Schottky diode array, in *Ultra-Wideband, Short-Pulse Electromagnetics,* ed. by Bertoni, A. L. et al. (Plenum)

Drazin P. G., and Johnson R. S. (1989) *Solitons: An Introduction.* (Cambridge University Press)

Drühl, K., Wenzel, R.G., and Carlsten, J. L. (1983) Observation of solitons in stimulated Raman scattering, Phys. Rev. Lett. **51**, 1171–1174

Drummond, P. D., Shelby, R. M., Friberg, S. R., and Yamamoto, Y. (1993) Quantum solitons in optical fibers, Nature **365**, 307–313

Dusuel, S., Michaux, P., and Remoissenet, M. (1998) From kinks to compacton-like kinks, Phys. Rev. E **57**, 2320–2326

Dysthe, K. B. (1979) Note on a modification to the nonlinear Schrödinger equation for application to deep water waves, Proc. R. Soc. London. A **369**, 105–114

Eilbeck, J. C. (1991) Numerical studies of solitons on lattices, in *Nonlinear Coherent Structures in Physics and Biology*, Proceedings of the 7th Interdisciplinary Workshop, Dijon 1991, ed. by Remoissenet, M., and Peyrard, M. (Springer)

Eilbeck, J.C., Lomdahl, P.S., and Scott, A. C. (1985) The discrete self trapping equation, Physica D **16**, 318–338

Eisenberg, H. S., and Silberberg, Y. (1998) Discrete spatial optical solitons in waveguide arrays, Phys. Rev. Letters **81**, 3383–3386

Elmore, W. C., and Heald, M. A. (1985) *Physics of Waves* (Dover)

Erneux, T., and Nicolis, G. (1993) Propagating waves in discrete bistable reaction-diffusion systems, Physica D **67**, 237–244

Fath, G. (1998) Propagating failure of traveling waves in a discrete bistable medium, Physica D **116**, 176–190

Feir, J. E. (1967) Discussion: some results from wave pulse experiments, Proc. R. Soc. London A **299**, 54–58

Fermi, E., Pasta, J. R., and Ulam, S. M. (1955) Studies of nonlinear problems, Los Alamos Sci. Lab. Rep., LA-1940

Feynman, R. P., Leighton, R. B., and Sands. M. (1965) The Schrödinger equation in a classical context: a seminar on superconductivity, in *The Feynman Lectures on Physics* Vol. 3, Sect. 21. 9 (Addison-Wesley) .

Fitzhugh, R. (1961) Impulse and physiological states in theoretical models of nerve membrane, Biophys. J. **1**, 445–466

Flach, S. (1994) Conditions on the existence of localized excitations in nonlinear discrete systems, Phys. Rev. E **50**, 3134–3142

Flach, S. (1995) Existence of nonlinear localized excitations in nonlinear Hamiltonian lattices, Phys. Rev. E **51**, 1503–1507

Flach, S., and Willis, C. R. (1998) Discrete Breathers, Phys. Rep. **295**, 181–264

Fogedby, H. C. (1998) Solitons and diffusive modes in the noiseless Burgers equation: stability analysis, Phys. Rev. E **57**, 2331–2337

Ford, J. J. (1961) Equipartition of energy for nonlinear systems, J. Math. Phys. **2**, 387–393

Frenkel, J., and Kontorova, T. (1939) On the theory of plastic deformation and twinning, J. Phys. **1**, 137–149

Fujmaki, A., Nakajima, K., and Sawada, Y. (1987) Spatio-temporal observation of the soliton–antisoliton collision in Josephson transmission lines, Phys. Rev. Lett. **59**, 2895–2198

Fulton, T. A., and Dynes, R. C. (1973) Single vortex propagation in Josephson tunnel junction, Solid State. Comm. **12**, 57–61

Gardner, C. S., Green., J. M., Kruskal, M. D., and Miura, R. M. (1967) Method for solving the Korteweg De Vries equation, Phys. Rev. Lett. **19**, 1095–1097

Gelfand, I. M., and Levitan, B. M. (1951) On the determination of a differential equation from its spectral function, Am. Math. Soc. Transl. **1**, 253–304

Gloge, D. (1979) The optical fiber as a transmission medium, Rep. Prog. Phys. **42**, 1777–1824

Gorbatcheva, O. B., and Ostrowsky, L. A. (1983) Nonlinear vector waves in a mechanical model of a molecular chain, Physica D **8**, 223–228

Green, D. E., and Ji, S. (1972) The electomechanochemical model of mitochondrial structure and function, in *The Molecular Basis of Electron Transport*, ed by Schultz, J., and Cameron, B.F. (Academic Press)

Grinrod, P. (1991) *Pattern and Waves* (Oxford)

Haken, H. (1978) *Synergetics. An Introduction*, 2nd Edtion (Springer)

Hammack, J. L. (1973) A note on Tsunamis: their generation and propagation in an ocean of uniform depth, J. Fluid. Mech. **60**, 769–800

Hammack, J. L., and Henderson, D.M. (1993) Resonant interactions among surface waves, Ann. Rev. Fluid Mech., **25**, 55–97

Hammack, J. L., and Segur, H. (1974) The Korteweg-de Vries equation and water waves. Part 2, Comparisons with experiments, J. Fluid. Mech. **65**, 289–314

Hasegawa, A. (1983) Amplification and reshaping of optical soliton in a glass fiber. 4. Use of stimulated Raman process, Opt. Lett. **8**, 650–654

Hasegawa, A. (1984) Generation of a train of soliton pulses by induced modulational instability in optical fibers, Opt. Lett. **9**, 288–290

Hasegawa, A. (1989) *Optical Solitons in Fibers* (Springer)

Hasegawa, A., and Brinkman, W. F. (1980) Tunable coherent IR and FIR sources utilizing modulational instability, IEEE J. Quant. Electron. QE-**16**, 694–697

Hasegawa, A., and Kodama, Y. (1995) *Solitons in Optical Communications* (Clarendon)

Hasegawa, A., and Kodama, Y. (1981) Signal transmission by optical solitons in monomode fiber, Proc. IEEE **69**, 1145–1150

Hasegawa, A., and Tappert, F. (1973) Transmission of stationnary nonlinear optical pulses in dispersive dielectric fibers. 1. Anomalous dispersion, Appl. Phys. Lett. **23**, 142–144

Hashizume, Y. (1985) Nonlinear pressure waves in a fluid-filled elastic tube, J. Phys. Soc. Japan **54**, 3305–3312

Hashizume, Y. (1988) Nonlinear pressure wave propagation in arteries, J. Phys. Soc. Japan **57**, 4160–4168

Haus, H. A., and Wong, W. S. (1996) Solitons in optical communications, Rev. Mod. Phys. **68**, 423–444

Hirose, A., and Lonngreen, K. E. (1985) *Introduction to Wave Phenomena* (Wiley)

Hirota, R. (1971) Exact solution of the Korteweg-de-Vries equation for multiple collisions of solitons, Phys. Rev. Lett. **27**, 1192–1194

Hirota, R. (1973) Exact envelope soliton solutions of a nonlinear wave equation, J. Math. Phys. **14**, 805–809

Hirota, R. (1976) Direct methods of finding exact solutions of nonlinear evolution equations, in *Backlünd Transformation and Its Relation to the Inverse Scattering Problem*, ed. by Miura, R. M. (Springer)

Hirota, R., and Suzuki, K. (1973) Theoretical and experimental studies of lattice solitons in nonlinear lumped networks, Proc. IEEE **61**, 1483–1490

Hochstrasser, D., Mertens, F. G., and Büttner. H. (1988) An iterative method for the calculation of narrow solitary excitations on atomic chains, Physica D **35**, 259–266

Hodgkin, A. L., and Huxley, A. F. (1952) A quantitative description of membrane current and its application to conduction and excitation in nerve, J. Physiol. (London) **117**, 500–544

Holtstein, T. (1959) Studies of polaron motion. Part I. The molecular crystal model. Part II. The "small" polaron, Ann. Phys. (NY) **8**, 325–389

Hopf, E. (1950) The partial differential equation $u_t + uu_{xx} = \mu u_{xx}$, Comm. Pure Appl. Math. **3**, 201–230

Hyman, J. M., McLaughlin, D. W., and Scott, A. C. (1981) On Davydov's alpha-helix soliton, Physica D **3**, 23–44

Infeld, E., and Rowlands, G. (1990) *Nonlinear Waves, Solitons and Chaos* (Cambridge University Press)

Jackson, E. A. (1963) Nonlinear coupled oscillators. I. Perturbation theory: ergodic problems, J. Math. Phys. **4**, 551–558

Jackson, E. A. (1992) *Perspectives of Nonlinear Dynamics* (Cambridge University Press)

Jäger, D. (1985) Characteristics of travelling waves along the nonlinear transmission lines for monolithic integrated circuits: a review, Int. J. Electron. **58**, 649–669

Jäger, D., Block, M., Kaiser, D., Welters, A., and Von Wendorff, W. (1991) Wave propagation phenomena and microwave-optical interaction in coplanar lines on semiconductor substrate, J. Electr.Waves and Appl. **5**, 337–351

Jakubowski, M. H., Steiglitz, K., and Squier, R. (1997) Information transfer between solitary waves in the saturable Schrödinger equation, Phys. Rev. E **56**, 7267–7272

Jensen, M. H., Per Bak, P., and Popielewicz, A. (1983) Pinning-free soliton lattices and bifurcation in a discrete double-well model: exact results, J. Phys. A: Math. Gen. **16**, 4369–4375

Josephson, B. D. (1962) Possible new effects in superconductor tunneling, Phys. Lett. **1**, 251–253

Josephson, B. D. (1964) Coupled superconductors, Rev. Mod. Phys. **36**, 216–220

Kalinikos, B. A., Kovishov, N. G., and Patton, C. E. (1997) Decay free microwave magnetic envelope soliton pulse trains in ytrium iron garnet thin films, Phys. Rev. Lett. **78**, 2827–2830

Kalinikos, B. A., Kovishov, N. G., and Patton, C. E. (1998) Self-generation of microwave magnetic envelope soliton trains in ytrium iron garnet thin films, Phys. Rev. Lett. **80**, 4301–4304

Kalinikos, B. A., Kovshikov, N. G., and Slavin, A. N. (1988) Envelope solitons and modulational instability of dipole-exchange magnetization waves in ytrium iron garnet films, Sov. Phys. JETP. **67**, 303–312

Kalinikos, B. A., Kovshikov, N. G., and Slavin, A. N. (1990) Spin wave envelope solitons in thin ferromagnetic films, J. Appl. Phys. **67**, 5633–5638

Kalinikos, B.A., Kovshikov, N. G., and Slavin, A. N. (1983) Nonlinear envelope spin waves in ferromagnets, JETP Lett. **38**, 413–420

Karpman, V. I. (1967) Self-modulation of nonlinear plane waves in dispersive media, Zh. Eksp. Teor. Fiz. Pis'ma, **6**, 829–832 [(1967) JETP Lett, **6**, 277–279

Karpman, V. I. (1975) *Nonlinear Waves in Dispersive Media* (Pitman)

Keener, J. P. (1980) Waves in excitable media, SIAM J. Appl. Math. **39**, 528–548

Keener, J. P. (1987) Propagation and its failure in coupled systems of discrete excitable cells, SIAM J. Appl. Math. **47**, 566–572

Kemeny, G., and Goklany, I.M. (1973) Polarons and conformons, J. Theor. Biol, **40**, 107–123

Kinsman, B. (1965) *Wind Waves* (Prentice Hall)

Kiselev, S. A., Bickam, S. R., and Sievers, A. J. (1995) Properties of intrinsic localized modes in one-dimensional lattices, Comment. Cond. Mat. Phys. **17**, 135–73

Kisvelev, S. A. (1990) Stationary vibrational modes of a chain of particles interacting via an even order potential, Physics. Lett. A **148**, 95–97

Kivshar, Y. S. (1993) Dark solitons in nonlinear optics, IEEE J. Quant. Elec. QE-**29**, 250–263

Kivshar, Y. S., and Luther-Davies, B. (1998) Optical dark solitons: physics and applications, Phys. Rep. **298**, 81–155

Kivshar, Y. S., and Peyrard, M. (1992) Modulational instabilities in discrete lattices, Phys. Rev. A **46**, 3198–3205

Kivshar, Y. S., and Salerno, M. (1994) Modulational instabilities in the discrete deformable nonlinear Schrödinger equation, Phys. Rev. E **49**, 3543–3546

Kivshar, Y. S., Champneys, A. R., Cai, D., and Bishop, A. R. (1998) Multiple states of intrinsic localized modes, Phys. Rev. B **58**, 5423–5428

Kodama, Y., and Ablowitz, M. J. (1980) Perturbations of solitons and solitary waves, Stud. Appl. Math. **64**, 225–245

Kodama, Y., and Hasegawa, A. (1987) Nonlinear pulse propagation in a monomode dielectric guide, IEEE J. Quant. Electron QE-**23**, 510–524

Kofane, T., Michaux, B., and Remoissenet, M. (1988) Theoretical and experimental studies of diatomic lattice solitons using an electrical transmission line, J. Phys. C: Solid State Phys. **21**, 1395–1412

Konwent. H., Machnikowski. P., and Radosz. A. (1996) Dynamics of a hydrogen-bonded linear chain with a new type of one-particle potential, J. Phys. C: Condens. Matter **8**, 4325–4338

Korteweg, D. J., and De Vries, G. (1895) On the change of form of long waves advancing in a rectangular channel, and on a new type of long stationary waves, Phil. Mag. **39** (5), 442–443

Kosevich, A. M., and Kovalev, A.S. (1974) Selflocalization of vibrations in a one-dimensional anharmonic chain, Zh. Eksp. Teor. Fiz. **67**, 1793-1798 [Sov Phys. JETP **40**, 891–896]

Kovshikov, N.G., Kalinikos, B.A., Patton, C.E., Wright, E. S., and Nash, J.M. (1996) Formation, propagation, reflection, and collision of microwave envelope solitons in ytrium iron garnet films, Phys. Rev. B **54**, 15210–15223

Krumhansl, J. A., and Schrieffer, J. R. (1975) Dynamics and statistical mechanics of a one-dimensional model Hamiltonian for structural phase transitions, Phys. Rev. B **11**, 3535–3545

Kundu, P. K. (1990) *Fluid Mechanics* (Academic)

Kuwabara, G., Hasegawa, T. and Kono, K. (1986) Water waves in a ripple tank, Am. J. Phys. **54**, 1002–1007

Lai, R., and Sievers, A. J. (1999) Nonlinear nanoscale localization of magnetic excitations in atomic lattices, Phys. Rep. **314**, 147–236

Lai, Y., and Haus, H. A. (1989a) Quantum theory of optical solitons in fibers. I. Time dependent Hartree approximation, Phys. Rev. A **40**, 844–853

Lai, Y., and Haus, H. A. (1989b) Quantum theory of optical solitons in fibers. II. Exact solution, Phys. Rev. A **40**, 854–866

Lake, B. M., Yuen, H. C., Rungaldier, H., and Ferguson, W. E. (1977) Nonlinear deep water waves: theory and experiment. 2. Evolution of a continuous wave train, J. Fluid. Mech. **83**, 49–74

Lamb, G. L. (1971) Analytical descriptions of ultrashort optical pulse propagation in a resonant medium, Rev. Mod. Phys. **43**, 99–124

Lamb, G. L. (1980) *Elements of Soliton Theory* (Wiley)

Lamb, H. (1984) *Hydrodynamics* (Dover)

Landau, L. (1933) Über die bewegung der elektronen in kristallgitter, Phys. Z. Sowietuninon **3**, 664–665

Landau, L. D., and Lifshitz, E. (1954) *Fluid Mechanics* (Pergamon)

Landau, L. D., and Lifshitz, E. (1958) *Quantum Mechanics*, 2nd Edition (Pergamon)

Landauer, R. (1960) Shock waves in nonlinear transmission lines and their effect on parametric amplification, IBM J. Res. Devel. **4**, 391–401

Lang, S. (1973) *Elliptic Functions* (Addison-Wesley)

Lax, P. D. (1968) Integrals of nonlinear equations of evolution and solitary waves, Commun. Pure Appl. Math. **21**, 467–490

Lazaridi, A. N., and Nesterenko, V. V. (1985) Observation of a new type of solitary waves in a one-dimensional medium, J. Appl. Mech. Technol, **26**, 405-408

Leblond, P. H., and Mysak, P. A. (1978) *Waves in the Ocean* (Elsevier)

Leon, J., and Mikhailov, A. V. (1999) Soliton generation in SRS, Phys. Lett A **253**, 33–37

Levring, O. A., Perdersen, N. F., and Samulsen, M. R. (1982) Fluxon motion in long overlap and inline Josephson junctions, Appl. Phys. Lett. **40**, 846–847

Lighthill, J. (1978) *Waves in Fluids* (Cambridge University Press)

Lighthill, M. J. (1965) Contribution to the theory of waves in nonlinear dispersive systems, J. Inst. Math. Appl. **1**, 269–306

Likharev, K . K. (1986) *Dynamics of Josephson Junctions and Circuits* (Gordon and Breach)

Lomdahl, P. S. (1985) Solitons in Josephson junctions: an overview, J. Stat. Phys. **39**, 551–561

Lomdahl, P. S., Soerensen, O. H., and Christiansen, P. L. (1982) Soliton excitations in Josephson tunnel junctions, Phys. Rev. B **25**, 5737–5748

Longuet-Higgins, M. S. (1980) Modulation of the amplitude of steep wind waves, J. Fluid. Mech. **99**, 705–713

Lonngren, K. E. (1978) Observation of solitons on nonlinear dispersive transmission lines, in *Solitons in Action*, ed. by Lonngreen, K. E., and Scott, A. C. (Academic Press)

Lonngren, K. E., Andersen, D. R., and Robinson, J. P. (1991) A soliton secure lock and key, Optical computing and processing **4**, 349–351

Lonngren, K., and Scott, A. C. (Eds) (1978) *Solitons In Action* (Academic Press)

Ludu, A., and Draayer, J. P. (1998) Nonlinear modes of liquid drops as solitary waves, Phys. Rev. Lett **80**, 2125–2128

Lukomskii, V. P. (1978) Nonlinear magnetostatic waves in ferromagnetic plates, Ukrainskii. Fiz. Zh. **23**, 134–139

Lynch, D. K. (1982) Tidal bores, Scientific American **247**, 134–143

MacKay R. S., and Aubry. S. (1994) Proof of existence of breathers for time-reversible or Hamiltonian networks of weakly coupled oscillators, Nonlinearity **7**, 1623–1643

Magyari, E. (1982) Travelling kinks in Schlögl's second model for non-equilibrium phase transitions, J. Phys. Math. Gen. **15**, L139–L142

Main, I. G. (1993) *Vibrations and Waves in Physics* (Cambridge University Press)

Makhankov, V. G. (1990) *Soliton Phenomenology* (Kluwer)

Malfliet, W. (1998) Remark on shock wave in a nonlinear LC circuit, J. Phys. Soc. Japan **67**, 701

Marchenko, V. A. (1955) On the reconstruction of the potential energy from phases of the scattered waves, Dokl. Acad. Nauk. SSR **104**, 695–698

Marquié, P., Bilbault, J. M., and Remoissenet, M. (1994) Generation of envelope and hole solitons in an experimental transmission line, Phys. Rev. E **49**, 828–835

Marquié, P., Bilbault, J. M., and Remoissenet, M. (1995) Observation of nonlinear localized modes in an electrical lattice, Phys. Rev. E **51**, 6127–6133

Marquié, P., Binczak, S., Comte, J. C., Michaux, B., and Bilbault, J. M. (1998) Diffusion effects in a nonlinear electrical lattice, Phys. Rev. E **57**, 6075–6078

Matsuda, A. (1986) Observation of fluxon–antifluxon collision in a Josephson transmission line, Phys. Rev. B **34**, 3127–3135

Matsuda, A., and Uheara, S. (1982) Observation of fluxon propagation on Josephson transmission line, Appl. Phys. Lett. **41**, 770–772

Mc Donald, D. A. (1974) *Blood Flow in Arteries* (Edward Arnold)

Mc Laughlin, D. W., and Scott, A. C. (1978) Perturbation analysis of fluxon motion, Phys. Rev. A. **18**, 1652–1680

McKinstrie, C. J., and Luther, G. C. (1990) The modulational instability of colinear waves, Physica Scripta **30**, 31–40

Melville, W. K. (1982) The instability and breaking of deep-water waves, J. Fluid. Mech. **115**, 165–185

Menyuk, C. R. (1987) Nonlinear pulse propagation in birefringent optical fibers, IEEE J. Quant. Electron QE-**23**, 174–176

Meron, E. (1991) Pattern formation in excitable media, Phys. Rep, **218**, 1-66

Millot, G., Seve, E., and Wabnitz, S. (1997) Polarization symmetry breaking and pulse train generation from the modulation of light waves, Phys. Rev. Lett. **79**, 661-664

Mills, D. L. (1991) *Nonlinear Optics* (Springer)

Mizumura, M., and Noguchi, A. (1975) Modulational instability and envelope soliton in a nonlinear transmission line, Elec. and Comm. in Japan **58-A**, 9–10

Mollenauer, L. F., and Smith, K. (1988) Demonstration of soliton transmission over more than 4000 km in fiber with loss periodically compensated by Raman gain, Opt. Lett. **13**, 675–677

Mollenauer, L. F., and Stolen, R. H. (1982) Solitons in optical fibers, Fiberoptic Technology, April, 193–198

Mollenauer, L. F., and Stolen, R. H., and Islam, M. N. (1985) Experimental demonstration of soliton propagation in long fibers: loss compensated by Raman gain, Opt. Lett. **10**, 229–231

Mollenauer, L. F., Stolen, R. H., and Gordon, J. P. (1980) Experimental observation of picosecond pulse narowing and solitons in optical fibers, Phys. Rev. Lett. **45**, 1095–1098

Murdoch, S. G., Leonhardt, R., and Harvey, J. D. (1995) Polarization modulational instability in weakly birefringent fibers, Opt. Lett. **20**, 866–868

Muroya, K., Saitoh, N., and Watanabe, S. (1982) Experiment on lattice solitons by an nonlinear LC circuit, observation of a dark soliton, J. Phys. Soc. Japan **51**, 1024–1029

Murray, J. D. (1993) *Mathematical Biology*, 2nd Edition (Springer)

Nagashima, H., and Amagishi, H. (1979) Experiment on solitons in the dissipative Toda lattice using nonlinear transmission lines, J. Phys. Soc. Japan **47**, 2021–2027

Nagumo, J. S., Arimoto, S., and Yoshizawa, S. (1962) An active pulse transmission line simulating nerve axon, Proc. IRE **50**, 2061–2071

Nagumo, S., Yoshizawa., and Arimoto, S. (1965) Bistable transmission lines, IEEE Trans. Circuit Theory CT-**12**, 400–412

Nakajima, K., Yamashita, T., and Onodera, Y. (1974) Mechanical analogue of active Josephson transmission line, Phys. Rev B **45**, 3141–3145

Nesterenko, V. F. (1984) Propagation of nonlinear compression pulses in granular media, J. Appl. Mech. Tech. Phys, **5**, 733-743

Nesterenko, V. F. (1994) Solitary waves in discrete media with anomalous compressibility and similar to "sonic vacuum", J. Phys. IV **55**, C-8, 729–734

Newell, A. C. (1978) Nonlinear tunelling, J. Math. Phys. **19**, 1126–1133

Newell, A. C. (1985) *Solitons in Mathematics and Physics* (SIAM)

Newell, A. C., and Moloney, J. V. (1991) *Nonlinear Optics* (Addison-Wesley)

Nitta, J., Matsuda., A., and Kawakami, T. (1984) Propagation properties of fluxons in a well damped Josephson transmission line, Phys. Rev E **55**, 2758–2762

Novikov, S. P., Manakov, S. V., Pitaevski, L. P., and Zakharov, V. E. (1984) *Theory of Solitons* (Plenum)

Olsen, M., Smith, H., and Scott, A. C. (1984) Solitons in a wave tank, Am. J. Phys. **54**, 826–830

Osborne, A. R., and Petti, M. (1994) Laboratory generated shallow-water surface waves: analysis using the periodic, inverse scattering transform, Phys. Fluids **6**, 1727–1744

Osborne, A. R., Segre, E., Boffeta, G., and Cavaleri, L. (1991) Solsis states in shallow-water ocean surface waves, Phys. Rev. Lett. **64**, 1733–1736

Ostrovskii, L. A. (1966) Propagation of wave packets and space-time self focusing in a nonlinear medium, J. Exp. Theor. Phys. **51**, 1189–1194. [(1964) Sov. Phys. JETP. **24**, 797–800]

Ostrovskii, L. A., Papko, V. V., and Pelinovskii, E. N. (1974) Solitary electro-magnetic waves in nonlinear lines, Radiophysics and Quant. Electron **15**, 438–446

Ostrovskii, L.A. (1963) Electromagnetic waves in nonlinear media with dispersion, Zh. Tek. Fiz, **33**, 905-908 [(1964) Sov. Phys. Tech. Phys. **8**, 679–681]

Ovschinnikov, A.A. (1970) Localized long-lived vibrational states in molecular crystals, Sov. Phys. JETP **30**, 147–150 [(1969) Zh. Eksp. Fiz. **57**, 263–270]

Page, J. B. (1990) Asymptotic solution for localized vibrational modes in strongly anharmonic periodic systems, Phys. Rev B **41**, 7835–7838

Paquerot, J. F., and Remoissenet, M. (1994) Dynamics of nonlinear blood pressure waves in large arteries, Phys. Lett. A **194**, 77–82

Parmentier (1978) Fluxons in long Josephson Junctions, in *Solitons in Action*, ed by Lonngreen, K., and Scott, A. C. (Academic Press)

Parmentier R. D. (1967) Analysis of neuristor waveforms, Proc. IEEE Vol 12 1497–1498

Parmentier, R. D., and Pedersen, N.F. (Eds.) (1994) *Nonlinear Superconducting Devices and High-T_c Materials*, Proceedings of the international conference, Capri, Italy, 1994 (World Scientific)

Patton, C. E., Kabos, P., Xia, H., Kolodin, P. A., Zhang, H.Y., Staudinger, R., Kalinikos, B.A., and Kovshikov, N. G. (1998) Microwave magnetic envelope solitons in thin ferrite films, in Proceedings of the 4th International Symposium on Physics of Magnetic Materials, Sendai Japan 1998, J. Magn. Mat. Soc. of Japan **23**, No 1–2

Pauwellussen, J. P. (1982) One way of traffic of pulses in a neuron, J. Math. Biol **15**, 151–171

Pedersen, N. F. (1986) Solitons in Josephson transmission lines, in *Solitons*, ed. by Trullinger, S. E., Zakharov, V. E., and Prokovsky, V. L. (North-Holland)

Pedersen, N. F., and Welner, D. (1984) Comparison between experiments and perturbation theory for solitons in Josephson junctions, Phys. Rev. B **29**, 2551–2258

Pedley, T. J. (1980) *The Fluid Mechanics of Blood Vessels* (Cambridge University Press)

Peierls, R. E. (1961) *Quantum Theory of Solids* (Wiley)

Pekar, S.I. (1954) *Untersuchungen über die electronentheorie der kristalle*, (Akademie Verlag)

Peregrine, D.H. (1983) Water waves, nonlinear Schrödinger equations and their solutions, J. Austral. Math. Soc. Ser. B **25**, 16–43

Perring, J. K., and Skyrme, T. H. R . (1962) A model unified field equation, Nucl. Phys. **31**, 550–555

Peters, R.D. (1995) Chaotic pendulum based on torsion and gravity in opposition, Am. J. Phys. **63**, 1128–1136

Peterson, G. E. (1984) Electrical transmission lines as models for soliton propagation in materials : Elementary Aspects of Video Solitons, AT&T Bell Lab. Tech. J. **63**, 901–919

Peyrard, M. (1998) The pathway to energy localization in nonlinear lattices, Physica D **119**, 184–199

Peyrard, M., and Kruskal. M. (1984) Kinks dynamics in the highly discrete Sine–Gordon system, Physica D **14**, 88–102

Peyrard. M; and Remoissenet, M. (1982) Soliton like excitations in a one-dimensional atomic chain with a nonlinear deformable substrate potential, Phys. Rev. B **26**, 2886–2899

Philipps, O. M. (1960) Theoretical and experimental studies of gravity waves interactions, Proc. R. Soc. London. Ser. A **299**, 104–119

Philipps, O. M. (1981) Waves interaction, the evolution of an idea, J. Fluid. Mech. **106**, 215–227

Pierce, R. D., and Knobloch, E. (1995) Asymptotically exact Zakharov equations and the stability of water waves with bimodal spectra, Physica D **81**, 341–373

Pippard, A. B. (1979) *The Physics of Vibrations* (Cambridge University Press)

Pitois, S., Millot, G., and Wabnitz, S. (1998) Polarization domain wall solitons with counterpropagating laser beams, Phys. Rev. Lett. **81**, 1409–1412

Rajaraman, R. (1982) *Solitons and Instantons* (North-Holland)

Ramo, S., Whinnery, J. R., and Van Duzer, T. (1965) *Fields and Waves for Communication Electronics* (Wiley)

Rasmussen, K.φ., Bishop, A. R., and Grφenbech-Jensen, N. (1998) Creation and annihilation of intrinsic localized excitations, Phys. Rev. E **58**, R40–R43

Rayleigh, Lord (1876) On waves, Phil. Mag. (5), **1**, 257–279 .

Remoissenet, M. (1986) Low-amplitude breather and envelope solitons in quasi-one-dimensional physical models, Phys. Rev. E **33**, 2386–2392

Remoissenet, M. (1998) In preparation for Phys. Rev E

Remoissenet, M., and Michaux, B. (1990) Electrical transmission lines and soliton propagation in physical systems, in *Continuum Models and Discrete Systems*, Proceedings of the CMDS6 Conference, Dijon, 1989, ed. by Maugin, G. (Longman)

Remoissenet, M., and Peyrard. M. (1981) A new simple model of a kink bearing Hamiltonian, J. Phys. C: Solid State Phys. **14**, L481–L485

Rinzel, J. (1990) Mechanisms for nonuniform propagation along excitable cables, Ann. NY Acad. Sc. **591**, 51–61

Rodwell, M. J., Allen, S. T., Yu, R.Y., Case, M.G., Bhattacharya, U., Reddy, M., Carman, E., Kamegawa, M., Konishi, Y., Pusl, J., and Pullela, R. (1994) Active and nonlinear wave propagation devices in ultrafast electronics and optoelectronics, Proc IEEE **82**, 1037–1058

Rosenau, P. (1987) Dynamics of dense lattices, Phys. Rev. B **36**, 5868–5876

Rosenau P., and Hyman, J. K. (1993) Compactons: solitons with finite wavelength, Phys. Rev. Lett. **70**, 564–567

Rosenau, P. (1994) Nonlinear Dispersion and compact structures, Phys. Rev. Lett. **73**, 1737–1741

Rosenau, P. (1996) On solitons, compactons, and Lagrange maps, Phys. Lett. A **211**, 265–275

Rosenau, P. (1997) On nonanalytic solitary waves formed by a nonlinear dispersion, Phys. Lett. A **230**, 305–318

Rothenberg, J. E. (1990) Modulational instability for normal dispersion, Phys. Rev. A **42**, 682–685

Rubinstein, J. (1970) Sine–Gordon equation, J. Math. Phys. **11**, 258–266

Russell, F. M., and Collins, D. R. (1995) Lattice-solitons and nonlinear phenomena in track formation, Rad. Meas. **25**, 67–70

Russell, F. M., Zolotaryuk, Y., Eilbeck, J.C., and Dauxois, T. (1997) Moving breathers in a chain of magnetic pendulums, Phys. Rev. E **55**, 6304–6308

Russell, J. S. (1844) Report on Waves, Rep. 14th Meet. Bristish. Assoc. Adv. Sci. (John Murray) pp. 311–390

Salerno, M. (1992) Quantum deformations of the discrete nonlinear Schrödinger equation, Phys. Rev. A **46**, 6856–6859

Satsuma, J., and Yajima, S. (1974) Initial value problems of one dimensional self-modulation of nonlinear waves in dispersive media, Prog. Theor. Phys. Suppl. **55**, 284–306

Schlögl, F. (1972) Chemical reaction models for non-equilibrium phase transitions, Z. Phys. **253**, 147–161

Schlögl, F., and Berry, R. S. (1980) Small roughness fluctutation in the layer between two phases, Phys. Rev. A **21**, 2078–2081

Schmidt, H. (1979) Exact solutions in the discrete case for solitons propagating in a chain of harmonically coupled particles lying in double minimum potential wells, Phys. Rev. B **20**, 4397–4405

Scott, A. C. (1964) Distributed device applications of the superconducting tunnel junction, Solid State Elec. **7**, 137–146

Scott, A. C. (1965) Analysis of nonlinear distributed systems, Trans. IRE CT-**9**, 284–286

Scott, A. C. (1969) A nonlinear Klein–Gordon equation, Am. J. Phys. **37**, 52–61

Scott, A. C. (1970) *Active and Nonlinear Wave Propagation in Electronics* (Wiley Interscience)

Scott, A. C. (1985) Solitons in biological molecules, Comments Mol. Cell. Biophys. **3** (1), 15–37

Scott, A. C. (1992) Davydov's soliton, Phys. Rep. **217**, 1–67

Scott, A. C., Chu, F. T. F., and Mc Laughlin, D. W. (1973) The soliton: a new concept in applied science, Proc. IEEE **61**, 1443–1483

Seeger, A., and Schiller, P. (1966) Kinks and dislocation lines and their effects on internal friction in crystals, in *Physical Acoustics*, ed. by Mason, P. A. W., Vol 3 (Academic Press) pp.361–495

Seeger, A., Donth, H., and Kochendörfer, A. (1953) Theorie der Versetzungen in eindimensionalen Atomreihen. III, Z. Phys. **134**, 173–193

Segev, M., and Stegeman, G. (1998) Self-trapping of optical beams: spatial solitons, Physics Today **51** (8), 42–48

Segur, H. (1982) Solitons and the inverse scattering transform, in *Topics in Ocean Physics,* Course LXXX, ed. by Osborne, A. R. (North-Holland)

Sepulchre, J. A., and MacKay, R. S. (1998) Discrete breathers in disordered media, Physica D **113**, 342–345

Sievers, A. J., and Takeno, S. (1988) Intrinsic localized modes in anharmonic crystals, Phys. Rev. Lett. **61**, 970–973

Singer, A. C. (1996) Signal processing and communications with solitons, PhD Thesis, MIT, USA

Slavin, A. N., Kalinikos, B. A., and Kovshikov, N. G. (1994a) Spin wave envelope solitons in magnetic films, in *Nonlinear Phenomena and Chaos in Magnetic Films*, ed. by Wigen, P. E (World Scientific)

Slavin, A. N., Kalinikos, B.A., and Kovshikov, N.G. (1994b) Nonlinear dynamics of propagating spin waves in magnetic films, in *Linear and Nonlinear Spin Waves in Magnetic Films and Superlattices*, ed. by Cottam, M.G (World Scientific)

Soto-Crespo, J. M., Akmediev, N., and Ankiewicz, A. (1995) Stationary solitonlike pulses in birefringent optical fibers, Phys. Rev. E **51**, 3547–3555

Stokes, G. G. (1847) On the theory of oscillatory waves, Camb. Trans. **8**, 441–473

Su, M. Y. (1982) Evolution of groups of gravity waves with moderate to high steepness, Phys. Fluids. **25**, 2167–2174

Swihart, J. C. (1961) Field solution for a thin film superconducting strip-line, J. Appl. Phys. **32**, 461–469

Tai, K., Hasegawa, A., and Tomita, A. (1986a) Observation of modulation instability in optical fibers, Phys. Rev. Lett. **56**, 135–138

Tai, K., Tomita, A., Jewell, J. L., and Hasegawa, A. (1986b) Generation of subpicosecond solitonlike optical pulses at 0.3 THz repetition rate by induced modulational instability, Appl. Phys. Lett. **49**, 236–238

Takagi, K. (1983) The power spectrum for a white noise passed through a nonlinear transmission line, Jpn. J. Appl. Phys. **22**, 1466

Takeno, S. (1992) Theory of stationary anharmonic localized modes in solids, J. Phys. Japan. **61**, 2821–2834

Taniuti, T., and Wei, C. C. (1968) Reductive perturbation method in nonlinear wave propagation, J. Phys. Soc. Jap. **21**, 209–212

The Open University Course Team. (1989) *Waves, Tides and Shallow-Water Processes* (Pergamon)

Toda, M. (1967) Vibrations of a chain with nonlinear interactions, J. Phys. Soc. Japan. **22**, 431–436

Toda, M. (1970) Waves in nonlinear lattices, Prog. Theor. Phys. Japan. Suppl. **45**, 174–201

Toda, M. (1978) *Theory of Nonlinear Lattices* (Springer)

Turing, A. M. (1952) The chemical basis for morphogenesis, Phil. Trans. Roy. Soc. London. B **237**, 37-72

Ursell, F. (1953) The long wave paradox in the theory of gravity waves, Proc. Camb. Phil. Soc. **49**, 685–694

Ustinov, A. V. (1998) Solitons in Josephson junctions, Physica, D **123**, 315–329

Ustinov, A.V., Cirillo, M., and Malomed, B. A. (1993) Fluxon dynamics in one-dimensional Josephson-junction arrays, Phys. Rev. B **47**, 8357–8360

Van der Weide, D. W. (1994) Delta-doped Schottky diode nonlinear transmission lines for 480-fs, 3.5-V transients, Appl. Phys. Lett. 881–883

Van der Zandt, H. S. J., Barahona, M., Duwel, A. E., Trias, E., Orlando, T. P., Watanabe, S., and Strogatz, S. (1998) Dynamics of one-dimensional Josephson-junction arrays, Physica D **119**, 219–226

Van der Zandt, H. S. J., Orlando, T.P., Watanabe, S., and Strogatz, S. H. (1995) Kink propagation in a highly discrete system: observation of phase locking to linear waves, Phys. Rev. Lett. **74**, 174–177

Volkenstein, M. V. (1972) The conformon, J. Theor. Biol. **34**, 193-195

Walls, D. F. (1983) Squeezed states of light, Nature **306**, 141–146

Watanabe, S., Ishiwata, S., M., Kawamura, K., and Oh, H. G. (1997a) Higher order solution of nonlinear waves. II. Shock wave described by Burgers equation, J. Phys. Soc. Japan **66**, 984–987

Watanabe, S., Kawaguchi, M., Kawamura, K., Ishiwata, S., Ohta, Y., and Oh, H. G. (1997b) Shock wave in a nonlinear LC circuit, J. Phys. Soc. Japan. **66**, 1231–1233

Watanabe, S., Miyarkawa, M., Tada, M.(1978) Asymptotic behaviour of collisionless shock in nonlinear LC circuit, J. Phys. Soc. Japan. **66**, 2030–2034

Weiner, A. M. (1992) Dark optical solitons, in *Optical Solitons: Theory and Experiments,* ed. by Taylor, J. R. (Cambridge University Press)

Werner, M. J. (1998) Quantum soliton generation using an interferometer, Phys. Rev. Lett. **81**, 4132–4135

Whitham, G. B. (1965) A general approach to linear and nonlinear dispersive waves using a Lagrangian, J. Fluid. Mech. **22**, 273–283

Witham, G. B. (1974) *Linear and Nonlinear Waves* (Wiley)

Yagi, T., and Noguchi, A. (1976) Experimental studies on modulational instability by using nonlinear transmission lines, Elec. and Commun. in Japan **59A**, 1–6

Yagi, T., and Noguchi, A. (1977) Gyromagnetic nonlinear element and its application as a pulse-shaping transmission line, Electron. Lett. **13**, 683–685

Yariv, A. (1989) *Quantum Electronics* (Wiley)

Yomosa, S. (1987) Solitary waves in large blood vessels, J. Phys. Soc. Japan **56**, 506–520

Yoshimura, K., and Watanabe, S. (1991) Envelope soliton as an intrinsic localized mode in a one-dimensional lattice, J. Phys. Soc. Japan **60**, 82–87

Yuen, H. C., and Lake, B. C. (1975) Nonlinear deep water waves: theory and experiments, Phys. Fluids **18**, 956–960; Nonlinear concepts applied to deep-water waves, in *Solitons in Action*, ed. by Lonngren, K., and Scott, A. C. (Academic Press)

Yuen, H. C., and Lake, B. M. (1975) Nonlinear deep water waves: Theory and experiment, Phys. Fluids **18**, 958–960

Zabusky, N. J. (1973) Solitons and energy transport in nonlinear lattices, J. Comput. Phys. Commun. **5**, 1–10

Zabusky, N., and Kruskal, M. (1965) Interaction of solitons in a collisionless plasma and the recurrence of initial states, Phys. Rev. Lett. **15**, 240–243

Zakharov, V. E. (1968) Stability of periodic waves of finite amplitude on the surface of a deep fluid, J. Appl. Mech. Tech. Phys. **9**, 86–94

Zakharov, V. E. (1980) The inverse scattering method, in *Solitons*, ed by Bullough, R. K. and Caudrey, P. (Springer)

Zakharov, V. E., and Shabat, A. B. (1972) Exact theory of two-dimensional self-focusing and one dimensional self-modulation of waves in nonlinear media, Sov. Phys. JETP **34**, 62–69

Zhakarov, V. E., and Shabat, A. B. (1973) Interaction between solitons in a stable medium, Sov. Phys. JETP **37**, 823–828

Zhang, H.Y., Kabos, P., Xia, H., Staudinger, R. A., Kolodin, P. A., and Patton, C. E. (1998) Modeling of microwave magnetic envelope solitons in thin ferrite films through the nonlinear Schrödinger equation, J. Appl. Phys. **84**, 3776–3785

Zolotaryuk, Y. (1998) Nonlinear dynamics of molecular and atomic chains, PhD Thesis, Heriott-Watt University, Edinburgh, Scotland.

Zvezdin. A. K., and Popkov. (1983) On the nonlinear theroy of magnetostatic spin waves, Sov. Phys. JETP **57**, 350–362

Subject Index

Druck: Mercedes-Druck, Berlin
Verarbeitung: Stürtz AG, Würzburg